IET TRANSPORTATION SERIES 32

Cable Based and Wireless Charging Systems for Electric Vehicles

Other volumes in this series:

Volume 1 **Clean Mobility and Intelligent Transport Systems** M. Fiorini and J-C. Lin (Editors)
Volume 2 **Energy Systems for Electric and Hybrid Vehicles** K.T. Chau (Editor)
Volume 5 **Sliding Mode Control of Vehicle Dynamics** A. Ferrara (Editor)
Volume 6 **Low Carbon Mobility for Future Cities: Principles and Applications** H. Dia (Editor)
Volume 7 **Evaluation of Intelligent Road Transportation Systems: Methods and Results** M. Lu (Editor)
Volume 8 **Road Pricing: Technologies, economics and acceptability** J. Walker (Editor)
Volume 9 **Autonomous Decentralized Systems and their Applications in Transport and Infrastructure** K. Mori (Editor)
Volume 11 **Navigation and Control of Autonomous Marine Vehicles** S. Sharma and B. Subudhi (Editors)
Volume 12 **EMC and Functional Safety of Automotive Electronics** K. Borgeest
Volume 16 **ICT for Electric Vehicle Integration with the Smart Grid** N. Kishor and J. Fraile-Ardanuy (Editors)
Volume 17 **Smart Sensing for Traffic Monitoring** Nobuyuki Ozaki (Editor)
Volume 18 **Collection and Delivery of Traffic and Travel Information** P. Burton and A. Stevens (Editors)
Volume 24 **Driving Simulators for the Evaluation of Human-Machine Interfaces in Assisted and Automated Vehicles** T. Ito and T. Hirose (Editors)
Volume 25 **Cooperative Intelligent Transport Systems: Towards high-level automated driving** M. Lu (Editor)
Volume 26 **Traffic Information and Control** Ruimin Li and Zhengbing He (Editors)
Volume 30 **ICT Solutions and Digitalisation in Ports and Shipping** M. Fiorini and N. Gupta
Volume 38 **The Electric Car** M.H. Westbrook
Volume 45 **Propulsion Systems for Hybrid Vehicles** J. Miller
Volume 79 **Vehicle-to-Grid: Linking Electric Vehicles to the Smart Grid** J. Lu and J. Hossain (Editors)

Cable Based and Wireless Charging Systems for Electric Vehicles

Technology and control, management and grid integration

Edited by
Rajiv Singh, Sanjeevikumar Padmanaban,
Sanjeet Dwivedi, Marta Molinas and Frede Blaabjerg

The Institution of Engineering and Technology

Published by The Institution of Engineering and Technology, London, United Kingdom

The Institution of Engineering and Technology is registered as a Charity in England & Wales (no. 211014) and Scotland (no. SC038698).

First published 2021

The Institution of Engineering and Technology
Michael Faraday House
Six Hills Way, Stevenage
Herts, SG1 2AY, United Kingdom

www.theiet.org

British Library Cataloguing in Publication Data
A catalogue record for this product is available from the British Library

ISBN 978-1-83953-178-1 (hardback)
ISBN 978-1-83953-179-8 (PDF)

Typeset in India by Exeter Premedia Services Private Limited
Printed in the UK by CPI Group (UK) Ltd, Croydon

Contents

Preface

Introduction

The history of EVs dates back to the year 1830; these earliest types used non-rechargeable batteries. The 19th century, however, witnessed their commercial production on a mass scale. Their popularity relatively declined after 1910 compared to IC engine vehicles due to the limitations of the battery storage capacity, causing limited travel and problems associated with refueling or recharging. During the late 20th century, the technological advancements in battery technology, chargers, electric drives, and controllers coupled with increasing environmental concerns resulted in the acceptability of EVs in the modern transportation system. The EVs are designed & manufactured in many forms and hence can be classified into six different types based on propulsion systems and the kind of energy storage components being used. *Battery Electric Vehicles (BEVs)* amongst the oldest and traditional forms. The second type is a hybrid system containing an IC engine and a Hybrid Electric Vehicle battery *(HEV)*. The third type uses fuel cells or metal-air batteries as replaceable fuels. The fourth type receives operating energy from the permanent power supply lines. EVs may be working using energy received from solar insolation, falling under the fifth category. In contrast, the sixth type may use super-capacitors and fly-wheels as additional storage besides primary battery storage.

Batteries being an essential component, are considered being the heart of EVs. They pump energy into the drive system and other features of EVs leading to their proper functioning. Maintaining battery health, therefore, becomes the primary issue for running most components in EVs satisfactorily. Dissuading failures by maintaining good battery health can be achieved by using appropriate charging mechanisms. Continuous use of improper charging methods may be dangerous for battery health in the long run. Therefore, intelligent and precise control of current and voltage fed into the battery becomes the primary objective of reliable EV chargers. Many smart chargers and battery management systems (BMS) have been reported in the literature. Cable-based and wire-less chargers employing intelligent control algorithms have been introduced by engineers & scientists through several articles and research papers. Commercially available EV chargers are being manufactured by several companies such as Delta, ABB, Schneider Electric, etc., nowadays. According to the Global outlook EV report 2020 published by IEA, in 2010, the total global electric car stock was approx. 0.02 million, which increased to 4.79 million by 2019. Also, nearly 7.3 million EV chargers were installed worldwide by the end-of-year 2019. The significant rise in the number of EVs and charging infrastructures worldwide causes the compilation of research articles, technical reports, and books for proper dissemination of knowledge and technical developments in

EVs and chargers. This book attempts the editors to bring before the readers the latest technological advancements in EV charging infrastructure. We hope the readers will come across a completely different experience compared to the existent literature while going through the book.

Key Features:

- It provides a better overview of the recent trends with the EV chargers.
- Besides giving basic understanding also provides in-depth concept about EV chargers.
- A better explanation is provided by incorporating relevant case studies in many chapters.
- Experimental and simulation results are provided for better insight.
- Lucid content delivery.

Content and coverage

The book introduces various configurations of wired & wireless chargers and strategies for charging the battery in EVs. It gives the readers a complete overview of the multiple technologies adopted for EV chargers and makes the book more meaningful for its readers by adopting a case study-based approach for analysis in many chapters.

The chapter-wise contents of the book are briefly described below:

Chapter-1: In this chapter, charging methods of EVs such as conductive or wireless charging have been discussed in detail. Also, recent advancements in static and dynamic wireless chargers are presented. Moreover, *standards and regulations* for conductive and wireless charging of EVs are also discussed in detail.

Chapter-2: Fast inductive charging technologies allow the wireless exchange of power more significant between the grid and EVs. This chapter presents a methodology for evaluating the impact of fast static and dynamic inductive charging on the distribution grid. The simulation-based case study has been given for the benefit of the readers.

Chapter-3: The chapter presents basic principles of wire-less EV chargers and the main challenges involved in the system design and control. Future trends have also been reviewed in the chapter to provide the readers an insight into the prospects of future research in this field.

Chapter-4: Dynamic inductive charging has been explained in detail in this chapter, and methods for achieving optimized power transfer at a high efficiency have been validated through experimental results.

Chapter-5: The chapter explains the various configurations of converters used for EV applications, their essential features, and control mechanisms. The topologies have been compared in terms of steady-state analysis, small-signal modeling, and passive component design. A case study of a bi-directional quadratic boost converter is also presented with the help of simulation and experimental results.

Chapter-6: The effect of using solar photovoltaic (PV) and energy storage units for alleviating the impact of high penetration of EVs at all the charging levels from level 1 to level 3 in EV charging system is discussed in this chapter.

Chapter-7: In this chapter, factors affecting the Li-ion battery life-time have been discussed. Also, different battery thermal management strategies for enhancing the battery life-time have been evaluated. Finally, considering the life span of batteries optimal energy management strategy has been proposed for EVs in a parking lot to minimize their operational costs.

Chapter-8: At the outset of the chapter, state-of-the-art scheduling EV charging and optimization approaches for integrating EVs into micro-grids are presented. Also, scheduling strategies as optimization problems encompassing several forms of objective functions and their limitations are investigated. Finally, a real case study from Iran is presented and analyzed through simulation for micro-grids with high penetration of renewables.

Chapter-9: Different energy management strategies to optimally regulate the energy among different components of micro-grids have been discussed. The role of energy management systems (EMS) in micro-grids containing renewables and EVs have been outlined by introducing the objective functions, constraints, and uncertainties.

Chapter-10: This chapter contributes to the technologies and various standards associated with the latest EVs and EV service (supply/charging) equipment (EVSE) along with associated infrastructure for its erection. The procedure for planning & installing charging infrastructure is also discussed to provide better insight into the field engineers.

Chapter-11: Thermal models for high-power semi-conductor modules have been developed systematically. Also, the thermal reliability study of power electronic converters used in EVs has been carried out in detail in this chapter.

Finally, we believe the book will meet the readers' expectations and will be helpful for the researchers working in the domain of EV chargers or willing to adopt it as one area of their affinity. It is also expected that this book will ignite a spark within the new researchers and industry experts to foray into the area of EV chargers for research activities. Moreover, we welcome valuable comments and feedback from the readers and guide us in improving our efforts for any such future endeavors.
At last, we would like to thank almighty God, our family, friends, and society for continuously motivating us throughout this project.

Editors

About the Editors

Rajiv Singh is an assistant professor at the Department of Electrical Engineering, College of Technology, G.B. Pant University of Agriculture & Technology, India

Sanjeevikumar Padmanaban is an assistant professor at the Department of Energy Technology, the Faculty of Engineering and Science, Esbjerg Energy Section, Aalborg University, Denmark

Sanjeet Dwivedi is a senior R&D engineer in control engineering, application, and motor control and a scrum master at Danfoss Power Electronics, Denmark. He is a fellow of the IET (UK), the IETE, and the SMIEEE (USA). He has published more than 40 research articles and four international patents.

Marta Molinas is a professor at the Department of Engineering Cybernetics, Faculty of Information Technology & Electrical Engineering, Norwegian University of Science and Technology, Norway

Frede Blaabjerg is a full professor at Aalborg University's Centre of Reliable Power Electronics, Denmark. He is the recipient of numerous prestigious awards in power engineering from the IEEE eg the 2020 IEEE Edison Medal. In 2019-2020 he has President of the IEEE Power Electronics Society. He was nominated in 2014-2020 by Thomson Reuters as one of the 250 most cited researchers in engineering in the world.

Chapter 1

Charging stations and standards

Sourabh Ghosh[1], Asheesh K Singh[1], Rakesh Maurya[2], Sri Niwas Singh[3], and Vassilios G Agelidis[4]

1.1 Introduction

The global call for green energy and to reduce the dependency on fossil fuel for public transportation requires new alternative energy. As a solution to this problem, the electrification of vehicles has replaced internal combustion engines and thus has the potential to reduce CO_2 emissions. Lower operating costs and better fuel economy with reduced carbon emissions are the reasons for higher preference of electric vehicles (EVs). Additionally, high penetration of EVs can facilitate grid integration and participate in load balancing and revenue generation. However, the development of EVs is facing many barriers such as immature battery technology, scarce charging infrastructure, standards, and regulations.

The battery pack of an EV can be charged at standstill or while driving using either the conductive or inductive power transfer (IPT) method. The conductive charging supports only standstill charging, whereas inductive charging supports both the standstill charging as well as charging while driving. There are a few limitations of conductive charging that can be overcome by using inductive charging. Many research organizations and governments are working together for the development of EV technology and policy formulation to make public transport sustainable in the near future.

1.2 Conductive charging of EVs

Conductive method of charging has been widely commercialized and is available in many countries. This method uses a standard cable for EV charging at standstill conditions. The users can charge their vehicles at home or in public spaces with a

[1]Electrical Engineering Department, Motilal Nehru National Institute of Technology, India
[2]Department of Electrical Engineering, Sardar Vallabhbhai National Institute of Technology, India
[3]Department of Electrical Engineering, Indian Institute of Technology Kanpur, India
[4]Department of Electrical Engineering, Technical University of Denmark, Denmark

dedicated charging station. However, different EV manufacturers follow different standards, which makes uniform global commercialization difficult.

1.2.1 EV charging infrastructure

For successful implementation of EV technology in future transportation, reliable operation of charging support is mandatory. This subsection includes various challenges associated with EV charging infrastructure.

1.2.1.1 Charging levels and accessories

The market penetration of EVs is greatly decided by EV charging infrastructure. Based on the type of power used, there can be two types of EV charging, namely AC-based charging and DC-based charging. Electricity in the grid is alternating by nature; however, in a battery, it is DC in nature. Hence, to charge an EV AC power from the grid is fed to the EV via the vehicle inlet using the charging cable and connector through the process of power conversion from AC to DC. Thus, the whole system that supplies electrical power to the battery pack of an EV from the electrical utility is called electric vehicle supply equipment (EVSE) and is commonly known as the charging station. The location of EVSE can be either on-board or off-board. The size and weight of the on-board charger are dependent on the maximum charging power requirement. Typically, on-board chargers can have a maximum power of up to 22 kW. If the charging power has to be raised, then on-board chargers' size and weight will increase. However, due to the restrictions on size and weight inside the vehicle, the on-board chargers are limited in size and weight as well. Thus, the solution to increasing the charging power is to move the AC/DC converter outside the vehicle and place it in the off-board charger. This is referred to as DC charging where the off-board charger converts power from AC to DC and this DC power is directly fed into the vehicle via a charging inlet. With no size and weight restrictions, maximum charging power can reach the level of hundreds of kilowatts. Thus, we can have either AC charging or DC charging depending on the load requirement.

Charging power depends on the load requirement for EV charging. Hence, EV charging power is classified into three different levels based on the load requirements. AC charging can be categorized into Level 1, Level 2, and Level 3 charging. A private or home charging station can be of Level 1 or Level 2 while installation and commissioning of Level 3 charging station involve separate power equipment, require consent from electricity providers, and are generally built in public spaces. DC charging compared with AC charging, generally, is relatively faster and provides high charging capacity but it requires an off-board charging facility. The main difference between charging levels is the input voltage of EVSE. Level 1 uses 120/230 V single-phase supply with 16 A of limited current supply, Level 2 uses 240 V single-phase supply with 80 A of limited current supply, Level 3 uses a regular three-phase supply and DC fast chargers use DC voltages up to 600 V or more.

The scarcity of charging infrastructure in public space forces users to charge their vehicles at home using a Level 1 charger, which takes a long time to full charge thus cannot be a regular charging option for EV users. This type of charging is suitable for residential spaces, but while planning a new charging installation, a Level 2 charging infrastructure will be cheaper to install as will provide higher charging power capacity as compared to Level 1 charging. Level 2 charging is also a solution for residential and commercial charging of EVs. While offering greater power delivery than its Level 1 counterpart, they also offer supplementary features that are not provided by Level 1 charging systems.

Networked chargers and non-networked chargers are two categories of Level 2 chargers. Networked chargers come with an additional feature of adjustable power that is not available with non-networked chargers. Due to reduced features, non-networked chargers are cheaper and may be sufficient for single-user EV charging. Non-networked chargers are applicable for installation at multi-unit dwellings (MUDs) like apartments, or at public spaces that take power from the individual's subpanels. In such cases, the power used for EV charging will be metered as per the individual's electricity bill, thus avoiding the necessity of separate metered chargers. However, networked chargers are applicable for use at single-owner residences, or at commercial sites, where payment is required for EV charging, or at MUDs, where multiple residents pool the total charging cost. Networked Level 2 chargers have features that include the capability to concede various formats of payment, remote access via cellular or Wi-Fi connection, load balancing across multiple EV chargers, etc. Thus, balancing, monitoring, and payment of electricity usage bills across single or multiple chargers with multiple drivers need networked Level 2 chargers. Networked chargers can also limit EV charging to specific hours in a day, which allows the user to charge the EV when electricity is at the cheapest rate.

DC off-board chargers are categorized under Level 3 of EV charging. They are often used to extend driving ranges for EV users for long-distance trips who do not have the support of a household charging facility. To increase power output even more, DC off-board chargers are employed as the components' size and weight will increase enough to accommodate the charger inside EV. Level 3 charging is provided by a three-phase AC supply and converted into DC power for fast charging of EVs. At present, DC fast chargers require a supply of more than 480 V and 100 A to deliver 50–60 kW to full charge an EV. However, the latest generations of DC fast chargers are gathering attraction of EV users and can supply up to 350 kW of power. A summary of the comparative analysis of various AC and DC charging power levels is given in Table 1.1 [2].

AC and DC charging is characterized by the connector types, ports, and charging power levels. Every country follows its own charging standard and the major difference lies in the ports and connector types used. Level 1 and Level 2 chargers are mostly designed for private charging while Level 3 off-board DC fast chargers are designed for public place charging. Chargers with Level 2 specifications are designed for outdoor and indoor use conforming to National Electrical Manufacturers Association (NEMA) defined standards such as NEMA 6P, NEMA 4x, and NEMA 3R.

DC off-board fast chargers allow three different types of connector specifications: Combo Charging System (CCS), CHAdeMO, and Tesla Supercharger. The CHAdeMO standard was co-developed by Tokyo Electric Power Company and

Table 1.1 Standards for cable-based charging of EVs

Power levels	Charger location	Power and current
Society of Automotive Engineering (SAE) standards		
Level 1: AC charging 230 V (EU) and 120 V (USA)	Single-phase On-board charger	1.4 kW with 12 A 1.9 kW with 20 A
Level 2: AC charging 400 V (EU) and 240 V (USA)	Single-phase/three-phase On-board charger	4 kW with 17 A 8 kW with 32 A 19.2 kW with 80 A
Level 3: AC charging 208–600 V	Three-phase Off-board charger	50 kW 100 kW
Level 1: DC charging 200–450 V	Off-board charger	40 kW with 80 A
Level 2: DC charging 200–450 V	Off-board charger	90 kW with 200 A
Level 3: DC charging 200–600 V	Off-board charger	240 kW with 400 A
International Electrotechnical Commission (IEC) standards		
Level 1: AC charging	Single-phase On-board charger	4–7.5 kW with 16 A
Level 2: AC charging	Single-phase/three-phase On-board charger	8–15 kW with 32 A
Level 3: AC charging	Single-phase/three-phase On-board/off-board charger	60–120 kW with 250 A
DC Fast charging	Off-board charger	1 000–2 000 kW with 400 A
CHAdeMO standard		
DC Fast charging	Off-board charger	62.5 kW with 125 A

Source: [1].

other Japanese automakers. The SAE established their own DC fast-charging standard CCS instead of adopting CHAdeMO. Likewise, Tesla developed their own proprietary fast-charging system branded as Tesla Supercharger and China framed their own GB/T fast-charging standard. CCS standard uses Combo 1 and Combo 2 connectors with additional DC contacts to supply power up to 350 kW. A revised CHAdeMO specification uses 1 000 V, 400 A DC to supply power up to 400 kW. Tesla Supercharger uses a three-phase supply to convert it to a 480 V DC for fast charging of EVs.

The USA and Japan use both AC and DC charging of EV facilitated by using Level 1 and Level 2 charger ports and connectors that conform to SAE1772 standard. Tesla uses its proprietary connector and port both for fast charging. Tesla has also introduced an adapter that can convert SAEJ1772 connectors into Tesla

Table 1.2 Standards of connectors and ports employed at different countries

	USA	**Japan**	**EU**	**China**
Single-phase/ three-phase AC charging	Level 1, Level 2 Single-phase SAE1772	Level 1, Level 2 Single-phase SAE1772	Level 1, Level 2 Single-phase/ three-phase IEC62196	Level 1, Level 2 Single-phase/ three-phase IEC62196
DC fast charging/ AC–DC combo	SAE1772 Combo	CHAdeMO DC fast charger	IEC62196 hybrid combo	GB/T 20234.3 DC fast charger

Source: [3].

connectors so that Tesla supercharging stations can be directly used. In Europe, a different model of charging connector is used called "combo" which uses additional pins for DC charging into existing connectors for AC charging. Level 1 and Level 2 charging ports and connectors are used in Europe that conform to IEC62196 standard. With respect to EV manufacturers, European and American EV manufacturers have adopted Combo while Japanese EV manufacturers have adopted CHAdeMO, and Chinese EV manufacturers have adopted GB/T standard. Therefore, it can be seen that EV charging equipment manufacturers have not agreed upon a single international standard. Organizations and governments are making efforts to develop uniformity in the charging standards and frame a universal solution for EV charging. Table 1.2 lists a few standards of connectors and ports employed in different countries by different manufacturers.

1.2.1.2 Power converters

The growth in EV charging technology based on power electronics converter topologies contributes towards many significant benefits, including peak load shaving, load balancing, revenue generation, and maximum power tracking of renewable energy resources. Despite many advancements in EV technology, there are still some technological barriers including charging infrastructure, power converter topologies for battery chargers, range anxiety, power quality, and battery life degradation issues. These issues must be overcome for the widespread adoption of EVs [4]. Power electronic converters are the main interface between the EVSE and the battery pack; therefore, it is necessary to develop low-cost and highly reliable power converters for advancement in charging infrastructure of EVs. Therefore, several power electronic converters are discussed here, considering various important requirements including the capability of bidirectional power flow, high power density, and high efficiency [5].

The improvement of power quality and power factor can be achieved by observing less distorted utility current drawn by EV chargers. Various converters topologies are employed to EV charging systems for power conversion from AC to DC,

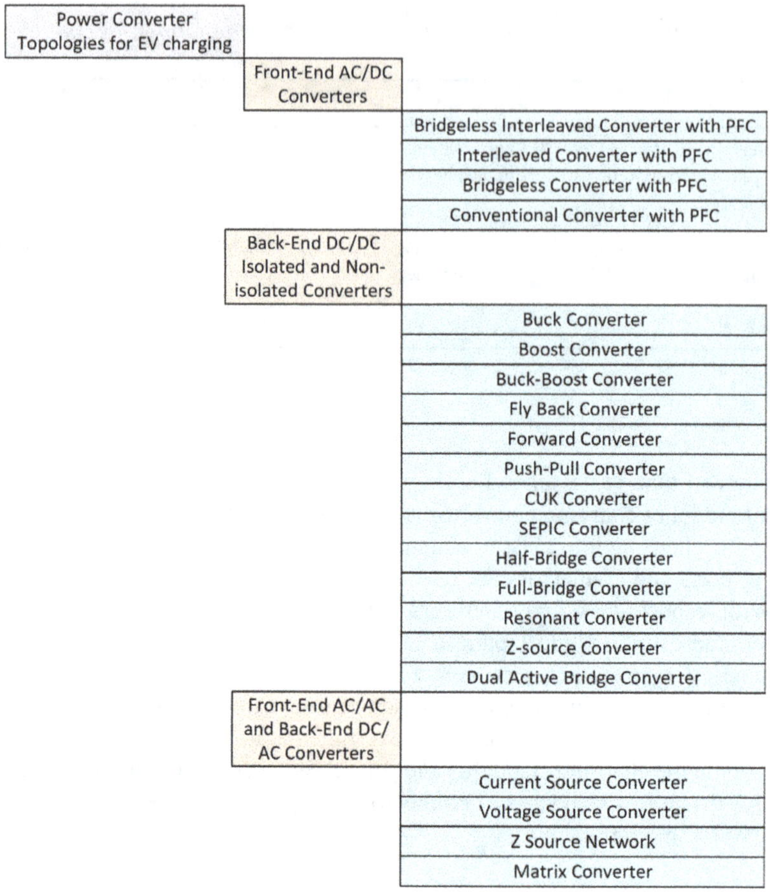

Figure 1.1 Classification of power converters for EV charging solution [1]

AC to AC, DC to DC, and DC to AC based on application requirements as depicted in Figure 1.1. Among them, those converters having input supply as AC can inject and generate current harmonics. Therefore, there is a necessity of implementing power factor correction (PFC) techniques to eliminate the issue of current harmonics. The PFC techniques using phase-locked loops (PLL) ensure the input currents to be sinusoidal and in phase with their respective phase voltages. These converters have two separate converter stages, AC to DC and DC to DC. A front-end converter is implemented with PFC to achieve an input power factor close to unity, while a back-end converter is used to achieve desired DC voltage/current to charge the battery of an EV.

A few popular topologies of front-end AC/DC converters are three-phase six-switch AC/DC converter, three-phase converter using multiple single-phase AC/DC converters, three-phase four-switch AC/DC converter with split DC link, and Vienna rectifier. The operation of the Vienna rectifier permits bidirectional power flow as the topology consists of 18 diodes and 3 bidirectional switches. Reduced number of switches and bidirectional power flow are admirable results but at the cost of increased complexity in control methods. The control methods that are extensively used are voltage control, current control, and power control. With increasing requirements of power in EV charging system, interleaved techniques are employed in multi-phase boost converters with PFC. The benefits of using the interleaved boost converters in high-power applications include reduced size of inductor, minimized ripples of the output capacitor, reduced conduction losses, better performance, and higher efficiency. However, there are a few drawbacks including complex control method, reverse recovery losses of boost diodes, and heating issues for input diode bridge. Therefore, soft-switching techniques are implemented for faster recovery diodes.

The DC/DC power converter provides regulated output voltage/current from the DC voltage output of the front-end converter and makes it suitable for EV battery charging. The converter topologies of back-end DC/DC converters can be categorized into isolated and non-isolated configurations depending on the presence of galvanic isolation between the input side and the output circuitry. These back-end converters possess energy storage passive elements for power conversion. The increase in switching frequency allows reduction in size and weight of passive elements. But the increase in on-off time and switching losses constraint the switching frequency. Thus, to increase the switching frequency, resonant circuits and soft-switching methods are widely adopted. Two soft-switching techniques that are most widely adopted are (a) zero current switching (ZCS) and (b) zero voltage switching (ZVS).

Generally, for low-power-level applications, the most widely used isolated topology is the fly-back converter. However, for medium power level applications, forward converters are often used. For high-power level applications, half-bridge, full-bridge, and push-pull converters are widely used. Underutilization of the transformer and the necessity of supplementary capacitors due to high input and output ripple currents are the drawbacks of the fly-back converter topology. The implementation of the forward converters can be helpful in overcoming the limitations of the fly-back converter; however, the maximum duty ratio is limited during the transformer flux reset process, as peak flux during transient states and startup can increase to very high levels [6]. The full utilization of the transformer core, no need to reset the transformer, and higher duty ratio are a few advantages of implementing push-pull and half-bridge converters. However, the full-bridge topology has the advantages of best core utilization, switch voltage less than the input, and much less current for a given power as compared to half-bridge topology, which consequently increases its efficiency at high load currents. In contrast, half-bridge topology uses a single inductor and active components with lesser ratings, as compared to CUK and SEPIC converters, which leads to less switching and conduction losses. The best way to minimize switching and conduction losses is the implementation of different soft-switching resonant methods. Different resonant topologies include series resonant converters (SRC), parallel resonant converters (PRC), series-parallel resonant converters (SPRC), and LLC. When compared with conventional resonant topologies such as SRC, PRC, and SPRC, the efficiency of an LLC resonant converter topology in constant voltage and constant current (CV and CC) charging modes is high, as circulating energy and conduction losses are little in both modes of EV battery charging. Also, for isolated bidirectional chargers, the dual active bridge (DAB) converter topologies are appropriate for the interface between an EV battery and a DC link. Therefore, a careful selection of power converter topology must be done for EV charging [1].

1.2.1.3 Communication and protection

A charging system for EV consists of AC/DC power electronic converter, cable, connector, charging port, a charge controller on the supply-side, and a battery management system on the battery side. Thus, a fast and reliable communication system is responsible for charging, protection, and control of the charging process. The charging current can be adjusted, and a complete shutdown of the system is also possible in case of overcharging, complete draining, short circuit, or overvoltage of the battery. It also ensures that the temperature and balancing between the battery cells are optimal for the long life of the battery. Depending on the charging strategy, there can be four modes of communication, control, and protection, as introduced in the international standard IEC61851-1:

Mode 1: It is the simplest mode of EV charging via a regular household outlet. There will be no communication between the EV and the supply equipment. The safe mode of operation of a charging point relies on the implementation of suitable protective equipment on the supply side, i.e., a CB or a fuse to protect against overcurrent, proper connection of earthing, and a residual current circuit breaker (RCCB) device to shut down the charging if a leakage current greater than a specific value is detected. Without the proper connection of earthing, an unsafe operating condition could occur with an earth fault within the vehicle; therefore, Mode 1 has been illegalized in many countries while a few countries allow Mode 1 charging at private spaces only.

Mode 2: The implementation of standardized socket-outlets for connection of charging the EV to the utility provides supplementary protection by using additional control box with a control pilot conductor built-in with the charging cable is responsible for communication, control, and protection. Mode 2 charging was introduced under the standards of the National Electrical Code (NEC) to ensure the protection and safety of the user in all charging conditions. The control box protects the cable and the EV battery but not the supply equipment. This is the main limitation of Mode 2 because the supply equipment is the main component liable to be damaged during charging.

Mode 3: This mode involves both slow and semi-fast charging of EV via a dedicated outlet. The charging cable uses plugs on both sides to connect between EV and supply equipment. The charging station will be responsible for the communication, control, and protection of the EV charging process. Mode 3 charging uses additional protection called control pilot, which verifies the proper connection of EV, verifies protective device integrity, and adjusts the charging rate. Mode 3 is widely used for public charging stations and can also facilitate grid integration.

Mode 4: This mode utilizes an off-board charger for the indirect connection of the EV to the utility. Mode 4 is especially recommended for DC fast charging of an EV. As the location of the DC fast charger is off-board, a communication link is needed for feedback of the voltage and state of charge (SoC) of the battery to provide it with the right amount of current supply that is provided by the control pilot conductor. The control pilot conductor and the charging cable are permanently connected to the off-board charger. The functionalities like communication, control, and protection are also built into the off-board charging station.

1.2.2 Integration of EV with power grid

The commercialization of EVs has increased recently and created many opportunities as well as challenges for the power sector. These EVs can be treated as distributed energy resources (DERs) when not in use, i.e., at standstill mode. Thus, a linkage between EVs and the grid can open up new possibilities for power management. Integration of EVs with the grid causes many positive as well as negative impacts on the grid, which are discussed here.

1.2.2.1 Positive impacts

Power management: The large-scale power demand for EV battery charging can put an additional burden on the electricity grid and also generate harmonic distortion. However, using scheduled charging or discharging to meet peak load demand, better performance and efficiency can be achieved on the grid side.

Improvement of power quality: In recent times, market penetration of renewable energy sources and EVs have increased, and they act as DERs. Uncontrolled grid integration of DERs causes voltage surges that can be solved by properly managed integration of EVs. Other possible advantages of EV grid integration are the elimination of voltage flickers and reactive power injection; harmonics injection by uncontrolled DERs can be lowered and the imbalance of voltage can be solved by sharing power flow through each phase [7].

Regulation: By varying the switching and control scheme of EV power converters, the voltage can be regulated by supplying or absorbing reactive power. Since EVs are capable of supplying real power to the grid, the frequency can also be regulated by correcting the grid frequency deviation. EV batteries act as power flow buffer by storing excess power in them. Therefore, stability can be improved for isolated electric networks or microgrids.

Renewable energy: There is always an uncertainty in power generation using renewable energy sources due to their dependency on nature's conditions. Using EVs, the uncertainty of power from renewable energy sources can be reduced and can act as a renewable energy buffer [8].

1.2.2.2 Negative impacts

Increase in load demand: As the charging technology evolves, there will be large-scale deployment of EVs in the market. This can put an additional burden on the electricity grid due to large-scale power demand for EV battery charging. Most of the EV users will charge their EV battery at home because of the current lack of public charging stations. Thus, there will be a huge increase in current flow through the transmission lines and distribution feeders. The present status of utility infrastructure may not be able to deliver such a huge load requirement for uncontrolled charging of EVs, which may lead to grid collapse [9]. Therefore, the existing power infrastructure components need to be redesigned to handle the additional load requirement and to avoid component overloading.

Unbalanced voltage and phase: If a large number of EVs are charged using the same phase of utility, then a phase voltage unbalance will be created. This unbalance in voltage is created due to the current unbalance in that phase. One of the main power quality effects caused by EVs is the unbalance in voltage profile of the low-voltage distribution system. Also, due to the grid integration of EVs, there will be a voltage deviation and drop at the interconnection point of the EV charger and grid [10].

Injection of harmonics: The components used in EV chargers use power electronic components that inherently generate harmonic distortion during power conversion. The contribution of harmonic distortion by power converters increases with more penetration of EVs [11].

Power loss: The additional burden of delivering active power demand to EV chargers would lead to increased flow of current in the power system components. This will cause a huge amount of power loss and heat dissipation in the power distribution system. The power loss is even higher at peak load hours. The required power to charge a large number of EVs can be shared among power substations and can act as one solution to the problem. However, recently, a few new techniques such as static load management, dynamic load management, and time-based load management have been introduced. Static load management uniformly distributes a charging power pre-defined for each charging station. In contrast, the total available charging capacity is adjusted according to the instantaneous power consumption in the entire building using dynamic load management. The available charging current is divided into each charging station according to the required travel time of the EV and their rated charging capacity [12].

Stability: EV charging stations act as non-linear loads and draw a huge amount of current in a small time duration, which leads to instability of the power system. Thus, the power system becomes sensitive to even small disturbances and it takes more time to return to a new steady state. This problem increases by manyfold with more penetration of EVs. However, if power flow is managed properly, then EVs can make the power system more stable [13].

Table 1.3 *Summary of a few international standards and guidelines for conductive charging of EVs*

Organization	Standards	Description
IEC	IEC61851	Cable-based charging systems
	IEC62196	Plugs, sockets, and couplers for EV applications
SAE	SAEJ2293	General requirements for cable-based and inductive charging of EV
	SAEJ1772	Ratings of power equipment used for EV charging
	SAEJ2931	Communication network requirement for charging of EV in smart grid
IEEE	IEEE1547	Interconnection of DERs with electric utility
UL	UL1741	Power conversion and its protection specifications
	UL1741SA	Inverter operation for optimal grid stability
NFPA	NFPA 70	Electrical equipment wiring and safety
NEC	NEC 625	Off-board charging systems
	NEC 626	Specifications for the electrical equipment

Source: [3].

1.2.3 International standards and regulations

The advancements in EVs have opened new prospects both in transport and power sector. To develop a uniform charging infrastructure worldwide, it is necessary to standardize each aspect of EVs. Table 1.3 summarizes a few international standards that govern various aspects of EVs. The standards related to EVs can be categorized into three sections: EV charging components, grid integration of EV, and safety standards. International Organization for Standardization (ISO) works on standardizing whole EVs, and the other organizations work on elementary specifications. EV charging or discharging comes under grid integration standards. When integrated with the power system, EVs act as DERs and the standards for integration of DERs with grid also apply on EVs. The Underwriters Laboratories (UL) and Institute of Electrical and Electronics Engineers (IEEE) work on standardizing grid integration of EVs. Organizations such as National Electrical Code (NEC) and National Fire Protection Association (NFPA) work mainly on safety standards [14]. The standards formed by these organizations are discussed in the subsections below.

1.2.3.1 Charging of EVs

IEC61851 extends to cable-based charging of EV and applies to both off-board and on-board components for charging EVs with supply voltages up to 1 500 V DC and 1 000 V AC. Under section IEC61851-1, three possible cases to charge EVs are covered, namely Cases A, B, and C. When the cable is always connected with the EV, it comes under Case A. In contrast, when the cable is attached neither to EV nor to the charging station, it comes under Case B. Otherwise, when the cable is always connected with the charging station, it comes under Case

Table 1.4 Summary of current and voltage levels in various standards

Standards	Source	Mode/Level	Voltage (V)	Phase	Max current (A)
IEC62196	AC	Mode 1	120	Single	16
		Mode 2	240	Single	32
		Mode 3	250	Single	32–250
	DC	Mode 4	600	DC	400
IEC61851	AC	Mode 1	120	Single	16
		Mode 2	240	Single	80
	DC	Mode 4	200–450	DC	80
SAEJ1772	AC	Level 1	120	Single	16
		Level 2	240	Single	32–80
	DC	Level 1	200–450	DC	80
		Level 2	200–450	DC	200

C. Electromagnetic compatibility (EMC) requirements for on-board conductive charger and for off-board conductive charger are given by IEC61851-21-1 and IEC61851-21-2, respectively. IEC61851-23 covers the required specifications of DC fast-charging stations and IEC61851-24 addresses the communication and control standards for charging of EV.

IEC62196 standard is divided into subsections where IEC62196-1 extends to plugs, sockets, and vehicle connectors that are used for cable-based charging of EVs, and IEC62196-2 addresses the types of main connecting systems that are known as type 1, type 2, and type 3. IEC62196-3 extends to plugs, sockets, and vehicle connectors for DC fast charging.

SAEJ2293 addresses the requirements of off-board and on-board charging components and equipment. SAEJ2293-1 covers the general requirements and system architecture for conductive AC and DC as well as for inductive charging. On the other hand, SAEJ2293-2 addresses the communication standards and architecture for charging of EVs.

SAEJ1772 covers the ratings of power equipment used for AC and DC charging of EVs including charging voltage rating, protective devices rating, etc. The charging rate of the battery is limited by many factors including the composition of battery, power converter topology, and charging power level. A summary of current and voltage levels for IEC62196, SAEJ1772, and IEC61851 is presented in Table 1.4.

SAEJ2931 addresses the requirements for communication between EVs, supply equipment, electric utility, and smart meters. In a smart grid environment, to set up network architecture for communication during EV charging, the SAEJ2931 should be followed. SAEJ2931/1 subsection addresses the network architecture for digital communication. SAEJ2931/2-4 specifies media access control (MAC) and implementation of physical layer using SAEJ1772 pilot wire [15].

1.2.3.2 Grid integration

IEEE1547 standardizes the interconnection of DERs with electric grids. It is applicable to DERs as well as EVs and covers all the relevant requirements for improved performance, operation, maintenance, testing, and safety for integration of DERs, and highlights the interconnection of DERs with primary and secondary power distribution networks [16]. IEEE1547 comprises a few subsections that cover the design, operation, and integration of DERs with the grid. IEEE P1547.1 provides a detailed procedure to test for confirming that grid integration requirements and equipment ratings follow the IEEE1547 standard. The required technical background and specifications are provided by IEEE P1547.2 to make the standard user-friendly. It also points out the interconnection issues that may occur. IEEE P1547.3 offers guidelines for interoperability among DERs for data exchange, monitoring, and control. IEEE P1547.4 covers the features of electrical islands when power is not available from the utility.

UL also published many standards to address various aspects of integration of DERs with a grid. One of them is UL1741, which covers the protection specifications and power conversion requirements applicable for the integration of DERs with the grid. UL1741SA is a supplementary draft standard for UL1741, which specifies the desired inverter operation for optimal stability of the grid. It also covers modern inverter topologies and the procedures for testing which is necessary for the integration of DERs with a grid.

1.2.3.3 Safety

The safety and security measures are the important aspects of grid integration and charging of EV, which are being led by the NEC and NFPA. NFPA is a global leader in the sector of electrical fire and safety. NFPA released standard NFPA 70, which addresses EV and its grid integration and specifies the instructions on wiring and safety of electrical equipment on the user side. NFPA 70 specifies the requirements needed for the interconnection of DERs and EVs to the grid, component ratings, appropriate wiring techniques, and emergency provisions.

NEC 625 standardizes the off-board charging systems and covers the specifications and installation instructions for the components used in cable-based and wireless charging conductors. NEC 626 standardizes the specifications and installation instructions for the components used to charge trucks.

1.3 Inductive charging of EVs

Recently wireless charging of consumer electronics devices has become a reality, and now it is not necessary to connect a charging cable to charge a device. This technology can be extended to charge batteries of EVs as well. In this section, key methods of wireless charging of EVs are discussed along with the relevant international standards. We will also learn about the underlying concept behind wireless power transfer, important components of a wireless charging system, and a few advantages and disadvantages of wireless charging.

1.3.1 Need for inductive charging of EV

The advantages of such systems over conventional plug-in charging systems are that we no longer need to add charging poles to the streets and hence aesthetically more compatible especially for traditional cities. Also, the inductive charge pads in the ground can be surface-mounted or buried. Further, they are safe and convenient as EVs can be charged at the comfort of parking without plugging any cable into the vehicle, and possibilities of sparking and tripping on the cable are avoided. This technology can be extended to charge a vehicle while driving; thus, on-board battery weight can be greatly reduced as well as eliminating range anxiety of EV users. Inductive charging systems for EVs are already commercially available today and this is only expected to rise in the future.

1.3.2 Modes of IPT

In an IPT system, power is transferred from a source to a load without the need of connecting conducting cables between them. Different types of wireless power transfer techniques are possible based on an energy carrier medium, and they can be mainly classified as:

1. microwave
2. laser
3. acoustic
4. capacitive
5. inductive.

Depending on the energy carrier medium, power can be transmitted wirelessly over a range of distances from small to large, with variable power transfer capability and efficiency. IPT is a form of wireless power transfer that relies on high-frequency magnetic fields, which typically uses frequency in the kHz range and capable to transfer power in the kW range with an efficiency of 75–92%. On the other hand, capacitive power transfer (CPT) relies on high-frequency electric fields, which typically uses frequency in kHz to MHz range but it is capable of transferring power only up to a few watts with an efficiency of 50–80%. Microwave-based power transfer uses frequency in GHz range for wireless power transfer in space applications. Other modes of WPT include laser-based power transfer that uses electromagnetic (EM) radiation at very high frequency and acoustic-based power transfer that uses frequency in the kHz range [17]. IPT is the most popular method of wireless power transfer used for charging batteries of EVs due to its high power transfer capability.

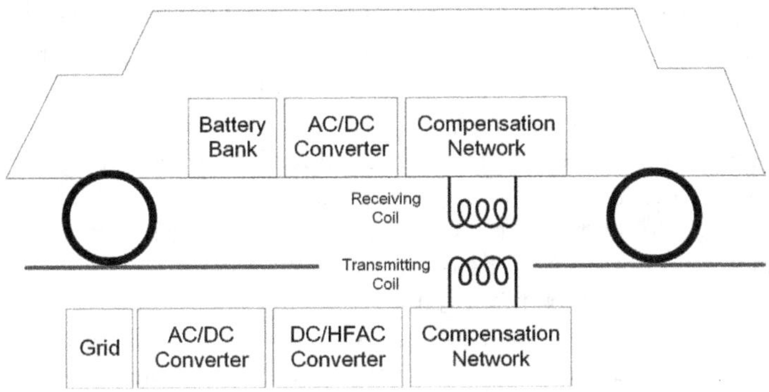

Figure 1.2 Inductive charging system for an EV

1.3.3 Operating principle of IPT

The underlying principle of IPT is governed by Faraday's law and Ampere Maxwell's law, which are mentioned below for steady-state condition:

$$\nabla \times \vec{H} = \vec{J} + \frac{\partial \vec{D}}{\partial t} \tag{1.1}$$

$$\nabla \times \vec{E} = -\frac{\partial \vec{B}}{\partial t} \tag{1.2}$$

Power delivery is a sequential process as explained below:

1. Dynamic magnetic flux is produced by the transmitter coil according to Ampere's law.
2. Generated magnetic flux induces an electromotive force (EMF) in the receiver coil according to Faraday's law.
3. EMF in the receiver coil circulates a current producing a dynamic magnetic flux that induces a counter EMF in the transmitter coil.
4. Power flow oscillates between transmitter and receiver coil, thus effective power flow is decided by phase angle between EMF of both coils.

It can be seen from Figure 1.2 that the working principle of IPT resembles with transformer but contains large leakage inductance on both sides. These leakage inductances are an effect of leakage magnetic flux due to power transfer over an air gap, thus giving rise to weak coupling. Therefore, to compensate for these leakage inductances, capacitive compensation networks must be incorporated into IPT systems. Equivalent circuit models for coil along with compensation networks can be analyzed as either transformer model, mutual coupling model, or T-model. These models are valid in steady state as well as during transient analysis.

1.3.3.1 Resonant IPT

It is one of the widely used variants of IPT systems. Resonant IPT (RIPT) requires a careful design of power converter and wireless transmission coils. Main AC voltage is first converted to high frequency AC and then applied to the transmitting coil. The receiving coil gets power via dynamic magnetic fields. To charge the EV's battery bank, the received power is then converted to DC using a converter and filter circuitry. Also, to create resonance and to enhance the performance of the system, additional compensation networks are installed for both transmitting and receiving coils. Hence, on the transmitting side, the main components will be the grid connection, power factor corrected rectifier, high frequency (HF) inverter, primary side compensation network, and the transmitting coil. Similarly, on the receiving side, the main components will be the receiving coil, secondary compensation network, HF rectifier with filter, and a battery set. Both subsystems are separated by an air gap and they should have a communication link to ensure a closed-loop operation.

The purpose of using compensation network is to match the resonant frequencies on both sides of the system. Operation of IPT at resonant frequency will improve performance and reduce losses. The resonant frequency will also determine the required switching frequency of HF inverters. Generally, the resonant frequencies for RIPT systems are within a range of 20–100 kHz. The flux produced at this frequency deteriorates the coefficient of coupling (k) and it varies between 0.2 and 0.3. However, if both the coils are strongly coupled then the coefficient of coupling will improve. It should be noted that at higher switching frequencies, converter losses increase due to skin and proximity effect in coil wires. This highlights the need for soft-switching techniques like ZCS/ZVS for the power converters. Soft switching of converters reduces switching losses and voltage stress across switches even at high frequency of operation. It should be noted that many international organizations such as the IEEE, SAE, and IEC along with research labs and governments have standardized the power levels, operating frequency, electromagnetic interference (EMI), and leakage flux levels from wireless charging systems to enable commercialization. Based on application, inductive charging systems can further be categorized into two methods to wirelessly charge the battery pack of EV, and they are static and dynamic/quasi-dynamic inductive charging systems.

1.3.4 Static inductive charging

1.3.4.1 Overview

EVs can be charged wirelessly with the ease of "park and charge" using static inductive charging where the transmitter coil is installed on the ground and the receiver coil is placed beneath the chassis of the EV. The transmitter coil system is often regarded as ground assembly (GA) whereas the receiver coil system is regarded as vehicle assembly (VA). To be ready to transmit ample power over long distances through an air gap, the system should operate at the resonant frequency with zero phase angle (ZPA) between grid current and voltage. To achieve resonant conditions, additional reactive elements are linked with both the coils in series or/and in parallel configuration. An appropriate selection of compensation topology helps in achieving soft switching on the transmitting side power converter. Thus, the VA rating of the HF inverter can be reduced, and subsequently total cost will be reduced with increased power transfer capacity. On the receiving side, a similar

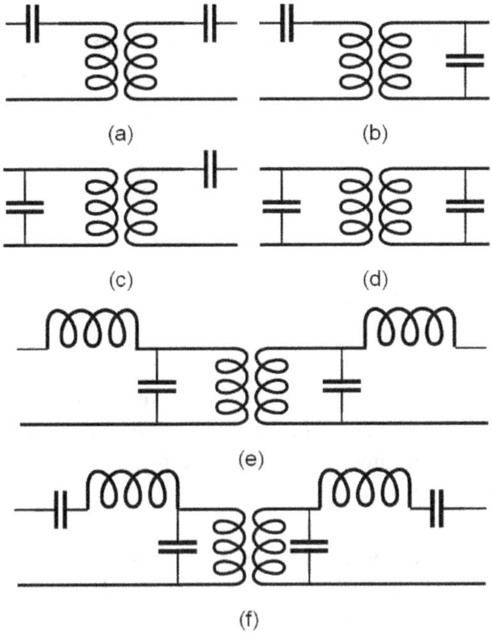

Figure 1.3 Conventional compensation topologies (a) SS, (b) SP, (c) PS, (d) PP, (e) LCL, and (f) CLCL

compensation topology is used to neutralize the effect of leakage inductance and to match the load impedance with source impedance for maximum power transfer. The most widely used compensation topologies for IPT systems consist of series or/and parallel-connected capacitors. This gives rise to six different compensation topologies, namely: (a) series-series (SS) (b) series-parallel (SP) (c) parallel-series (PS) (d) parallel-parallel (PP) (e) LCL, and (f) CLCL, as shown in Figure 1.3. The selection of proper compensation topology depends on the application type of the IPT system. As discussed earlier, careful mathematical modeling of the equivalent circuit can be done to analyze the relationship between parameters such as power, efficiency, load resistance, operating frequency, and coefficient of coupling [18].

Static inductive charging systems can replace the conventional plug-in chargers and solve the issues of electric shock and protection hazards. It follows the principle of operation of WPT to charge battery bank using three-stage power conversion, i.e., AC/DC, DC/HFAC, and HFAC/DC. Wireless communication networks are employed to exchange information from the other coil side to eliminate protection hazards and for a better battery management system. The amount of power transmitted and charging time depends on the coil structures, operating frequency, source voltage level, and coefficient of coupling between coils. A few manufacturers have built static inductive charging systems that vary in pricing due to variation in charging levels as per SAE standards. Static inductive charging systems can be installed in private or public spaces, and thus, a set of standards and guidelines must be followed to avoid human exposure to time-varying electric and magnetic fields. A carefully designed and constructed wireless charging system will have very few health and safety concerns. However, an external factor such as the sudden entrance of a metallic object in the space between the charging pads will lead to large eddy current losses in that object. This condition will create high temperature and fire between the coils; therefore, power transfer should be interrupted immediately on detection of a foreign object.

1.3.4.2 Design challenges

1.3.4.2.1 Charging pad design

To achieve maximum power transfer with low EMI, receiver and transmitter pads are made up of three components: coil, EM shielding material, and supportive and protective layers. Until now, a collection of various planar coils has been developed to enhance the performance and to reduce misalignment issues between both coils. Pads are mainly divided into two types: (a) polarized pads (PPs) and (b) non-polarized pads (NPPs).

NPPs produce vertical components of flux using single coil, whereas PPs use multiple coils to produce vertical as well as horizontal components of flux. Planar NPPs can take shapes like circular, square, rectangular, and polygonal. Planar circular NPP coils are widely used in IPT systems because there are no sharp edges and eddy current is kept minimum. Flux distribution can be controlled by the proper design of coil diameter. The magnetic flux distribution can be expanded by increasing the center diameter of the coil, which will help in tolerating misalignment of coils. However, circular coils can produce magnetic flux up to a limited height, which again depends on coil diameter. Planar square and rectangular coils are suitable for array structure due to perfectly aligned coil sides. But the presence of sharp corners increases eddy currents, and subsequently coil inductance is increased. Thus, coil structures with sharp edges and corners are not preferred for high-power IPT applications. Similar issues are there with polygonal coil structures, where maximum power is transmitted when coils are centrally aligned but provide a significant reduction in power transfer as misalignment increases. To counter the poor performance of NPPs during horizontal misalignment, PPs were developed by placing multiple coil structures in a variety of arrangements. Examples of PPs include solenoidal coil, bipolar coil, double D (DD), double D quadrature (DDQ), and quad D quadrature (QDQ), as shown in Figure 1.4 [20].

Magnetic flux will be distributed in space between the coils during wireless transfer of power and a large share of it will be wasted as the system will be loosely coupled. In IPT systems, large leakage magnetic fluxes are produced, which may also cross the maximum allowable limits as mentioned in relevant guidelines and may be harmful to the personnel engaged in wireless charging. So, it is necessary to limit the magnetic flux within coil structures to enhance the performance of system and to prevent health issues of nearby public. Mutual inductance can also be improved significantly by employing properly designed magnetic ferrite cores. Conventional ferrite shapes can be circular, square, and rectangular. U-core, EE-core, and pot

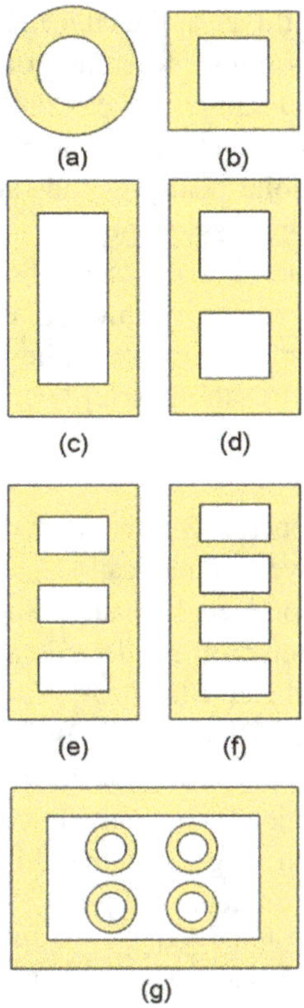

Figure 1.4 Top view of coil shapes (a) circular, (b) square, (c) rectangular, (d) DD, (e) bi-polar, (f) DDQ, and (g) QDQ [19]

ferrite shapes offer a high coefficient of coupling, but they are not applicable due to height constraints, because vehicles have limited ground clearance. Thus, to limit the magnetic flux within coil structures higher permeability materials are the first choice for RIPT application because of applicability and availability [19].

1.3.4.2.2 Misalignment issues

If the transmitter and receiver coils are not perfectly aligned with each other, then resonant operating frequency and the induced voltage drops also change. The misalignment or lateral displacement of the coils affects the power transfer capability, soft-switching condition, and operating power factor. Therefore, to avoid misalignment issues alternative coil structures can be developed.

By winding coils around a flat shape ferrite plate, a solenoidal coil is created to produce polarized magnetic fluxes on both sides of the coupler. Such polarized fluxes are greater than the fluxes produced by NPPs. DD PPs are created from two coils either square or rectangular in shape that generate magnetic flux only in one direction with minimal leakage fluxes at the edge. This pad is suitable for static and dynamic applications due to its higher tolerance of misalignment. The DDQ coil is the next generation of the DD pad which can produce a flux twice the height as compared to the circular pad. The DDQ pad is applicable in single- or three-phase power supply IPT systems. Bipolar charging pads are constructed from more than one similar size coils and require less copper for the construction as compared to the DDQ pad but show a reduction in coefficient of coupling with angular misalignment between the coils. QDQ pads utilize more than two circular and square coils to create charging pads and offer significantly higher coefficient of coupling and can transfer sufficient power during misalignment [21].

1.3.4.2.3 Communication link

A few control techniques cannot operate without a fast communication interface between both coil sides. It makes an effective communication interface very crucial for power flow control. Key parameter information includes battery SoC, power requirement, misalignment, ground clearance, and foreign object detection. Wireless communication interface uses technologies such as cellular, Bluetooth, Wi-Fi, Zigbee, and Radio Frequency Identification (RFID). Sometimes, Bluetooth or Wi-Fi is not well suited for wireless communication network when duplex data transfer is a key requirement. In dynamic IPT systems, synchronized operation with a moving vehicle at high speed requires low latency, increased communication range, and the ability to communicate with multiple vehicles simultaneously. The guidelines given by SAE show dedicated short-range communication (DSRC), Wi-Fi cellular, satellite, RFID, and World Interoperability for Microwave Access (WiMAX) to be suitable for wireless power transfer communication interface. However, a few studies concluded that cellular and DSRC communication are the most suitable alternatives. DSRC offers very low latency and maintains high rate of data with limited coverage, on the other hand, cellular communication also maintains high rate of data with large coverage but with higher latency as compared to DSRC. Another alternative would be 5G, as it offers fast data transfer across multiple subsystems with very low latency.

1.3.4.2.4 Control system

To control the power flow through the IPT system, three types of control techniques can be employed: (a) transmitter-side control (b) receiver-side control, and (c) transmitter and receiver both combined. When there exists only one receiving coil, the current and frequency on the transmitting side can be adjusted to regulate the load power on the receiving side. Transmitter-side control requires a fast exchange of information between transmitting and receiving sides, but this control strategy provides higher efficiencies at small load resistances. The use of fast communication link provides information about battery state of charge, battery voltage, and battery current, which is used by transmitter-side controller, ultimately helps in removal of complex electronics from receiver side and cost reduction of the system. If multiple coils are connected on the receiving side, then receiver-side control will be useful where an active rectifier for each receiving subsystem adjusts its switching to draw the required power and maintain constant supply current.

For HF inverter control, there exist two key techniques: (a) frequency control and (b) phase shift control. In phase shift control, the difference of phase angle between HF inverter legs varies at a constant switching frequency. This technique controls the fundamental voltage output, but soft switching gets compromised because of the frequency bifurcation phenomenon. It leads to switching losses and non-inductive conditions. Despite these limitations, phase shift control is cheaper as communication link is not required between both coils, but this control technique is only applicable for full-bridge inverter [22]. In frequency control, the switching frequency varies at a constant phase shift between inverter legs. A set of switching states must be predefined so that the control action can vary the switching frequency to track ZVS by operating under inductive conditions. This requires a communication link between both coils as the actual resonant frequency must be measured in real time, which makes the system complex and costly [23]. The advantages of transmitter-side control and receiver-side control can be combined with HF inverter control, to achieve the highest efficiency.

1.3.4.3 Recent developments

Detection of a foreign object is an important accessorial system used in EV wireless charging to detect any between the charging pads. If any conducting object is detected then, the charging system shuts down immediately and stops the transfer of power. Thus, preventing heating of the intermediate object, which can cause safety and fire hazards. Detection of the incoming vehicle is a distinct form of detection of conducting object and is applicable where automatic switching of the transmitting pad is required depending on the presence of the vehicle receiving coil [24, 25].

During the charging process, the mutual inductance of coils and load resistance depend on SoC of battery pack change. This causes the efficiency of the WPT system to deteriorate below the acceptable range. So, the maximum efficiency of the system needs to be tracked with varying operating conditions. The most common method is to use a DC–DC converter and an active rectifier to the wireless charging system. By adjusting the phase shift angle and the output voltage of the active rectifier, maximum efficiency at a given operating point can be achieved.

1.3.5 Dynamic inductive charging

1.3.5.1 Overview

Static inductive charging systems take a long time to charge the battery pack of EV and also the driving ranges are limited. Also, to charge the vehicle, it must be parked at a fixed place. Dynamic inductive charging can provide an attractive solution to this problem. A user can charge his/her EV continuously while driving along the road, which results in the reduction of the size and weight of the EV batteries, thereby eliminating range anxiety and increasing convenience for EV users. This technique uses a coil fixed on the track surface that can transmit power wirelessly to the receiving coil which is fixed beneath the EV chassis. Dynamic inductive charging systems can be implemented by using two types of on-road coils: (a) a single-coil track formed using a coil that is longer than the receiver coil beneath the EV and

(b) segmented-coil track formed using multiple coils of similar size to the receiver coil beneath the EV. The on-road subsystem includes an AC/DC converter, an HF inverter, primary compensation network, and a power rail. The vehicle subsystem includes a pickup coil, secondary compensation network, an AC/DC converter, and a converter for charging of battery pack. The on-road subsystem should be able to withstand harsh weather conditions and should be economical enough to be installed over long distances, whereas the on-board vehicle subsystem should be small in size and lightweight so that it may be incorporated into the EV [26]. The control system is crucial for dynamic inductive charging systems as it must sense incoming EV for appropriate switching of the segmented on-road coils. Therefore, dynamic inductive charging systems must have a cheap and simple control strategy, larger lateral tolerance, and lower construction cost for commercial use. Since this technique of inductive charging is under research and development phase, not many standards and regulations have been formed. However, organizations such as SAE, IEEE, and ISO are working towards the formation of standards for dynamic inductive charging.

1.3.5.2 Design challenges

1.3.5.2.1 Segmentation issues

Similar to static inductive charging, the receiver coil is mounted beneath the EVs. When the EVs pass over the on-road subsystem, it cuts the magnetic flux and utilizes the energy to charge the battery bank using power converters. Overall battery size and weight can be reduced by frequent charging of EVs. For dynamic inductive charging, on-road subsystems need to be fixed on specific locations and routes. The power supply for dynamic charging is divided into centralized and individual power supply schemes [27]. In the centralized power supply scheme, power is supplied simultaneously to all the multiple small charging pads fixed on the road. When compared with segmented scheme, the centralized scheme has greater losses, lower efficiency, and higher installation and maintenance costs. Also, to improve the efficiency of the system, selective switching can be employed based on the vehicle detection over transmitter coils [28]. Based on the switching schemes of segmented power rails, there are three modes: (a) centralized switching (b) distributed switching, and (c) X-segmented switching. Centralized switching uses only one utility supply and all segments are connected individually to that source, thus using large cumulative length of power cables. On the other hand, distributed switching also uses one source but the switching is distributed among adjacent segments. This scheme also uses long power cables to connect to the distributed switch boxes. A long single power cable is used in X-segmented switching scheme as an auto compensation switch box is used [29]. Hence, to reduce the cumulative length of power cable as well as to have reliable switching, X-segmented switching scheme should be employed.

1.3.5.2.2 Compensation topology

It is inefficient to use a single coil on-road structure, because the whole on-road subsystem is always active and the receiver coil is coupled to a very small percentage of the on-road structure [30]. So, by using segmented coil track, the desired coil segment can be selected to excite by detecting the location of vehicle, thus increasing system efficiency and reducing leakage flux from the uncoupled segments of the coil [31]. Operating efficiency also depends up on proper selection of compensation of leakage inductance in dynamic charging environment. Due to its inherent advantages, LCC compensation provides good controllability and high efficiency in dynamic operating conditions, thus an LCC tuned topology can be used for dynamic EV charging applications [32]. LCC compensation topology allows the power between different segments to be distributed automatically when each segment of on-road coil can be driven by one supply [33]. By carefully analyzing the relationship between various parameters such as track length, power utilization, and speed of the vehicle, maximum efficiency can be obtained by ensuring optimal design of the length of the segmented coil accompanied with properly designed compensation topology.

1.3.5.3 Recent developments

The on-line electric vehicle (OLEV) project was started by Korea Advanced Institute of Science and Technology (KAIST) since 2009. Since then many developments have been witnessed by researchers [34]. Many generations of OLEV have been developed and deployed since 2010, started with the first-generation (1G) concept of OLEV, the second-generation (2G) OLEV buses, the third-generation (3G) OLEV passenger cars, and improved third-generation (3^{+}G) OLEV trains were developed and tested at KAIST. The development of fourth-generation (4G) OLEV has shown significant improvement in performance and the fifth-generation (5G) OLEV is under development and testing.

1.3.6 Bidirectional power flow

Depending on the power flow direction, conductive as well as wireless chargers can be categorized as bidirectional and unidirectional [35]. A wireless EV charger with a unidirectional power flow uses an uncontrolled AC/DC converter and a DC/DC converter for unidirectional EV battery charging process. The unidirectional converters are less complex to control as compare to bidirectional converters [36]. They reduce the degradation of battery life and have a fewer grid integration issues as compared to their bidirectional counterpart. However, most of the grid ancillary services cannot be provided by unidirectional chargers as the power flow is in one direction. In contrast, bidirectional chargers can provide various grid ancillary services as the power flow is in both directions, i.e., either charging or discharging of EVs. But, the frequent reversal of power back and forth from utility can lead to degradation of battery life. Moreover, the process is more complex with the grid stability and metering issues, which involve buying and selling electricity from utilities.

Bidirectional wireless power transfer depends on the proper selection of compensation topologies and power converters. A dual active bridge topology when complemented with symmetrical compensation topologies can facilitate bidirectional power flow. Recent developments have introduced LCL and CLCL topologies that show better performances as compared to conventional compensation topologies. An LCL compensation network uses a parallel-connected capacitor and a filter inductor in series with the coils. By using LCL compensation, the switching losses can be reduced. A major advantage of using LCL compensation is that it produces a constant current, which can be supplied to multiple receiver coils. Bidirectional power transfer is enabled by using LCL compensation on both sides of the IPT system but requires a complex control for synchronous switching.

Another approach to achieve bidirectional wireless power transfer is to use CLCL compensated transmitter and receiver coils. However, with increasing number of energy storage elements, system order increases and it becomes difficult to develop a precise mathematical model of the system. By selecting appropriate state variables, a dynamical equivalent circuit can be developed to study its transient and steady-state behavior [37]. These studies will ultimately help in controller design for bidirectional WPT system. Furthermore, to optimize the parameters of controller, any optimization technique can be used [38].

In the era of DERs, EVs can act as both energy source and energy storage. It is only possible with the development of bidirectional converters to realize G2V and V2G concepts. With the use of bidirectional converters, EVs also can act as energy storage and energy source, which can consequently reduce the load on the grid. During peak demand, EV batteries are discharged to supply power back to the grid otherwise EVs can be charged [39].

1.3.7 International standards and regulations

Since the inception of WPT for charging purposes, many international standards such as Qi and Rezence were formed for low-power consumer electronic devices. The power level required for EV charging is much higher than low-power applications, and therefore, different standards need to be formed. Standardization confirms a minimum quality of a product, confirms safe operating modes, and allows comparison between similar products from multiple manufacturers. Commercialization of wireless EV charging was hindered due to the absence of any international standard until 2016. SAE was the first organization to attempt standardization of WPT for EVs with the introduction of SAE J2954, which has been followed by numerous other standards, regulations, and guidelines, which are mentioned in Table 1.5.

1.3.7.1 General requirements

Recently revised SAE J2954 standard for EV charging earmarks a small band between 81.39 and 90 kHz for wireless charging of EVs with a center frequency of 85 kHz. SAE J2954 addresses only unidirectional power flow with surface mounted static inductive charging, from grid to vehicle (G2V) with allowable ground clearance between 100 and 250 mm. This standard does not address bidirectional dynamic inductive charging. In future, those EVs that can be charged wirelessly under SAE J2954 should also be able to be charged conductively by SAE J1772 plug-in chargers. This standard addresses home (private) charging as well as public wireless charging. Currently, the standard is limited to the three levels of power as 3.3, 7.7, and 11.1 kVA.

Table 1.5 Summary of a few international standards and guidelines for wireless charging of EVs

Organization	Standards	Detail
IEC [40]	IEC61851	General requirements of EV charging
	IEC62704	Approach for the implementation of finite difference time domain (FDTD) technique for SAR simulations
	IEC61980	Wireless charging of EVs
	IEC62196	Plugs, sockets, and connectors for EV charging
SAE [41]	J1773	Contactless charging of EV
	J2847	Communication between EV, grid, and off-board charger
	J2836	Use cases for communication
	J2954	WPT for EVs
Institute of Electrical and Electronic Engineers (IEEE) [42]	P2100.1	WPT and charging system standardization
	P1547	Grid connection of DERs
International Organization for Standardization (ISO) [43]	ISO14117	EMC test protocols for implantable medical devices
	ISO15118	Vehicle to grid communication interface
	ISO19363	Magnetic field safety requirements
International Commission on Non-Ionizing Radiation Protection (ICNIRP) [44]	ICNIRP1998	Exposure limit for frequency up to 300 GHz
	ICNIRP2010	Exposure limit for frequency up to 100 kHz

Similarly, IEC announced IEC 61980 standard, which consists of three parts, covering general requirements of EV WPT systems, specific communication requirements, and specific requirements for the magnetic field WPT systems. IEC 61980-1 currently addresses unidirectional charging of EV battery pack using static inductive charging with standard supply voltages per IEC 60038 up to 1 000 AC and up to 1 500 V DC. It covers both conductive and inductive modes of charging; however, safety aspects for maintenance or EMC requirements of WPT system are not discussed.

1.3.7.2 Health and safety

International Commission on Non-Ionizing Radiation Protection (ICNIRP) has two major guidelines for limiting RF exposure:

ICNIRP 1998: limiting exposure to dynamic electric, magnetic, and EM fields up to 300 GHz.

ICNIRP 2010: limiting exposure to dynamic electric and magnetic fields in the range of 1 Hz–100 kHz.

ICNIRP has issued a guideline under ICNIRP 1998 to mention basic restriction limits for Central Nervous System (CNS) under various frequency ranges. Restriction limits are expressed as current density (A/m^2) to prevent effects on the nervous system for frequency range of 1 Hz–10 MHz. For frequency range between 100 kHz and 10 GHz, restriction limits are expressed as specific absorption rate (SAR) to prevent localized and whole-body tissue heating. Between 10 MHz and 300 GHz restriction limits are expressed as a power density to prevent excessive heating in body tissue and surface. IEEE 1528.x and IEEE C95.3.1 have standardized the computational methods to assess human exposure against specified ICNIRP restriction limits. Recommended steps that should be followed to assess WPT system compliance with ICNIRP guidelines:

1. Model WPT system with coil geometry, operating frequency, and misalignment.
2. Characterize vehicle form factor and highest air gap between ground and vehicle assembly.
3. Perform uncertainty analysis, and calculate the combined extended uncertainty of measurement and simulation.
4. Assess human exposure with driver/passenger and bystander use cases.

As per ICNIRP 1998, restriction limits for WPT system at 85 kHz operating frequency is 170 A/m^2 averaged over 1 cm^2 area or 11.5 V/m averaged over 8 mm^3 volume. Also, interference limit for implanted medical devices (IMD) while using WPT system at 85 kHz operating frequency is 15 μT for driver/passenger and bystander cases. SAR limits are not framed for general public exposure while using WPT system at 85 kHz, but the limits are well defined for frequency above 100 kHz.

ICNIRP has also issued a guideline under ICNIRP 2010 for limiting exposure to dynamic electric and magnetic fields in the range of 1 Hz–100 kHz. The ICNIRP 2010 guidelines mention the limits in electric field E (V/m) for both the peripheral nervous system (PNS) and the central nervous system (CNS). It covers both direct effects resulting from direct interactions with the human body as well as indirect effects resulting from interaction with a conducting object placed in the field of effect. Effects of low-frequency EM fields exposure have been reviewed by the World Health Organization (WHO), ICNIRP, and the International Agency for Research on Cancer (IARC), who provide the scientific basis for upcoming guidelines. At low-frequency electric fields, the body acts as a good conductor, and the flux lines outside the body are nearly normal to the body surface. Oscillating charges are induced on the human body surfaces that produce current inside the body. At low-frequency magnetic fields, the permeability of body tissue is the same as that of air, which induces electric fields and associated current in the tissues. This risk is increased when dynamic WPT systems are used. Under the guidelines of ICNIRP 2010, reference levels for general public exposure to time-varying electric and magnetic fields for a frequency range of 3 kHz–10 MHz are given as 83 V/m for electric field strength, 21 A/m for magnetic field strength, and 27 μT for magnetic flux density. These reference levels may vary for occupational exposure. High exposure to dynamic magnetic and electric fields may prove to be carcinogenic and may be fatal for patients with medical implants. Therefore, it is necessary to strictly follow ICNIRP guidelines to prevent any interference with electronic devices, fires, and explosions resulting from ignition of flammable materials by sparks caused by induced fields.

It is understood that SAR restriction limits must be followed for frequency range between 100 kHz and 10 GHz. IEEE/IEC 62704-1 is an international standard for determining the peak SAR in the human body from Wireless Communications Devices in the range of 30 MHz–6 GHz. The scope of IEEE/IEC 62704-1 standard defines the approach to determine the peak spatial-average SAR in the human body by using the FDTD technique. It provides instructions to confirm the numerical model of the device under test and to evaluate its uncertainty when used in SAR simulations. IEEE/IEC 62704-1 does not mention any restriction limits of SAR as addressed by ICNIRP guidelines but only covers the methodology to assess SAR values under any given condition.

ISO 14117:2019 specifies test methodologies to assess the EMC of implantable medical devices (IMDs). Patients with IMDs are exposed to different sources of EM emitters on daily basis. Therefore, standard test methods help manufacturers to assess the EMC performance of a product and show that the product achieves the right level of EMC that patients might encounter. Interference of EM waves with IMDs depends on frequency, signal power, proximity to the patient, and exposure duration. WPT systems are also a source of EM waves and are categorized under low-frequency bands (<385 MHz), which risks EM interference between emitter and IMDs.

1.3.7.3 Communication links

IEC TS 61980-2 addresses the aspects of communication between EV and WPT systems during charging mode when connected to a standard supply as per IEC 60038 up to 1 000 AC and up to 1 500 V DC. This also applies when wireless power transfer equipment is supplied from buffer battery packs at standard supply voltages. IEC TS 61980-2 covers the required standards for operational and functional characteristics of the WPT communication subsystem. However, the required communication standards for dynamic inductive charging systems, two- and three-wheel vehicles, and bidirectional power flow are under consideration.

SAE J2847/6 demonstrates minimum requirements for communication between an EV and a WPT charging system. This document exploits the work of the SAE J2954 Alignment and Controls Sub-Team in the WPT and Alignment Task Force by extending a JSON-based message protocol originally developed to bench test WPT interoperability between manufacturer mismatched GA and VA systems. SAE J2847/6 extends the work by adding messages sufficient to indicate that proper coil alignment has been achieved, initialize the subsystems for wireless charging, ramp-up to full power, perform active WPT, and terminate the WPT session. It also provides guidance for the implementation of the JSON protocol as a reference for verification and validation of test plans by enabling the GA and VA to transmit a basic set of messages by using standard Wi-Fi IEEE 802.11n communications hardware.

SAE J2836/6 addresses the requirements for communication between vehicle assembly and ground assembly in support of detection of the charging pad, the charging process, and monitoring of the charging process as specified in SAE J2954. It provides standards for digital and analog communication over the wireless interface between EV and grid that are expected to be identical with V2G communications specified in SAE J2836/1.

A fast and reliable communication system is necessary for energy resource management to ensure recharge of EVs in the most economic and efficient way. It is also essential to develop reliable and convenient billing systems in order to cover the resulting micro-payments. Developments in communication network may provide contribution in the stabilization of the electrical grid as well as support supplementary services required to charge and run EVs efficiently and economically.

1.4 Conclusion

In this chapter, we discussed about charging methods of EVs that may be either conductive or wireless charging. Currently, most of the EV and charging equipment manufacturers are utilizing conductive charging method. This can be further categorized on the basis of load requirement and charger location. Accessories used for charging EVs follow different standards in different countries. However, a few international organizations are working towards a universal standard to make manufacturers follow the same standard globally, which will boost EV commercialization. An alternative for conductive charging is wireless charging of EVs, which has the potential to eliminate the shortcomings of conductive charging and can also provide additional advantages, which are thoroughly discussed. Recent developments in the field of static and dynamic wireless charging are presented. Standards and regulations for conductive and wireless charging of EVs are also discussed in detail.

References

[1] Habib S. 'Contemporary trends in power electronics converters for charging solutions of electric vehicles'. *CSEE Journal of Power and Energy Systems*. 2020, vol. 6(4), pp. 911–29.

[2] Khaligh A., D'Antonio M. 'Global trends in high-power on-board chargers for electric vehicles'. *IEEE Transactions on Vehicular Technology*. 2019, vol. 68(4), pp. 3306–24.

[3] Das H.S., Rahman M.M., Li S., Tan C.W. 'Electric vehicles Standards, charging infrastructure and impact on grid'. *Renewable and Sustainable Energy Reviews*. 2020, vol. 120, p. 109618.

[4] Singh B., Singh B.N., Chandra A., Al-Haddad K., Pandey A., Kothari D.P. 'A review of three-phase improved power quality AC-DC converters'. *IEEE Transactions on Industrial Electronics*. 2004, vol. 51(3), pp. 641–60.

[5] Tie S.F., Tan C.W. 'A review of energy sources and energy management system in electric vehicles'. *Renewable and Sustainable Energy Reviews*. 2013, vol. 20(4), pp. 82–102.

[6] Kumar S.L.V.S., Reddy N.R.S., Kumar M.V. 'Voltage mode control of integrated boost series parallel fly-back converter for energy storage applications'. *Energy Procedia*. 2017, vol. 117(12), pp. 62–70.

[7] Taghizadeh S., Hossain M.J., Lu J., Water W. 'A unified multi-functional onboard EV charger for power-quality control in household networks'. *Applied Energy*. 2018, vol. 215, pp. 186–201.

[8] Domínguez-Navarro J.A., Dufo-López R., Yusta-Loyo J.M., Artal-Sevil J.S., Bernal-Agustín J.L. 'Design of an electric vehicle fast-charging station with integration of renewable energy and storage systems'. *International Journal of Electrical Power & Energy Systems*. 2019, vol. 105(3), pp. 46–58.

[9] Hadley S.W., ATsvetkova A. 'Potential impacts of plug-in hybrid electric vehicles on regional power generation'. *The Electricity Journal*. 2009, vol. 22(10), pp. 56–68.

[10] Ul-Haq A., Cecati C., Strunz K., Abbasi E. 'Impact of electric vehicle charging on voltage unbalance in an urban distribution network'. *Intelligent Industrial Systems*. 2015, vol. 1, pp. 51–60.

[11] Pieltain Fernandez L., Gomez San Roman T., Cossent R., Mateo Domingo C., Frias P. 'Assessment of the impact of plug-in electric vehicles on distribution networks'. *IEEE Transactions on Power Systems*. 2011, vol. 26(1), pp. 206–13.

[12] Fernandes F., Morais H., Vale Z., Ramos C. 'Dynamic load management in a smart home to participate in demand response events'. *Energy and Buildings*. 2014, vol. 82(6), pp. 592–606.

[13] Wu D., Chau K.T., Liu C., Gao S., Li F. 'Transient stability analysis of SMES for smart grid with vehicle-to-grid operation'. *IEEE Transactions on Applied Superconductivity*. 2012, vol. 22(3), p. 5701105.

[14] Yilmaz M., Krein P.T. 'Review of battery charger topologies, charging power levels, and infrastructure for plug-in electric and hybrid vehicles'. *IEEE Transactions on Power Electronics*. 2013, vol. 28(5), pp. 2151–69.

[15] SAE International. *Digital Communications for Plug-in Electric Vehicles J2931/1_201412 [online]*. Available from https://www.sae.org/standards/content/j2931/1_201412/ [Accessed 6 Dec 2020].

[16] Basso T.S., DeBlasio R. 'IEEE 1547 series of standards: interconnection issues'. *IEEE Transactions on Power Electronics*. 2004, vol. 19(5), pp. 1159–62.

[17] Bi Z., Kan T., Mi C.C., Zhang Y., Zhao Z., Keoleian G.A. 'A review of wireless power transfer for electric vehicles: prospects to enhance sustainable mobility'. *Applied Energy*. 2016, vol. 179(4), pp. 413–25.

[18] Ahmad A., Alam M.S., Chabaan R. 'A comprehensive review of wireless charging technologies for electric vehicles'. *IEEE Transactions on Transportation Electrification*. 2018, vol. 4(1), pp. 38–63.

[19] Mohammad M., Choi S., Islam Z., Kwak S., Baek J. 'Core design and optimization for better misalignment tolerance and higher range of wireless charging of PHEV'. *IEEE Transactions on Transportation Electrification*. 2017, vol. 3(2), pp. 445–53.

[20] Xiang L., Li X., Tian J., Tian Y. 'A crossed DD geometry and its double-coil excitation method for electric vehicle dynamic wireless charging systems'. *IEEE Access*. 2018, vol. 6, pp. 45120–8.

[21] Kalwar K.A., Aamir M., Mekhilef S. 'A design method for developing a high misalignment tolerant wireless charging system for electric vehicles'. *Measurement*. 2018, vol. 118(3), pp. 237–45.

[22] Evangelou S.A., Rehman-Shaikh M.A. 'Hybrid electric vehicle fuel minimization by DC-DC converter dual-phase-shift control'. *Control Engineering Practice*. 2017, vol. 64(2), pp. 44–60.

[23] Nguyen B.X., Vilathgamuwa D.M., Foo G.H.B., *et al.* An efficiency optimization scheme for bidirectional inductive power transfer systems'. *IEEE Transactions on Power Electronics*. 2015, vol. 30(11), pp. 6310–19.

[24] Jeong S.Y., Kwak H.G., Jang G.C., Choi S.Y., Rim C.T. 'Dual-purpose nonoverlapping coil sets as metal object and vehicle position detections for wireless stationary EV chargers'. *IEEE Transactions on Power Electronics*. 2018, vol. 33(9), pp. 7387–97.

[25] Tian Y., Li Z., Lin Y., *et al.* Metal object detection for electric vehicle inductive power transfer systems based on hyperspectral imaging'. *Measurement*. 2021, vol. 168(83), p. 108493.

[26] Shin J., Shin S., Kim Y., *et al.* Design and implementation of shaped magnetic-resonance-based wireless power transfer system for roadway-powered moving electric vehicles'. *IEEE Transactions on Industrial Electronics*. 2014, vol. 61(3), pp. 1179–92.

[27] Jayathurathnage P.K.S., Alphones A., Vilathgamuwa D.M., Ong A. 'Optimum transmitter current distribution for dynamic wireless power transfer with segmented array'. *IEEE Transactions on Microwave Theory and Techniques*. 2018, vol. 66(1), pp. 346–56.

[28] Li X., Hu J., Wang H., Dai X., Sun Y. 'A new coupling structure and position detection method for segmented control dynamic wireless power transfer systems'. *IEEE Transactions on Power Electronics*. 2020, vol. 35(7), pp. 6741–5.

[29] Zhu Q., Guo Y., Li S., Wang L., Liao C. 'Segmental switching strategy with zero output pulsation for dynamic EV wireless charging system'. *IET Power Electronics*. 2019, vol. 12(6), pp. 1563–70.

[30] Guo Y., Wang L., Zhu Q., Liao C., Li F. 'Switch-on modeling and analysis of dynamic wireless charging system used for electric vehicles'. *IEEE Transactions on Industrial Electronics*. 2016, vol. 63(10), pp. 6568–79.

[31] Sampath J.P.K., Vilathgamuwa D.M., Alphones A. 'Efficiency enhancement for dynamic wireless power transfer system with segmented transmitter array'. *IEEE Transactions on Transportation Electrification*. 2016, vol. 2(1), pp. 76–85.

[32] Li W., Zhao H., Deng J., Li S., Mi C.C. 'Comparison study on SS and double-sided LCC compensation topologies for EV/PHEV wireless chargers'. *IEEE Transactions on Vehicular Technology*. 2016, vol. 65(6), pp. 4429–39.

[33] Feng H., Cai T., Duan S., Zhao J., Zhang X., Chen C. 'An LCC-compensated resonant converter optimized for robust reaction to large coupling variation in dynamic wireless power transfer'. *IEEE Transactions on Industrial Electronics*. 2016, vol. 63(10), pp. 6591–601.

[34] Jang Y.J., Suh E.S., Kim J.W. 'System architecture and mathematical models of electric transit bus system utilizing wireless power transfer technology'. *IEEE Systems Journal*. 2016, vol. 10(2), pp. 495–506.

[35] Lee J.-Y., Han B.-M. 'A bidirectional wireless power transfer EV charger using self-resonant PWM'. *IEEE Transactions on Power Electronics*. 2015, vol. 30(4), pp. 1784–7.

[36] Sortomme E., El-Sharkawi M.A. 'Optimal charging strategies for unidirectional vehicle-to-grid'. *IEEE Transactions on Smart Grid*. 2011, vol. 2(1), pp. 131–8.

[37] Thrimawithana D.J., Madawala U.K. 'A generalized steady-state model for bidirectional IPT systems'. *IEEE Transactions on Power Electronics*. 2013, vol. 28(10), pp. 4681–9.

[38] Neath M.J., Swain A.K., Madawala U.K., Thrimawithana D.J. 'An optimal PID controller for a bidirectional inductive power transfer system using multiobjective genetic algorithm'. *IEEE Transactions on Power Electronics*. 2014, vol. 29(3), pp. 1523–31.

[39] Wang X., Liang Q. 'Energy management strategy for plug-in hybrid electric vehicles via bidirectional vehicle-to-grid'. *IEEE Systems Journal*. 2017, vol. 11(3), pp. 1789–98.

[40] IEC. Available from https://standards.iteh.ai/catalog/standards/iec [Accessed 5 Dec 2020].

[41] SAE. Available from https://www.sae.org/standards/ [Accessed 5 Dec 2020].

[42] IEEE. Available from https://www.ieee.org/standards/index.html [Accessed 5 Dec 2020].

[43] ISO. Available from https://www.iso.org/standards.html [Accessed 5 Dec 2020].

[44] ICNIRP. Available from https://www.icnirp.org/en/activities/news/news-article/rf-guidelines-2020-published.html [Accessed 5 Dec 2020].

Chapter 2

Grid impact of static and dynamic inductive charging and its mitigation through effective management

Ioannis Karakitsios[1], Evangelos Karfopoulos[1], and Nikos Hatziargyriou[1]

2.1 Introduction

Increased needs for the decarbonization of transport have sparked increased interest in electric vehicles (EVs) in the past years. Inductive charging, which allows the wireless transfer of energy between the EV and the charging station, appears to be a very promising charging method, considerably simplifying the charging process and eliminating any risks associated with the use of cables. Apart from the ability to transfer energy in a contactless way while the EV is parked over the station (i.e. static inductive charging), energy can also be wirelessly transferred while the EV moves on the road (i.e. dynamic inductive charging).

Fast inductive charging technologies allow the wireless exchange of power greater than 20 kW between the electrical grid and an EV. Such a demand can introduce considerable modifications to a distribution network's load profile, while affecting its planning and operation. It is, therefore, necessary to evaluate the grid impact of fast inductive chargers while also identifying the maximum allowable number of chargers that can be installed within the grid, in order not to violate the network's technical constraints.

In this respect, a methodology for evaluating the impact of fast static and dynamic inductive charging on the distribution grid is analyzed in this chapter. This methodology is implemented in a realistic model of an urban medium voltage (MV) distribution network. Additionally, the ability of an urban distribution network to host inductive charging technologies is determined considering the adequacy of the network capacity (i.e. line thermal limits, transformers' nominal power).

[1]National Technical University of Athens, Greece

Moreover, the grid impact of fast inductive charging will be evaluated considering a near-future scenario where a load increment is noticed in the grid demand, while also taking into account different renewable energy source (RES) penetration scenarios. More specifically, scenarios with varying installed photovoltaics (PV) capacity will be examined, in order to observe whether the PV production can increase the number of inductive charging stations that can be installed in the grid.

It shall also be noted that in case of a mass inductive charging infrastructure deployment an energy management system (EMS) offering network support services to the system operators can be proven quite significant. In this respect, the ability of such an EMS to release the network from the high inductive charging consumption, while effectively managing grid violations under variable scenarios, will be evaluated.

2.2 Tool for estimating the demand for fast inductive charging stations

In order to reduce the amount of time an EV shall remain in a charging station in order to charge its battery, fast inductive charging stations allow significant amounts of power to be transferred from the electrical grid. Moreover, the ability of drivers to charge their EV wirelessly, especially in the case of dynamic inductive charging, indicates considerable differences among the charging patterns on inductive or conductive charging stations. In this respect, existing tools for the estimation of the conductive charging demand of EVs are not sufficient to evaluate the demand for inductive charging stations. Thus, it is important to define a demand estimation tool for fast static and dynamic inductive charging stations, taking into account the energy requirements of drivers willing to charge their EV in an inductive charging station, as well as the high energy amount of energy transferred between the grid and the vehicle in such applications.

2.2.1 Estimation tool for static inductive charging

Only a few actual applications of fast static inductive charging are noted today, while most of them mainly concern experimental or research-related implementations. However, the operational behavior of fast static inductive charging stations is expected to be similar to the relevant one of fast conductive charging stations [1]. In this respect, data for conductive charging stations can be evaluated in order to observe when an EV driver is expected to require fast-charging services during the day, as well as the duration of the respective charging activity. The relevant data have been obtained by the work performed in the European research project FastInCharge, as well as reference [1] by the authors. In particular, the hour of the day EV users is expected to use fast-charging stations, as well as the amount of time an EV is expected to remain on a station is depicted in Figures 2.1(a) and 2.1(b), respectively. Figure 2.1 also indicates the relevant distributions that have been defined according to Gaussian distributions, as further explained in [1].

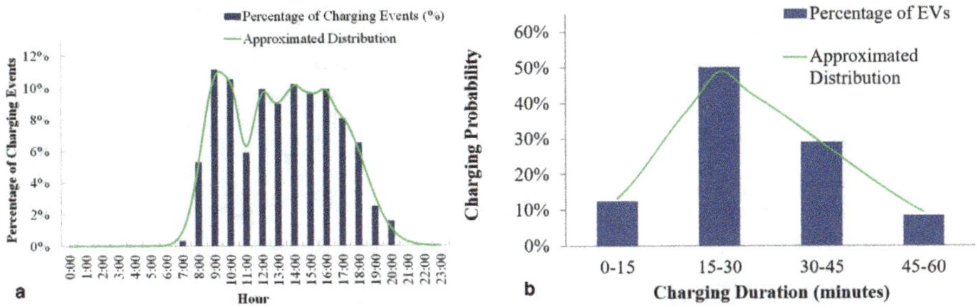

Figure 2.1   Charging probability indicating (a) the hour of the day EV users are expected to use fast-charging stations and (b) the amount of time an EV is expected to remain on a station [1]

In order to estimate the total EV demand for each day, the estimation tool that has been developed in [1] is taken into account. More specifically, this estimation tool considers the home charging demand, alongside the static inductive charging demand in order to achieve a more comprehensive analysis. The process for defining the home charging demand, as well as the demand for static inductive charging, is presented in Figures 2.2(a) and (b), respectively. The total EV demand comprises the energy requirements for home charging and fast static inductive charging.

Concerning home charging (Figure 2.2(a)), the estimation tool that has been taken into account is similar to the one discussed in [2]. In particular, the EVs are considered to charge at either Level-1 (3.7 kW) or Level-2 (11 kW). The EV fleet comprises two classes: L7e and M1, according to Directives 2002/24/EC[a] and 2007/46/EC.[b] The drivers are considered to charge their EV as soon as they return home after the last trip of the day. In order to calculate the amount of time required for charging, the battery consumption of each EV is taken into account in addition to the distance traveled by each EV during the day. Both the arrival time and the travel distance are usually randomly defined [1, 2]; thus, Monte Carlo is applied in order to estimate the daily charging profile, taking into account the stochastic nature of the parameters in the analysis.

In order to estimate the daily energy requirements in the case of static inductive charging stations, the process discussed in [1] and depicted in Figure 2.2(b) is considered. More specifically, the time within the day that an EV is expected to enter a static inductive charging station is randomly defined according to the distribution depicted in Figure 2.1(a). Similarly, the amount of time the EV will remain on the stations is also randomly defined according to the distribution in Figure 2.1(b). Following this procedure for all the EVs that are considered to charge in static inductive charging stations, the daily static inductive charging requirements can be defined.

[a]http://eur-lex.europa.eu/legal-content/EN/TXT/PDF/?uri=CELEX:32002L0024&from=EN
[b]http://eur-lex.europa.eu/legal-content/EN/TXT/PDF/?uri=CELEX:32007L0046&from=EN

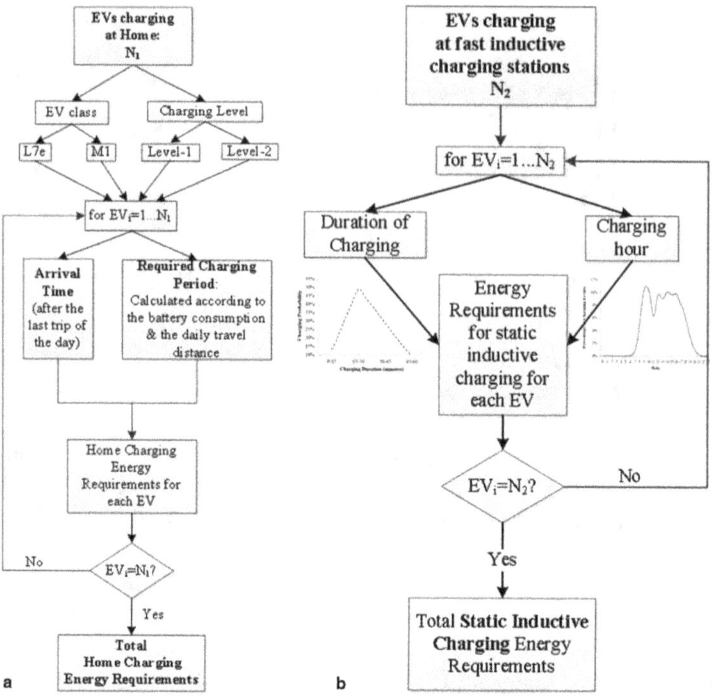

Figure 2.2 Estimation tool for (a) home charging energy requirement and (b) energy requirements of fast static inductive charging [1]

In order to estimate the number of charging stations required to fulfill the requirements of the relevant EV users, a process similar to the one depicted in Figure 2.2(b) can be considered [1]. More specifically, taking into account the distributions for the hour a vehicle is expected to enter a station, as well as the amount of time it will remain on the station, the number of vehicles charging each hour of the day can be defined. For instance, when considering that 300 EVs daily charge at static inductive charging stations, taking into account the distributions in Figure 2.2, the number of EVs simultaneously charging at static stations is depicted in Figure 2.3. In this case, it is evident in Figure 2.3 that a maximum number of 19 EVs are simultaneously charging at static stations, requiring a maximum of 19 stations in order to cover the relevant demand.

2.2.2 *Estimation tool for dynamic inductive charging*

In the case of dynamic inductive charging, the exact definition of the daily energy inductive charging demand requires many stochastic parameters. In particular, the trips that each EV user performs each day, the emergency charging needs, the traffic on every road, as well as many similar parameters are necessary to be defined. In addition, it shall be noted that more than one type of dynamic inductive charging stations may be installed: stations near a traffic light indicating small EV speeds,

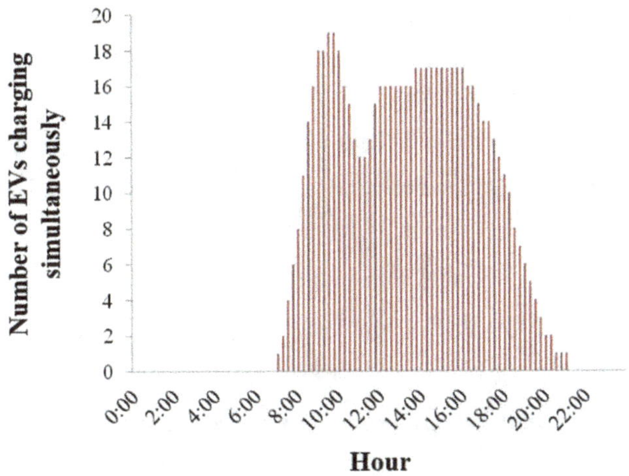

Figure 2.3 Number of vehicles charging at static stations each hour of the day, when considering that 300 EVs daily use inductive charging infrastructures

stations in highways allowing charging at very high EV speeds, etc. Each one of those stations will be linked with specific parameters, like the speed of the vehicle or the intervals between each charging event. It is evident that such a detailed analysis, requiring all the above-mentioned parameters, would be very difficult to be performed. So, a more simplistic approach, as described in [1], was considered in order to determine the maximum number of dynamic inductive charging stations that can be installed in an examined distribution grid, without noticing any grid-related issues due to the operation of the stations.

More specifically, the probability of a dynamic station operating to charge a passing EV ($P_{dynamic}$) is defined taking into account the traffic on the roads ($P_{traffic}$) [1]. However, EV users are not expected to always use dynamic chargers while moving on the roads. Instead, they will use the dynamic stations if a case for emergency charging arises: the battery's SoC is getting close to low levels, therefore the driver will seek a charging opportunity. Hence, the probability of charging during an emergency ($P_{charging}$) is defined according to the distribution depicted in Figure 2.1(a). The relevant Gaussian distribution is provided in [1].

The probability for the operation of a dynamic inductive charging station at each hour of the day 'h' can be defined as [1]:

$$P_{dynamic}(h) = P_{charging}(h) * P_{traffic}(h) \tag{2.1}$$

For instance, using real data from the Department for Transport in Great Britain[c] ($P_{traffic}$), as well as the emergency charging needs of the driver ($P_{charging}$), the probability

[c]https://www.gov.uk/government/collections/road-traffic-statistics

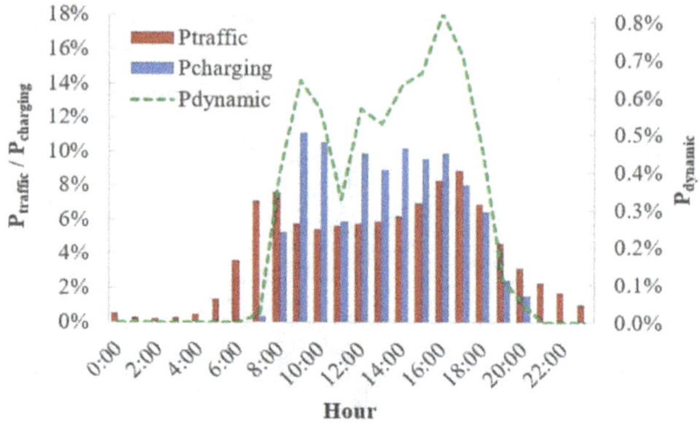

Figure 2.4 Traffic on the roads, probability for emergency charging and probability of a dynamic inductive charging station being occupied by a vehicle [1]

for a dynamic inductive charging station to be occupied with an EV ($P_{dynamic}$) is depicted in Figure 2.4.

When considering that a maximum number of stations, S_{max}, operate in a distribution network, the number of stations, S_h, operating each hour, h, of the day can be calculated as [1]:

$$S_h = \frac{P_{dynamic}(h)}{P_{dynamic}(h_{max})} * S_{max} \tag{2.2}$$

where h_{max} is the hour of the day the maximum charging probability (according to Figure 2.4) is noted.

In order to estimate the maximum number of stations that can operate without any grid violation, the process described in [1] and depicted in Figure 2.5 can be applied. In particular, at each iteration of the process, the number of stations that are considered to operate in the network is increased by one. The number of stations operating each hour of the day, as well as the relevant dynamic inductive charging demand can be defined according to (2.2).

The total EV charging demand comprises the dynamic inductive charging demand, the home charging demand and the static inductive charging demand. The EV charging energy requirements are, then, added to the grid demand in order to calculate the total network's demand.

A power flow analysis is performed in the next step in order to evaluate:

- The loading of lines and transformers (each line and transformer shall be loaded less than the relevant maximum thermal limit)
- The voltage values at the network's nodes: The voltage values shall not be greater or less than 5% of their nominal value
- The voltage deviation at the network nodes, as defined in (2.3), shall be below 3%.

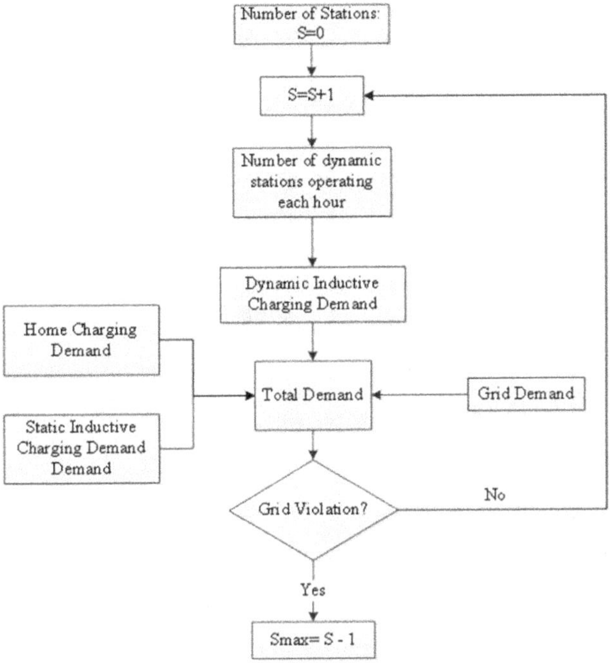

Figure 2.5 *Process defining the maximum number of dynamic inductive charging stations that can safely operate in a distribution network, according to [1]*

$$V_{dev} = 100 * \frac{V_{max} - V_{min}}{2 * V_{nom}} \% \tag{2.3}$$

The number of chargers is not increased further if any of the above constraints is violated, as depicted in Figure 2.5.

2.3 Impact of inductive charging on the distribution grid

In order to observe the impact of inductive charging on the grid, a single radial distribution line of the urban MV distribution grid operating in an area of North Greece has been considered. The studied distribution line comprises 35 MV/LV substations:

- Three substations of 160 kVA installed capacity
- One substation of 250 kVA
- Three substations of 400 kVA
- One substation of 500 kVA
- Twenty-two substations of 630 kVA
- One substation of 1 000 kVA and
- Four substations of 1 260 kVA.

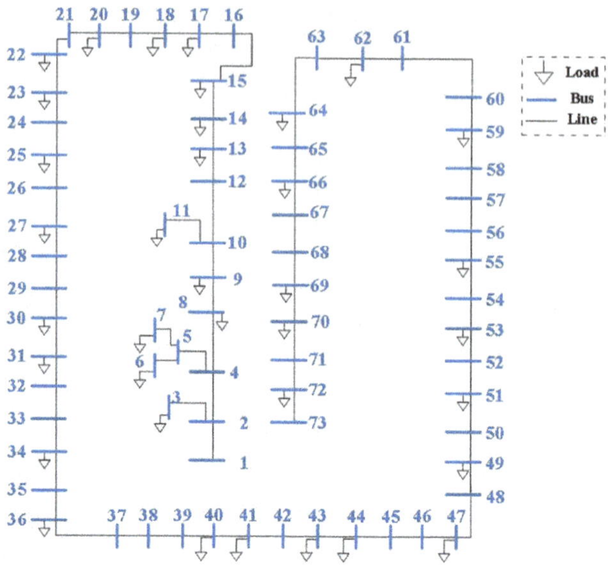

Figure 2.6 Single-phase simulation model of the examined line

The installed capacity of the central HV/MV substation is equal to 25 MVA. The single-phase simulation model of the distribution line is presented in Figure 2.6.

For the reference case represented as Scenario I in Table 2.1, no EVs are considered to charge in the examined network. In Scenario II, 1 000 EVs are considered to daily rely on home charging facilities in order to cover their energy needs. Two EV classes are taken into account in this case: type M1 commercial vehicles, as well as smaller L7e vehicles, with the percentages indicated in Table 2.2. The consumption for each type of vehicle is also depicted in Table 2.2.

The charging duration in the case of home charging is defined according to the battery consumption of the vehicle, which was calculated according to the distance traveled each day. In order to evaluate the daily energy needs of the EVs in

Table 2.1 Description of the scenarios under study

Scenarios	**Sub-case**	**Home charging (no. of users)**	**Fast inductive charging (no. of users)**	**Dynamic inductive charging**
Scenario I		0	0	–
Scenario II		1 000	0	–
Scenario III	A	1 000	100	–
	B	1 000	300	–
Scenario IV	A	1 000	100	✓
	B	1 000	300	✓

Table 2.2 Parameters for the evaluation of the home charging needs

	Category			
Parameter	**L7e**		**M1**	
Percentage of EVs	20		80	
Consumption (kWh/km)	0.13		0.16	
Charging probability for each level	*Level-1* 90%	*Level-2* 10%	*Level-1* 85%	*Level-2* 15%
Traveled distance (km)	*Mean value* 34.2		*Standard deviation* 5	
Arrival time	*The probability is defined according to the distribution in* Figure 2.7			

Scenario II, real data concerning the behavior of drivers in the city of Douai in France [3] is taken into account. According to this data, the daily travel distance was regarded as a stochastic parameter defined by a normal distribution with a mean value of 34.2 km and a standard deviation of 5 km. The EV drivers were considered to plug in their EV as soon as they return home after the last trip of the day. Taking into account real data for the city of Douai in France [3], the arrival time for the EVs is depicted in Figure 2.7. It is evident that most users are expected to be at home at 18:00 while a large number of drivers are also expected to return home between 19:00 and 20:00. Figure 2.7 also depicts the relevant approximated Gaussian distribution for the arrival time of the EVs.

Static Inductive Charging is considered in *Scenario III*. The output power of a static inductive charging station is considered equal to 30 kW, taking into account the inductive charging stations analyzed in [5]. The average efficiency for static and dynamic inductive charging stations is considered equal to 90% and 85%, respectively.

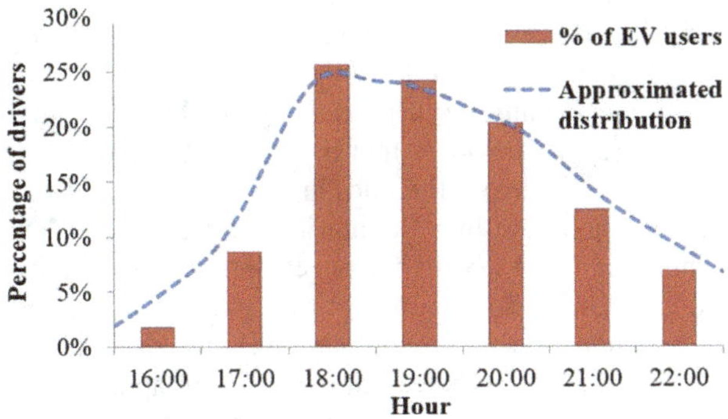

Figure 2.7 Arrival time for the EVs and the respective distribution [4], according to data provided in [3]

For the case of static inductive charging, two sub-cases, as mentioned below, are taken into account:

- *Scenario III-A*: 100 EV users are daily considered to use fast static inductive charging stations
- *Scenario III-B*: 300 EV users are daily considered to use fast static inductive charging stations

The hour each driver is expected to use a charging station, as well as the duration of charging, is defined according to the process described in Section 2.2 and the relevant distributions in Figure 2.1.

The impact of *Dynamic Inductive Charging* on the grid is considered in *Scenario IV*. Similar to static inductive charging the number of EV users charging at static inductive charging stations is considered equal to 100 and 300 in Scenarios IV-A and IV-B, respectively (Table 2.1).

In order to identify the number of chargers that can be installed in the distribution grid in each one of these two cases, the tool depicted in Figure 2.5 is taken into account.

The aim of the dynamic inductive charging analysis is to evaluate the grid's capability to host inductive charging stations. Hence, a worst-case scenario is considered for the placement of the charging stations within the grid [1]. In particular, the inductive charging stations are placed in the most loaded buses of the grid.

2.3.1 Impact of static inductive charging on the grid

Considering the *Reference Case Scenario* (*Scenario I*), where no EVs are considered in the grid, the summer load curve for the examined network is taken into account, as depicted in Figure 2.8.

In this case a peak load of around 8.7 MW is noted in the daily load curve between 13:00 and 14:00. In *Scenario II*, where 1 000 EVs are considered to daily charge at home, an increase is noted in the evening demand, as evident from Figure 2.8. In particular, the peak of the load is now noted between 18:45 and 19:00, due to the high home charging demand.

The maximum line loading, as a percentage of the lines' thermal limits, is depicted in Figure 2.9(a). It is evident that the home charging energy needs introduce an increment in the loading of the lines in this scenario.

In particular, the maximum line loading in this case (85.2% of the maximum thermal limit) is noted in line 12 (which connects buses 12 and 13 in Figure 2.6).

This line is quite close to the central HV/MV sub-station supplying the load of the majority of the system's buses. On the other hand, furthest line sections are less loaded, since they have significantly less amounts of demand.

The loading of the system's transformers in the case of *Scenario I* is depicted in Figure 2.9(b). In particular, the energy needs for home charging in *Scenario II* indicate an increment in the transformers' loading. However, the relevant variations

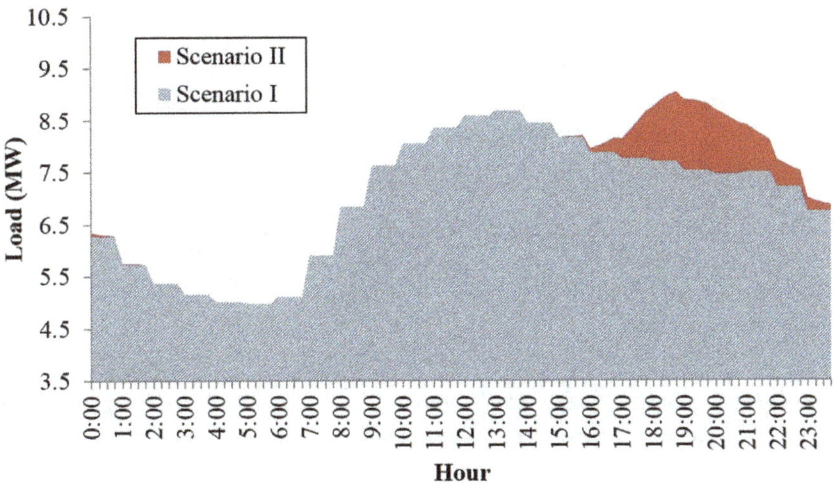

Figure 2.8 Load curves for Scenarios I and II

do not indicate any alarming increase in transformers' loading, which remains less than 41%, as evident in Figure 2.9(b).

The additional home charging demand in Scenario II also introduces an increase in the network's daily active losses. More specifically, an increase of 5.1% is noted in the system's losses, compared to the reference case scenario as depicted in Figure 2.10.

Concerning the voltage limits in the system's buses, the energy needs for home charging in Scenario II result in a decrease of the relevant voltage values; however, the relevant limits are always respected. More specifically, in Scenario II, the minimum voltage value of 0.96 p.u. and the maximum voltage deviation of 0.92% are noted in buses 72 and 73, which are the farthest buses from the HV/MV transformer.

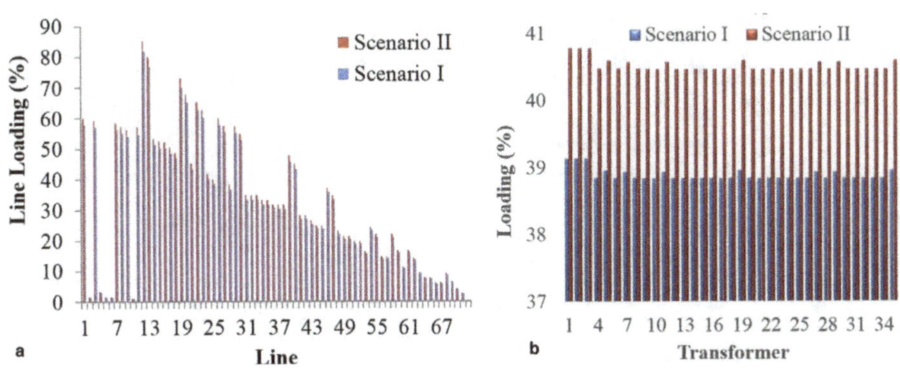

Figure 2.9 Scenarios I and II: (a) Line loading at the time when the peak load is noted and (b) maximum transformer loading as a percentage of the maximum thermal limit

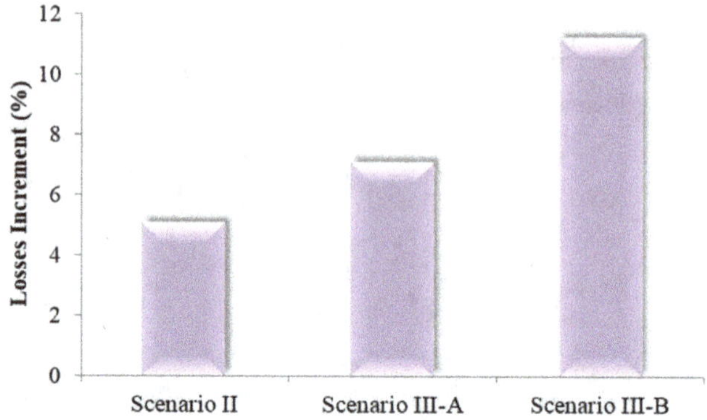

Figure 2.10 *Increment of the daily active losses, compared to the reference case scenario (Scenario I) for Scenarios II and III*

The methodology presented in Section 2.2.1 was applied in the examined network in order to estimate the energy demand of fast static inductive charging for *Scenarios III-A* and *III-B*. The results obtained by calculating the number of charging stations operating each hour of the day are depicted in Figure 2.11. It is evident that 6 and 19 static inductive charging stations are required in order to serve 100 and 300 EV users in Scenarios III-A and III-B, respectively.

The EV charging demands for Scenarios III-A and III-B are depicted in Figure 2.12. Fast inductive charging stations introduce an additional demand during the morning and middle-day hours. More specifically, a peak in the static inductive charging demand is noted between 9:30 and 10:00 when most charging stations are expected to operate, as indicated in Figure 2.11. At this specific time interval, the EV charging demand is increased by 0.2 and 0.63 MW for scenarios III-A and III-B,

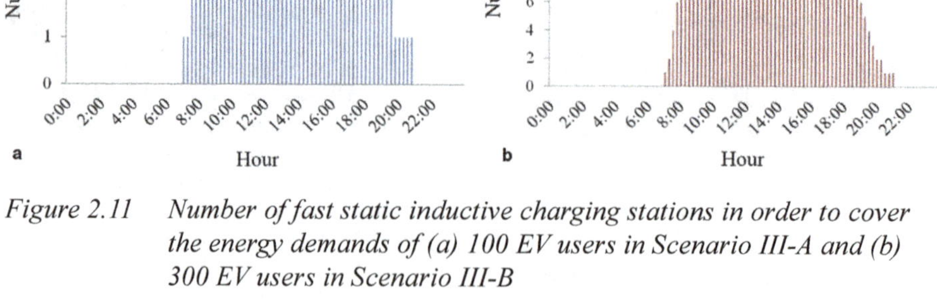

Figure 2.11 *Number of fast static inductive charging stations in order to cover the energy demands of (a) 100 EV users in Scenario III-A and (b) 300 EV users in Scenario III-B*

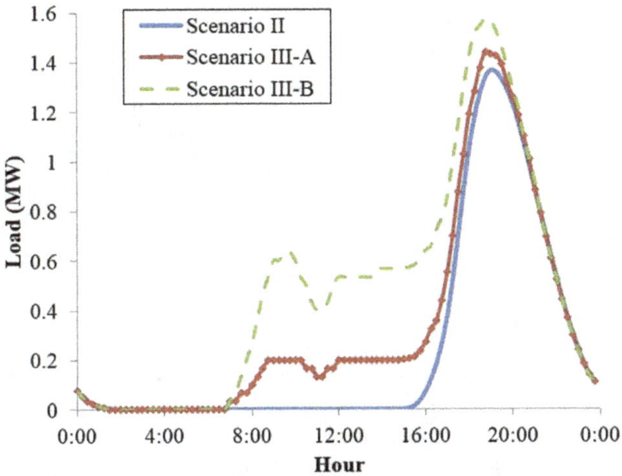

Figure 2.12 EV demand for Scenarios II, II-A and III-B

respectively. A small number of fast static inductive charging stations are also expected to operate during the evening hours as depicted in Figure 2.11. More specifically, during 18:45 and 19:00 an increase is evident in the relevant EV demand compared to Scenario II, which is equal to 6.1% and 15.2% for Scenarios III-A and III-B, respectively.

The network's load curve for the two sub-cases of Scenario III is indicated in Figure 2.13. Fast inductive charging stations do not introduce significant variations in the load of the system during the evening hours. More specifically, a slight increase of 1.11% and 2.58% (compared to Scenario II) is noted between 18:45 and 19:00 in Scenarios III-A and III-B, respectively. However, the operation of the fast inductive charging stations introduces a considerable load increase in the morning and mid-hours of the day. Due to the increased system demands as already observed

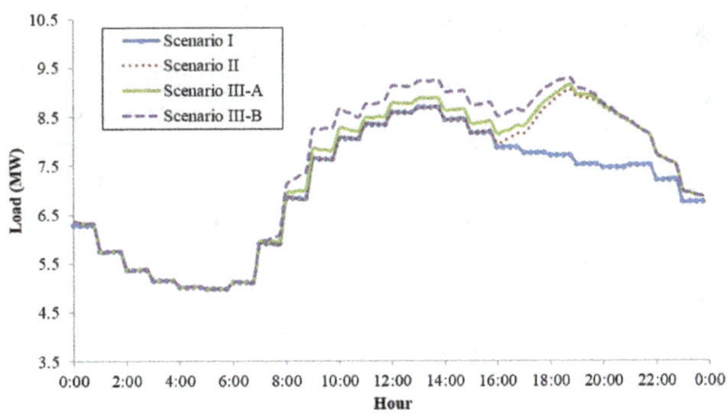

Figure 2.13 Load curve for Scenario III

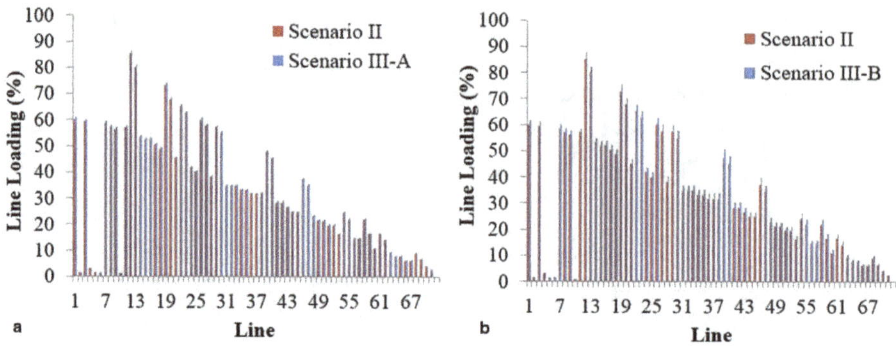

Figure 2.14 Maximum line loading for Scenarios (a) III-A and (b) III-B

in the reference case Scenario I during the mid-hours of the day, an additional peak is noted in the system's load curve at this time interval, as evident in Figure 2.13.

More specifically, when a small number of static inductive charging stations are considered to operate in the grid as represented in Scenario III-A, an increase of 2.3% is observed in the system's load during the time interval 13:00–14:00. However, the already high system demand at this specific hour results in a system load equal to 8.9 MW. An additional increment in the system's demand shall be expected when considering an increased static inductive charging deployment according to Scenario III-B. In this case, an increase of 6.5% is observed in the system's load between 13:45 and 14:00, resulting in a load equal to 9.25 MW.

Concerning the line loading in Scenario III, it is evident in Figure 2.14 that the most loaded line is line 12, connecting the buses 12 and 13 in the network depicted in Figure 2.6.

It shall be noted that compared to Scenario II, only a small increase is noted in the loading of line 12. However, the already increased loading of the specific line in Scenario II is now increased to 88.3% and 89.4% of its maximum thermal limit in Scenarios III-A and III-B, respectively.

The additional charging needs of the static inductive stations also increase the active power losses in the network by 7.12% and 11.25% compared to the reference case scenario (Scenario I) in Scenarios III-A and III-B, respectively (Figure 2.10).

Installing fast static inductive chargers in the network's buses increases the loading of the relevant transformers. However, the transformers' loading always remains within the respective safety limits. More specifically, the worst cases for the transformer loading noted in Scenarios III-A and III-B are 44.1% and 45.8% of the relevant maximum loading limit, respectively.

Similarly, the maximum/minimum values of the voltage are below the relevant safety limits in Scenario III. More specifically, the minimum voltage value noted in buses 72 and 73 in both sub-cases of Scenario III is equal to 0.96 p.u. An increase is observed in the voltage deviation of the buses as observed in Figure 2.15, due to the increased energy needs of fast static inductive charging stations. However, it is

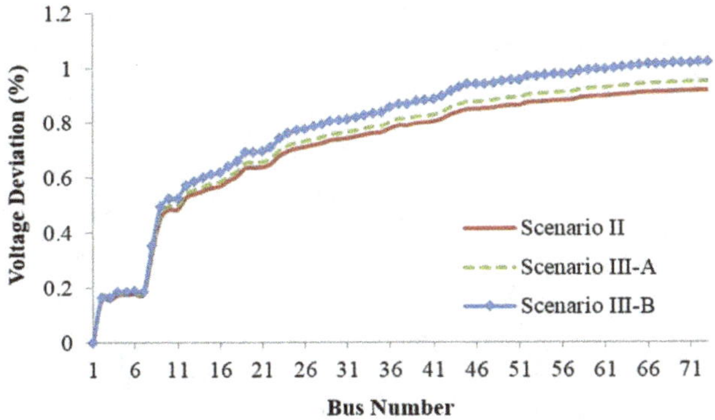

Figure 2.15 Voltage deviation for Scenarios III-A and III-B

evident in Figure 2.15 that the voltage deviations respect the relevant safety limit of 3% in both sub-cases of Scenario III.

2.3.2 Impact of dynamic inductive charging on the grid

The impact of dynamic inductive charging in the distribution network is examined in Scenario IV. For the two sub-cases of Scenario III, where 100 and 300 EV users are considered to daily rely on static inductive charging stations it was determined that 6 and 19 static stations are required, respectively, in order to fulfill the relevant energy needs.

The maximum number of dynamic inductive charging stations that can be installed in the grid, according to the tool described in Figure 2.5, is determined equal to 54 and 45 for Scenarios IV-A and IV-B, respectively, as indicated in Table 2.3.

Dynamic inductive charging introduces a considerable demand both in the morning and the evening hours, as depicted in Figure 2.16. More specifically, during 9:00 and 10:00 the dynamic inductive charging demand becomes quite high in Scenarios IV-A and IV-B and gets equal to 1.5 and 1.2 MW, respectively. Moreover, the EV demand remains quite high in the evening hours: during 17:00 and 18:00, the dynamic inductive charging demand is equal to 1.7 and 1.4 MW for both cases of Scenario IV. This dynamic inductive charging demand coincides with the already increased home charging demand, resulting in a significant peak in the total EV

Table 2.3 Number of static and inductive charging stations for the examined scenarios

	Static inductive charging stations	**Maximum number of dynamic inductive charging stations**	**Total number of inductive charging stations**
Scenario IV-A	6	54	60
Scenario IV-B	19	45	64

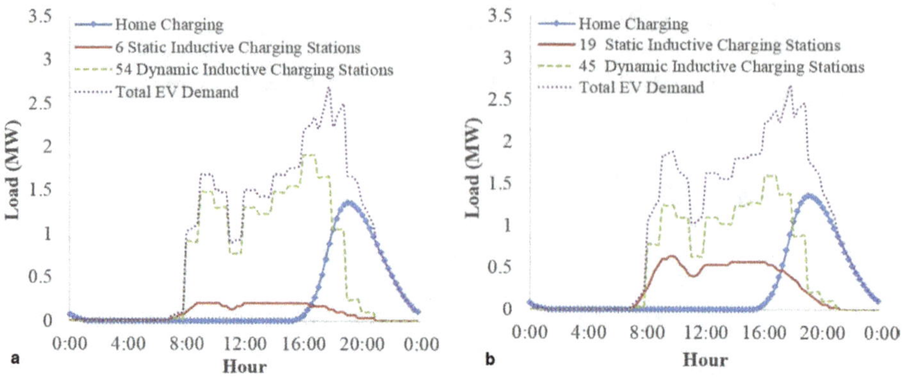

Figure 2.16 EV demand for Scenarios (a) IV-A and (b) IV-B

demand. More specifically, it is evident in Figure 2.16 that the total EV demand during 17:45 and 18:00 gets equal to around 2.7 MW for Scenarios IV-A and IV-B.

The load curves for both cases examined in Scenario IV are depicted in Figure 2.17. The considerably increased demand for dynamic inductive charging stations results in a significant system load for almost all of the day's hours. More specifically, during 12:00–19:00, the system's demand is higher than 9.9 MW for both cases of Scenario IV, while the demand gets equal to 10.4 MW during 17:45–18:00. At this specific time interval, a high system demand is observed due to the operation of both static and dynamic inductive charging stations, as well as the increased energy needs of home charging observed during the evening hours.

For both cases of Scenario IV, the dynamic inductive charging demand introduces a considerable loading in the system's lines. In fact, the thermal capacity limits of the lines are the constraints that define the maximum number of dynamic stations that can be installed in the grid. Additionally, the daily active losses of the

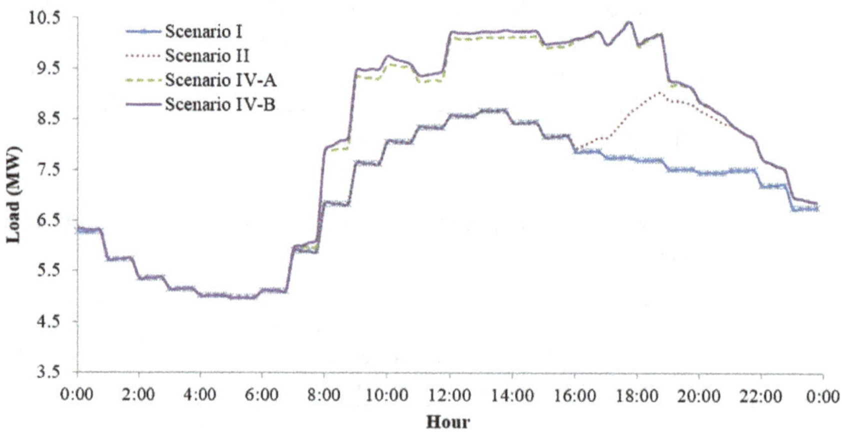

Figure 2.17 Load curves for the two sub-cases of Scenario IV

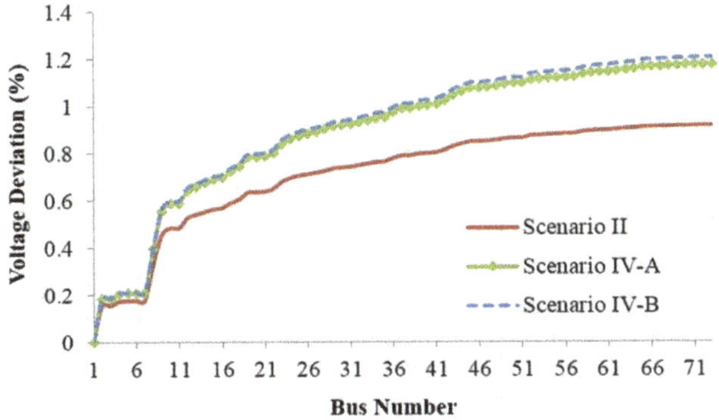

Figure 2.18 Voltage deviation for Scenario IV

system are increased by 24.5% and 26.2% for Scenarios IV-A and IV-B, respectively, as compared to Scenario I.

Moreover, it is evident in Figure 2.18 that the increased need for dynamic inductive charging stations increases the voltage deviation in the system's buses. However, the voltage deviation always remains well within the limit of 3%. Similarly, the limits for the minimum and maximum voltage values are not violated in any of the system's buses. A review of the results obtained from the examined scenarios is presented in Table 2.4.

2.4 RES and inductive charging

In case of a future load increase in the examined system, the system's hosting capacity for inductive charging infrastructure might be reduced. Similarly, the system might not be able to host the desired increased number of inductive charging stations, in case a mass adoption of fast inductive charging is noted, due to potential limits related to the already installed grid infrastructures. In order to avoid costly economical solutions, related to the upgrade of the existing grid infrastructure that will allow installing additional inductive charging stations, the installation of RESs within the grid shall be considered. In particular, the installation of PV within the examined network can be proven quite significant.

In order to examine the impact of PV installation on the number of chargers that can be installed in the examined network, the daily profile in Figure 2.19, concerning the PV production, is taken into account. For the examined network in Section 2.3 represented by Figure 2.6, it was determined that in the case of Scenario III-B, 19 static inductive charging stations are required in order to cover the energy needs of 300 EV users.

PVs with an installed capacity equal to 10 MW are considered to be installed in the examined network. Those PVs are considered to be installed in all the buses of the grid. The installation capacity of PVs in each bus is determined according to the percentage of the total load served by each bus. It is evident in Figure 2.20 that

Table 2.4 Summary results for the scenarios under study

Parameter		Scenario					
		I	**II**	**III-A**	**III-B**	**IV-A**	**IV-B**
Peak load (MW)	Total	8.68	9.05	9.15	9.28	10.45	10.43
	EV peak	0.00	1.34	1.44	1.57	2.69	3.17
	Peak hour	13:00–14:00	18:45–19:00	18:45–19:00	18:45–19:00	17:45–18:00	17:45–18:00
Daily active losses (MW)		17.6	18.5	18.8	19.6	21.9	22.2
Maximum voltage deviation (%)		0.92	0.92	0.95	1.02	1.17	1.21
Minimum voltage value (α.μ.)		0.958	0.958	0.957	0.956	0.953	0.952
Maximum line loading (%)		81.85	85.23	86.25	87.64	99.85	99.70

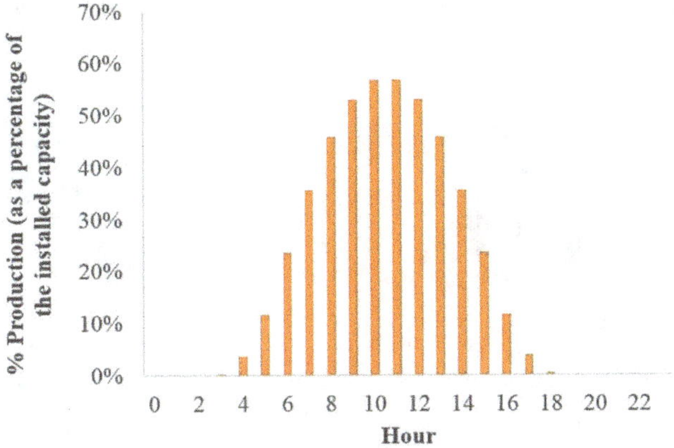

Figure 2.19 *Daily, hourly, profile of the PV production as a percentage of the installed PV capacity*

the installation of 10 MW PVs can cover a significant amount of both the initial system's load (reference case scenario in Figure 2.20) as well as a significant amount of the inductive charging demand.

Concerning the line loading, a decrease can be observed during the hours that the PV production remains at high levels, as evident in Figure 2.21(a). The significant reduction in the line loading during the morning and mid-day hours also results in a considerable decrease in the system's active losses, as depicted in Figure 2.21(b). In particular, the installed PV production manages to reduce the daily system's losses by 29%.

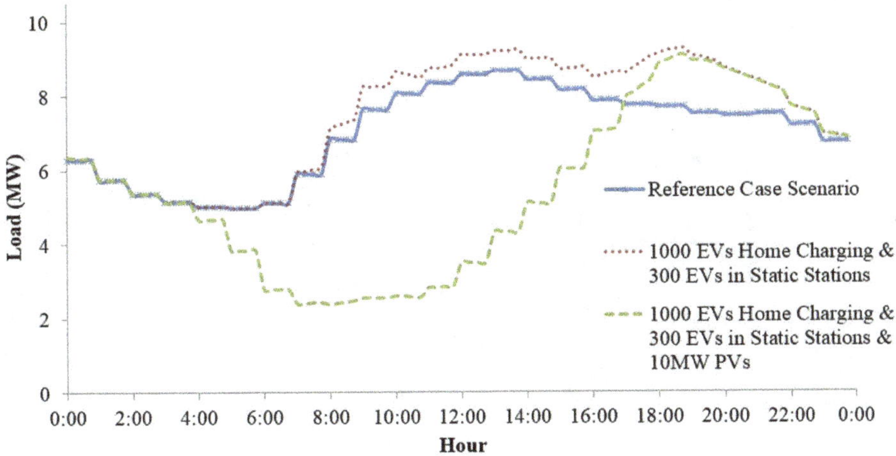

Figure 2.20 *Load curve when considering the installation of 10 MW PVs in the examined network for the scenario that 300 EV users daily use static inductive charging stations*

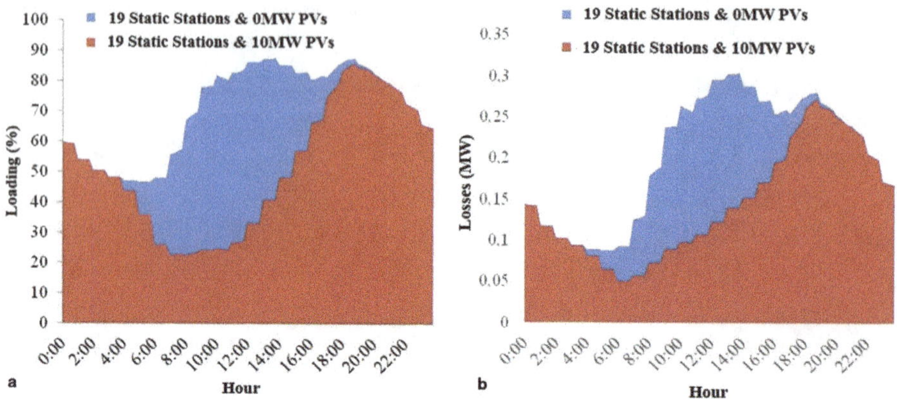

Figure 2.21 *(a) Loading of the network's most loaded line and (b) active losses during the day for the case where PVs of 10 MW installed capacity are considered in the scenario that 300 charging events take place in static inductive charging stations*

The application of the tool described in Figure 2.5 for the case where 300 EV users daily charge at static inductive charging stations (Scenario III-B) indicates that 45 charging stations can be installed in the examined network (Table 2.3). However, when considering the installation of 10 MW PVs within the network, the relevant maximum number of dynamic stations can be increased by around 27%, as evident in Figure 2.22.

The number of stations that the examined network can host may be further increased in case of a greater PV penetration. In particular, PVs of 20 MW installed capacity can increase the maximum number of stations that the grid can host by 47%, as depicted in Figure 2.22.

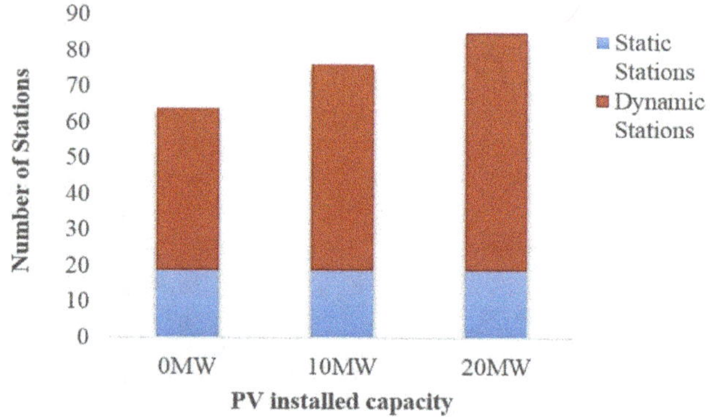

Figure 2.22 *Number of inductive charging stations that can be installed in the examined network when considering PVs of 10 MW or 20 MW installed capacity*

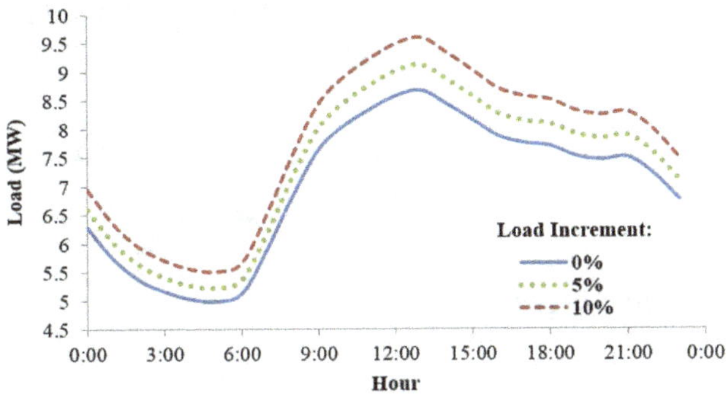

Figure 2.23 *Future load increment of 5% or 10% in the examined network*

Moreover, a PV penetration in the network can be proven quite significant in case of a potential future load increase. The tool described in Figure 2.5 has been applied for load-increment scenarios of 5% or 10% (Figure 2.23), in order to identify the maximum number of dynamic stations that can be installed in the examined network. The results are depicted in Figure 2.24.

In particular, considering Scenario III-B in Table 2.1, where 300 users daily charge their EV at static stations, a 5% load increase can result in a 24% reduction in the number of dynamic stations that can be installed in the grid. PVs of 10 MW installed capacity can avoid such a reduction in the number of inductive charging stations, while additional PVs within the grid (for instance, 20 MW installed capacity) can further increase the number of stations that can safely be installed (Figure 2.24).

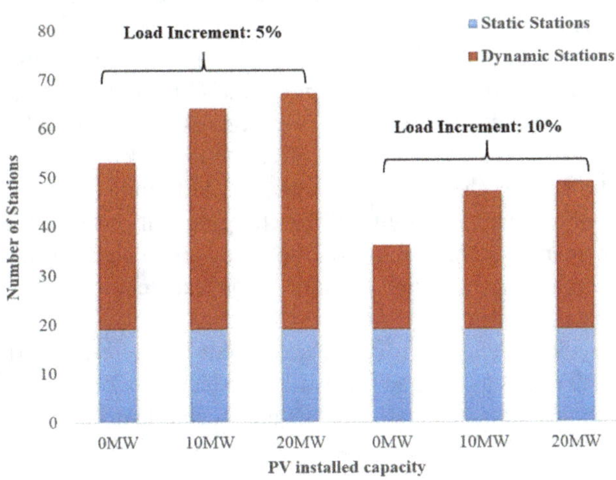

Figure 2.24 *Number of stations that can be installed in the examined network considering various cases of load increment and installed PV capacity*

A greater increase in the system's load by 10% can significantly decrease the number of stations that can be installed in the grid by 62%, as evident in Figure 2.24. However, PVs of 10 MW installed capacity may limit this reduction to 38%. A further increase in the maximum number of stations that can be hosted in the grid can be achieved when considering the installation of additional PVs, as can be seen in Figure 2.24.

2.5 EMS for inductive charging of EVs

Potential mass deployment of inductive charging infrastructures in a distribution network may result in the installation of more charging stations than the grid is able to tolerate with the existing grid facilities. In this case, in order to avoid an upgrade in the grid's infrastructure that will be able to cover the increased energy needs, an EMS may be developed.

Such an EMS is described in [6, 7]. This particular EMS takes into account the charging needs of EV users as well as constraints linked with the operation of the distribution grid. In particular, this system satisfies the following objectives [7]:

- **Monitor the stations' operation**: The consumption of the charging stations can be monitored in real-time by the operator of the charging stations. The monitor of the consumption allows billing purposes to be achieved, while also allowing demand flexibility programs to be offered in order to support the operation of the network.
- **Remotely selecting the maximum charging rate of the stations:** In case of emergency network operational conditions due to the increased need for fast inductive charging, the management system allows the remote control of the stations' maximum charging rate to be enabled.
- **User awareness of fast inductive charging stations**: The EMS allows the EV users to be aware of the location, the availability and the electricity cost of the charging stations. In particular, EV drivers can be aware of the location of the existing fast inductive charging infrastructures in order to select the most convenient place for charging their vehicle. Additionally, the EMS informs the EV drivers about the availability of the charging stations (i.e. free, busy, etc.). The management system also allows the EV users to be aware of specific pricing policies that may be adapted (aiming, for instance, to incentivize charging during off-peak hours).
- **Booking services:** The EV owners, taking into account their trip destination, their mobility needs as well as the available electricity prices, are able to book the most suitable charging station at the most convenient time.

An overview of the management system architecture is depicted in Figure 2.25. Dashed lines indicate interactions among different actors, while continuous lines depict the power flow paths. The following main components can be distinguished [7]:

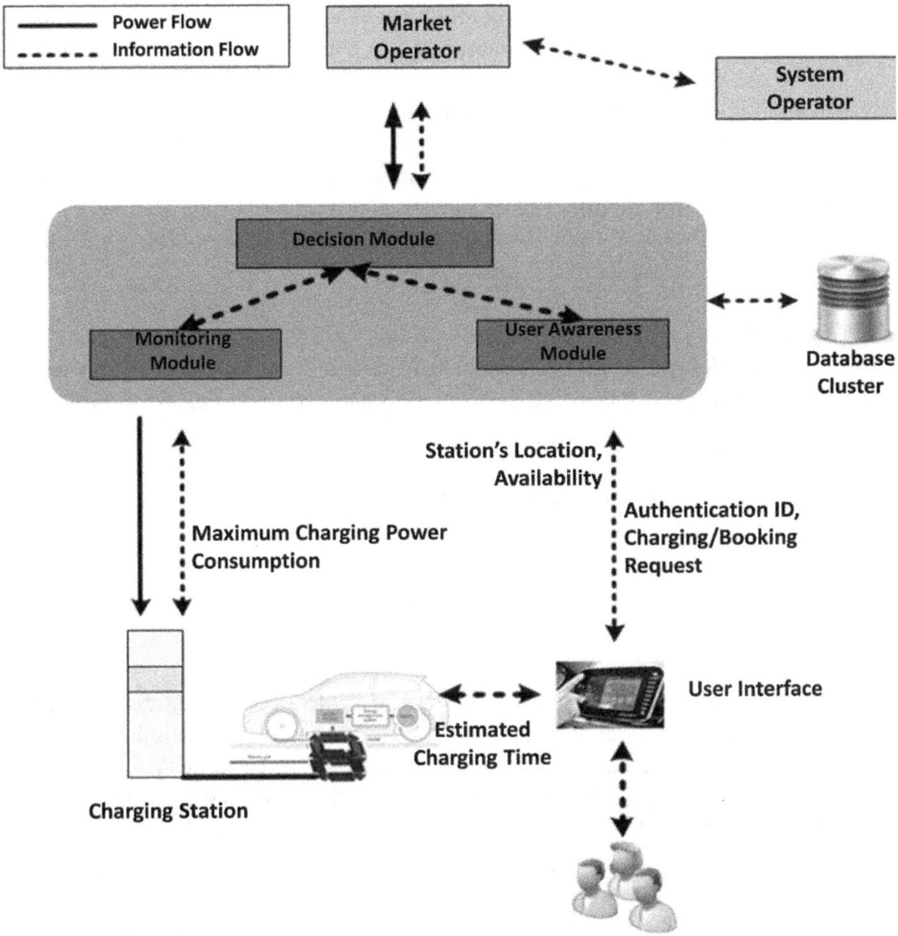

Figure 2.25 Architecture of the EMS [7]

- **User Awareness Module:** The EMS allows EV users to be aware of the charging stations' position, as well as their operational-status (free, busy or out-of-service). The management system also indicates the amount of time required for a charging station to be available in case it is not free at a specific time. Thus, the EV drivers can choose the most convenient charging station, taking into account their travel direction, their charging needs, as well as the location of the station.
- **Monitoring Module:** This module allows the interaction between the charging station and the EMS. In particular, it communicates with the metering infrastructure of the charging station in order to gather the necessary data indicating the energy consumption. Moreover, this module is responsible for the remote control of the maximum charging rate of the station.

- **Decision Module:** This module communicates with the wholesale market while also supplying the charging demand of EV users. In a simple case, constant price rates may be considered for charging at the stations. However, different pricing policies may be adopted, like multi-tariff or dynamic pricing. This module is also able to offer demand response services to the market operators. In particular, when network operational issues (equipment overloading, etc.) are noted, the decision module can reduce the charging rate of the stations in order to overcome the raised issues. Finally, the decision module processes the charging or booking requests of EV users.

Thus the EMS, apart from being able to facilitate the charging requests of the EV users, also allows the management of the stations in case grid-related issues are noticed. In this respect, the following demand response scenarios are examined in this section:

- 'Global'demand response services
- 'Local' demand response services at the substation level.

For each one of those scenarios two different cases are considered in order to avoid any grid violations:

- **Case 1:** In this case the EMS provides less power to all of the static and dynamic stations in order to overcome any type of violations noticed within the grid.
- **Case 2:** In this case, the EMS does not provide power to some specific stations (static or dynamic). All the other stations operate at their nominal power.

The network under study is the same distribution line that has been examined in Section 2.3 and depicted in Figure 2.6. In this case, a specific time interval is taken into account, rather than the whole day, considering a specific problem in the grid.

2.5.1 'Global' demand response services

Considering that the stations are distributed among the grid as depicted in Figure 2.26(a), the line connecting buses 12 and 13 is overloaded when the stations operate at their maximum charging power. The specific line is quite close to the central HV/MV substation, while all the stations are connected in buses served by this line. Thus, any decision taken by the EMS shall consider all the stations within the network.

The overloading of the line can be avoided when the EMS uses specific measures. In particular, when considering the reduction of the charging power of *all* the inductive charging stations to a percentage equal to 74% of their actual nominal power, the loading of the problematic line can be reduced to 90% of the line's maximum thermal limit.

Moreover, the overloading of the line can be overcome if the EMS does not provide power to specific stations, as indicated in Figure 2.26(b), while all the other

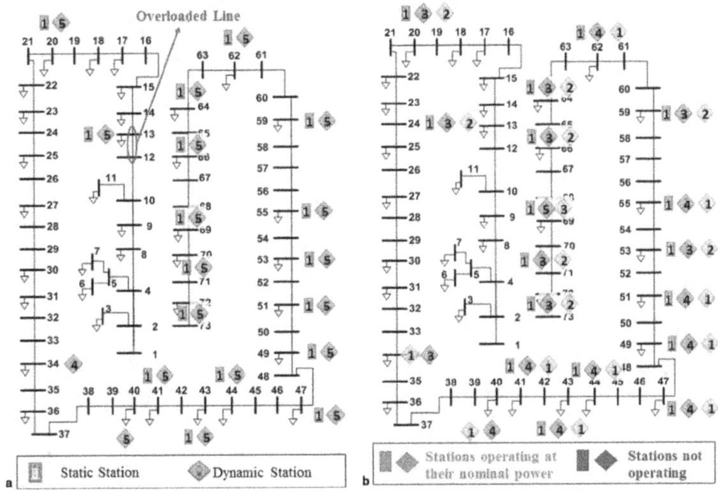

Figure 2.26 Distribution of the charging stations: (a) When all stations operate at full power, the noted line is overloaded. (b) The line overloading can be avoided by not allowing specific stations to operate.

stations operate at rated power. In particular, turning off the power to a small percentage of the dynamic stations (around 38% of the stations in Figure 2.26(a)) can significantly reduce the loading of the problematic line to less than 90%.

2.5.2 'Local' demand response services at the substation level

In the previous case, the stations were distributed among almost all of the system's buses. However, a large number of stations may be connected at specific buses of the grid, indicating potential locally related issues for those buses. For instance, in case the stations are distributed among the grid as depicted in Figure 2.27(a), none of the system's lines are overloaded. However, the increased number of stations connected to buses 17 and 49 results in the overloading of the relevant transformers at those buses. Apparently, any decision taken by the EMS in this case shall not consider all the stations in the grid but shall rather be limited to the stations connected at these transformers.

In particular, when the power provided to *all* the stations connected at those transformers was reduced to 85% of their nominal value, as depicted in Figure 2.27(b), the overloading of the transformers was avoided. It is also observed that in this case the rest of the system's inductive charging stations, connected in buses other than 17 and 49, operate at their nominal power.

The overloading of the transformers may also be avoided, if the EMS does not provide power to *specific stations* (Figure 2.27(c)). In particular, by turning off the power provided to two of the nine *dynamic* stations connected to each bus, the loading of the respective transformers can be reduced to lower than 90% of their thermal limit.

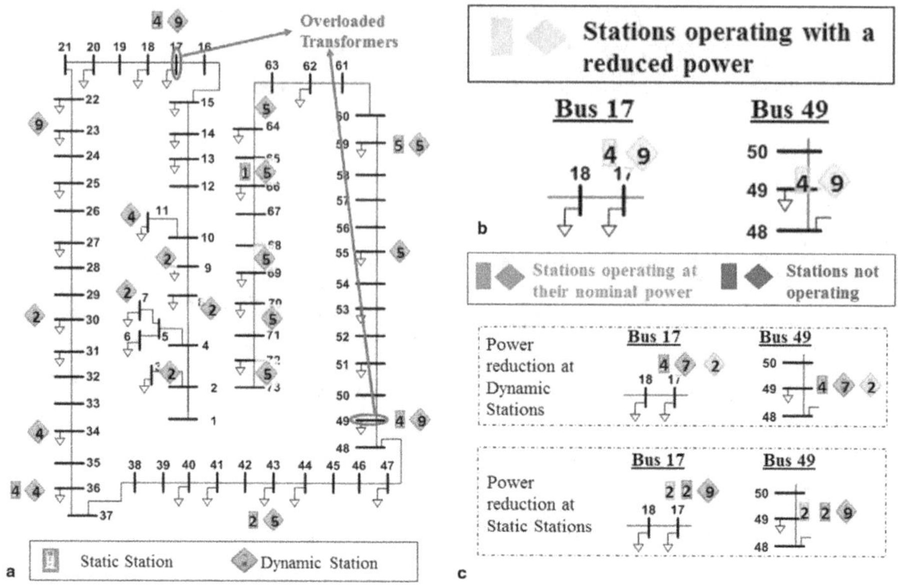

Figure 2.27 *(a) Distribution of the charging stations in the examined network and overloading of the transformers in buses 17 and 49, (b) reduction of the power of all the stations connected in the overloaded transformers and (c) turning off the power provided to specific stations (static or dynamic) in the problematic transformers*

Similarly, the EMS may resolve the overloading of the transformers by not providing power to specific *static* stations. More specifically, by turning off the power to two out of the four static stations connected at buses 17 and 49, the overloading of the transformers can be avoided.

2.6 Conclusions

The differences in the amount of power required for fast inductive charging applications compared to conventional conductive charging, as well as the different charging patterns distinguished among conductive and inductive charging (particularly in the case of dynamic stations) indicate that new tools shall be developed in order to estimate the inductive charging demand. The estimation tools analyzed in this chapter take into account both the high energy amounts as well as the charging patterns that are expected to arise in the case of static and dynamic fast inductive charging stations in order to evaluate the impact of those stations on the distribution grid.

In particular, the results of the analysis presented in this chapter indicate that static inductive charging affects the network load during the morning and mid-day hours. Moreover, dynamic inductive charging stations introduce a significant EV charging demand resulting in a high system load during most of the day's hours.

Both static and dynamic inductive charging do not appear to significantly impact the voltage values in the buses of the system. Similarly, although the inductive charging demand increases the loading of the transformers within the distribution network, the thermal limits of the transformers are not violated in any of the examined scenarios. On the other hand, the demand for inductive charging significantly increases the loading of the system's lines, which, in many cases gets very close to their thermal limit. Moreover, the increased inductive charging demand appears to significantly increase the power losses noted in the network.

Concerning the impact of RES production on the inductive charging demand, it is observed that the installation of PVs within a network can significantly increase the network's hosting capacity for inductive charging stations. Additionally, in case of a potential future load increase, the PV production could be proven quite beneficial. In particular, a load increase of 5% or 10% could significantly reduce the maximum number of inductive charging stations that may be installed in the grid. However, the installation of PVs in the network could assist in preventing such a reduction.

The estimation tools presented in this chapter can allow the system operator to identify the maximum number of inductive charging stations that can be installed in the network. However, in case of a mass inductive charging deployment scenario, the maximum allowable number of stations that can be installed in the grid may be exceeded. In this case, an upgrade of the system's infrastructure may be required in order for the system to host the additional charging infrastructure. Such costly measures could be avoided when considering the application of an EMS that will efficiently coordinate the operation of the stations. In this respect, the EMS presented in this chapter can adjust the charging rate of the stations in order to offer demand response services related to the whole grid or even specific buses of the grid. In particular, the results of the analysis presented in this chapter indicate that grid-related issues can be avoided either by reducing the power provided to the charging stations by a small percentage or by turning off the power provided to a small number of stations.

References

[1] Karakitsios I., Karfopoulos E., Hatziargyriou N. 'Impact of dynamic and static fast inductive charging of electric vehicles on the distribution network'. *Electric Power Systems Research*. 2016, vol. 140(5), pp. 107–15.

[2] Hatziargyriou N., Karfopoulos E., Tsatsakis K. 'The impact of EV charging on the system demand'. *Springer Electric Vehicle Integration into Modern Power Networks*; 2012.

[3] Syndicate Mixte du SCoT Grand Douaisis (France). *Rapport d'exploitation sur le territoire du SCOT - Enquête Ménages Déplacements Standard Certu du Grand Douaisis*. 2012.

[4] Karakitsios I., Karfopoulos E., Hatziargyriou N. *Report on the technical impact of the implementation*. FastInCharge Project - Deliverable D5.2; 2015.

[5] Madzharov N.D., Ilarionov R.T., Tonchev A.T. 'System for dynamic inductive power transfer'. *Indian journal of applied research.* 2014, vol. 3(7), pp. 173–6.
[6] Karfopoulos E., Karakitsios I., Safalidis G., Hatziargyriou N. *Demand management system report.* FastInCharge Project - Deliverable D3.2; 2014.
[7] Karfopoulos E., Hatzoplaki E., Karakitsios I., Hatziargyriou N., Safalidis G., Kamarinopoulos A. 'Energy management system for fast inductive charging network: The FastInCharge project'. Medpower Conference 2014; Athens; 2014. pp. 1–6.

Chapter 3

Wireless power transfer in EVs during motion

Ali Zakerian[1], Amir Babaki[1], and Ryosuke Ota[2]

3.1 Introduction

Transportation infrastructures have become a decisive component of our modern cities. Having been the second high energy-consuming sector and the largest air-polluting sector in the United States, transportation systems mostly have become more efficient and environment-friendly recently [1]. The use of electric vehicles (EVs) has helped in overcoming problems such as the emission of polluting greenhouse gases and limited efficient operating points generally encountered in internal combustion engine-based vehicles [2]. However, EVs are not free from deficiencies.

Comprising more than 40% of the total EV cost, batteries are the major issue in these systems [1]. Due to the limited energy storage capacity of batteries, EVs can travel limited distances after each charging. This problem would affect the consumer's comfort, which is regarded as range anxiety [3]. Moreover, using batteries of large capacities significantly increases the weight of EVs [4]. Also, disposing-off the batteries after the end of life generates soil and water contamination due to the presence of toxic chemicals and heavy elements within them, and recycling them becomes a costly affair [5]. While charging EVs by conventional systems using cables to connect them to the grid is limited to houses, urban parking lots, and charging stations, *wireless power transfer* (WPT) facilitates improved and easy accessibility of chargers [6]. This maintains the state of charge (SoC) within batteries thereby leading to lower average depth of discharge and extending their lifetime significantly [7, 8]. Moreover, consumer tendency gets improved by extended distance limitation and decreased range anxiety provided by wireless chargers [9]. Apart from that, the safety level in wireless charging is higher especially during wet weather conditions [10]. WPT systems can be exploited either as *stationary wireless chargers* (SWCs) or as *dynamic wireless chargers* (DWCs) [11]. The former refers to charging an EV battery while it is parked in a parking lot or stopped, for example, at a bus station or urban intersections. However, the latter implies charging an EV

[1]School of Electrical and Computer Engineering, College of Engineering, University of Tehran, Iran
[2]Department of Electrical Engineering, Faculty of Science and Technology, Tokyo University of Science, Japan

battery while it is moving [12]. It is noteworthy that the DWCs are used in order to feed the EV motor drive directly rather than just charging its battery [13].

Even though there are many common features and challenges between DWC and SWC, the major accentuation of this chapter is on the DWC. The chapter outline proceeds as follows. The basic theories and applications of WPT systems are reviewed consciously in Section 3.2. Mathematic modeling of the system is elaborated in Section 3.3. Diverse design schemes for system configuration and parameters are presented in Section 3.4. Section 3.5 represents the control challenges of DWC systems and some proposed methods in the literature. The future trends of the DWC systems are reviewed in Section 3.6 before presenting the conclusion in Section 3.7.

3.2 WPT systems: basic theories and applications

The idea of WPT conceived by Nikola Tesla during the nineteenth century provides advantages of wireless connectivity, safer connections, and broad accessibility to power grid. Thus, it has proven itself as promising technology engrossing the future of almost every industry [14]. This technology exhibited relatively slower growth during the twentieth century; it, however, gathered pace with the advent and inclusion of semiconductors and ultra-fast processor technology during the recent years. WPT technology has a very widespread scope and applications in areas such as bio-medical devices involving transcutaneous energy transfer [15–17], cell-phone and laptop chargers [18], industrial rotating structures [19], EVs [20], and maglev trains [21, 22].

The WPT methods can be categorized into three distinct approaches. The first one that is based on the electromagnetic waves fascinates low level of power to be transferred via long distances [6]. Using the strong electric field between plates of a capacitor constitutes the principle of the second type WPT method, which is called capacitance-coupled WPT [23]. This method enables power transfer to relatively shorter distances compared to the previous one. Despite its simplicity, the requirement of a large electric field results in its limited application [6]. The third method involving magnetic coupling between two windings uses inductive power transfer (IPT) in short air gaps [24]. Although the air gap between the windings mitigates the induced voltage, increasing the operating frequency will amplify the mutual inductance [25]. Being capable of transferring high rate of power through short air gaps, this method serves as the best option for wireless charging of EVs. Thus, the application of IPT in the charging of EVs is accounted for as the main subject of this chapter. It is noteworthy here that in the subsequent sections the term WPT has been used synonymously for IPT and the other two methods of WPT have not been given attention. Obviously, the two other methods of WPT will be dismissed in this chapter.

The two basic albeit principal laws of electricity hold the keys to understanding the operation of IPT systems. According to Ampere's circuital law, the high-frequency current passing through the transmitter coil will cause a high-frequency

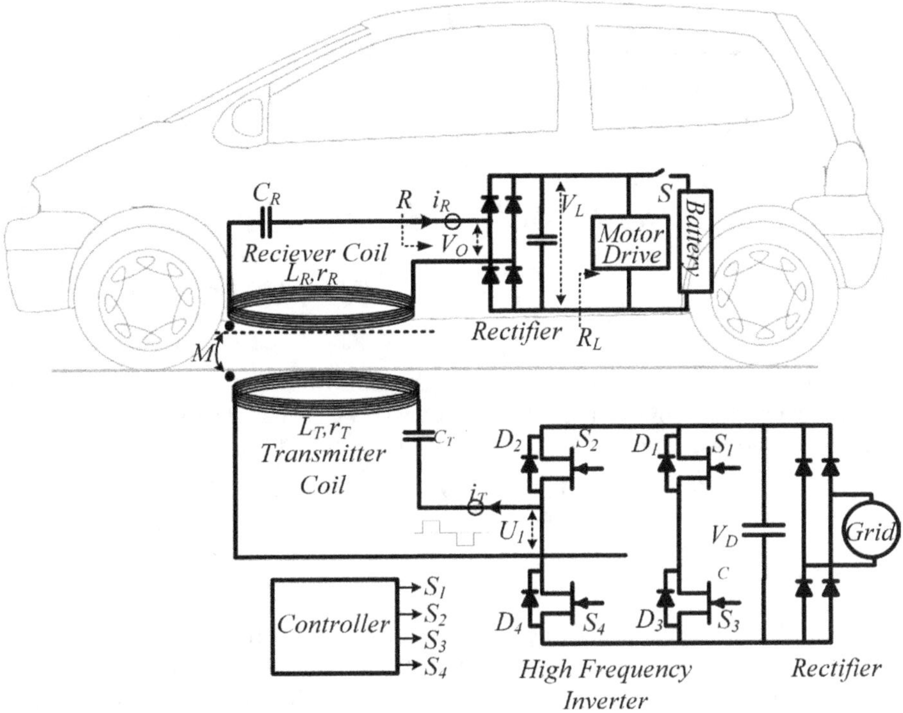

Figure 3.1 A schematic view of a WPT system for charging EVs [9]

magnetic field around it. Based on Faraday's law of induction, this high-frequency magnetic field induces a high-frequency voltage with considerable magnitude across the receiver coil. This enables a sufficient amount of power transfer through a relatively large air gap. A schematic view of WPT systems for charging EVs is depicted in Figure 3.1. The high-frequency input excitation voltage is provided by the transmitter-side H-bridge inverter. On the other hand, in order to make the received high-frequency voltage usable for charging, a rectifier is employed on the secondary side to convert the high-frequency AC voltage into the DC. As shown in Figure 3.1, due to the large inductance of the coils offered by high operating frequency, series compensation capacitors are used on both sides in order to cancel the reactive power of the system and improve both power transfer and efficiency.

3.3 System modeling

Developing a proper mathematical model of the WPT system is the first step to perceive how it operates. In this section, the system is modeled by an equivalent circuit depicted in Figure 3.2 and related equations using sinusoidal approximations.

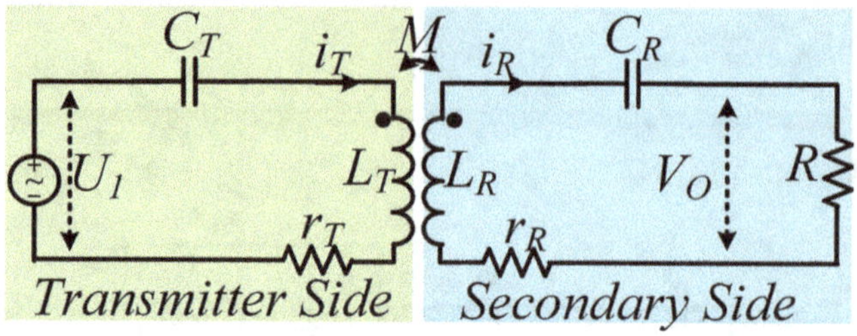

Figure 3.2 Equivalent circuit of a WPT system with series-series compensation capacitors [9]

The equivalent circuit represents a WPT system with series-series (SS) compensation capacitors. The Kirchhoff's Voltage Law (KVL) equations for transmitter and receiver-side loops, respectively, are given by (3.1) and (3.2).

$$\bar{U}_1 = \left[r_T + j\left(\omega L_T - \frac{1}{\omega C_T}\right)\right]\bar{i}_T - j\omega M\bar{i}_R, \tag{3.1}$$

$$j\omega M\bar{i}_T = \left[r_R + j\left(\omega L_R - \frac{1}{\omega C_R}\right)\right]\bar{i}_R + R\bar{i}_R, \tag{3.2}$$

where $\bar{U}_1$, θ_{i1}, and $\bar{i}_R$ are the phasors of fundamental harmonic of input square-wave voltage, transmitter-side current, and receiver-side current, respectively.

L_T, L_R, r_T, r_R, C_T, C_R, and ω are denoted as transmitter coil inductance, transmitter coil resistance, receiver coil inductance, receiver coil resistance, transmitter compensation capacitance, receiver compensation capacitance, and the operating angular frequency of the system, respectively. Moreover, R, the equivalent load resistance can be written as follows

$$R = \frac{8}{\pi^2}R_L, \tag{3.3}$$

where the load resistance R_L is given by (3.4).

$$R_L = \frac{V_L^2}{P_L}, \tag{3.4}$$

where V_L and P_L are load voltage and load power, respectively. M, denoted as the mutual inductance of transmitter and receiver windings, is given by (3.5).

$$M = K\sqrt{L_T L_R}, \tag{3.5}$$

where K is the coupling coefficient and is a function of the relative position of transmitter and receiver windings. The power transfer efficiency (η) can be obtained as the ratio of the load power, P_L, to the input power, P_{in}, as follows given by (3.6).

$$\eta = \frac{P_L}{P_{in}} = \frac{R}{R + r_R + r_T\left[|R + r_R + j\left(\omega L_R - 1/\omega C_R\right)|/\omega M\right]^2}. \tag{3.6}$$

It can be observed from (3.6), that operating frequency can be selected so as to maximize efficiency. Also it can be proved that the operating frequency is close to the resonant frequency of the receiver circuit. Overall, the resonant frequency is of great importance in WPT systems since its operation is closely related to the resonant converters. The resonant frequency of primary (transmitter) and secondary (receiver) sides is given by (3.7) and (3.8), respectively.

$$f_T = \frac{1}{2\pi\sqrt{L_T C_T}}. \tag{3.7}$$

$$f_R = \frac{1}{2\pi\sqrt{L_R C_R}}. \tag{3.8}$$

Moreover, the input power factor is an important criterion as it determines the power transfer capability of the system. The input power factor (PF) of the system is given by (3.9).

$$PF = \frac{P_{in}}{S_{in}} = \cos\left(\tan^{-1}\left(\frac{\mathrm{Im}\,(z_{in})}{\mathrm{Re}\,(z_{in})}\right)\right), \tag{3.9}$$

where z_{in}, the impedance seen from the inverter, is given by (3.10).

$$z_{in} = R_{in} + jX_{in} = r_T + j\left(\omega L_T - \frac{1}{\omega C_T}\right) + \frac{\omega^2 M^2}{R + r_R + j\left(\omega L_R - 1/\omega C_R\right)}. \tag{3.10}$$

From (3.9) and (3.10), it can be observed that if the resonant frequencies of both transmitter and receiver sides are the same as the operating frequency, X_{in} the imaginary part of z_{in} would be canceled and the *PF* would reach unity.

The load voltage of the system can be modeled using the sinusoidal approximation. The inverter output voltage can be stated by using the Fourier series as:

$$U(t) = \sum_{n=2k+1} U_n \sin(n\omega t), \tag{3.11}$$

where the Fourier series coefficient, U_n, is given by (3.12).

$$U_n = (-1)^{n+1} \frac{4}{n\pi} V_{DC} \sin\left(nD_i \frac{\pi}{2}\right). \tag{3.12}$$

On the other hand, the voltage across the input of the secondary-side diode bridge rectifier can be stated as follows.

$$V_O = \sum_{n=2k+1} V_n \sin(n\omega t), \tag{3.13}$$

where the Fourier series coefficient, V_n, is given by (3.14).

$$V_n = (-1)^{n+1} \frac{4}{n\pi} V_L. \tag{3.14}$$

From (3.1) and (3.2), the sinusoidal voltage across the input of the secondary-side diode bridge rectifier is given by (3.15).

$$V_O = \frac{U_1 \omega M R}{r_T(r_R + R) + (\omega M)^2} \sin(\omega t). \tag{3.15}$$

On the other hand, by using the sinusoidal approximation, the fundamental component of V, V_O, is given by (3.16).

$$V_O = \frac{4}{\pi} V_L \sin\left(\frac{\pi}{2}\right) \sin(\omega t). \tag{3.16}$$

Using (3.12), (3.15), and (3.16), the load voltage (V_L) is given by (3.17).

$$V_L = V_{max}\sin\left(\frac{\pi}{2}D_i\right), \tag{3.17}$$

where V_{max}, the maximum possible load voltage, is given by (3.18).

$$V_{max} = V_{DC}\frac{R\omega M}{r_T(R+r_R)+(\omega M)^2}. \tag{3.18}$$

3.4 Circuit and parameter design of the system

The design of a high-power wireless charging system incorporates several aspects. Shapes of the transmitter and receiver coils and the compensation circuits of both sides are the main features affecting system efficiency and power transfer. Moreover, proposing design algorithms for determining the system parameters is another approach to make the system more efficient. Thus, these three different design approaches are presented in this section. Moreover, some standards defined for the sake of coordination between different sections of relevant academic and industrial groups are discussed in this section as well. Moreover, the efficacy of soft-switching in the design stage of WPT systems is explored at the end of this section.

3.4.1 Standards for WPT system

In order to achieve proper coordination of researches and industries, the Society of Automotive Engineers (SAE) has defined *SAEj2954* standards for wireless charging of EVs. Meanwhile, the standards have been associated with SWC only and no standard regarding DWC has been presented until now. However, it is predicted that SAE would incorporate DWC in the near future [11].

SAEj2954 is composed of several issues regarding the efficiency and power rates, air gap, frequency range, and some practical instructions. In terms of system nominal VA, WPT systems are categorized into WPT1, WPT2, and WPT3 for 3.7, 7.7, and 11.1 kVA, apparent power values, respectively. In all of the three categories, the efficiency must be higher than 85% and 80% for aligned and misaligned conditions, respectively [26]. Moreover, a transmitter sub-system with a specific power value must be able to support receiver sub-systems of vehicles with lower power values. Yet, for the vehicles with higher power values, a transmitter must provide its nominal power. In terms of the air gap, the Z_1, Z_2, and Z_3 categories have been defined respectively in the standard for 10–15, 14–21, and 17–15 cm distance between the coils. Furthermore, a transmitter sub-system designed for a specific air gap category must be able to support the receiver sub-systems constructed for lower air gap class too [26].

Regarding the defined categories for power and air gap, several instructions for system configuration, circuit parameters, and coil parameters have been presented in the SAEj2954 [27]. Moreover, 85 kHz frequency has been defined as the nominal frequency of SWC systems for all of the mentioned categories. Also, the standard allows for frequency tuning in the range of 79–90 kHz [9].

Due to the possible effects of the high-frequency magnetic fields used in WPT systems on pedestrians and other creatures, instructions must be provided for industrial and academic activists. International Commission on Non-Ionizing Radiation Protection (ICNIRP) standard has been presented to confine the high-frequency magnetic field of such systems [11, 28]. For instance, the magnetic field peak value must not exceed 38.2 μT [11].

3.4.2 Types of transmitter and receiver coils

The design of transmitter and receiver coils of WPT systems has been a major issue since the early stages of this idea. It is desired that the mutual inductance of designed coils must be high enough for high-power WPT applications. Although using large size coils will lead to large mutual inductance, involving extremely large-sized coils exaggerate the difficulty in implementation and usage. Apart from that, the variations of coupling coefficient and mutual inductance highly depend on the design of coils. In other words, it is desired that mutual inductance does not vary significantly in the case of coil misalignment.

As basic structures, circular and rectangular coils have been proposed in several literature [29]. These basic structures are easy to analyze and manufacture rather than more complicated ones. While the rectangular coils involve a larger area as compared with the circular one, their magnetic field would not be as symmetrical as that of the circular one [30]. Using Bessel-Fourier transformation and Dual-Fourier transformation, the analytical models of planar spiral circular and rectangular coils have been presented in the literature [30]. The self-inductance and mutual inductance of these coils have been analyzed by considering the relative permeability of substrates and line spacing of coils [30]. The variations of mutual inductance due to misalignment have been studied as well [30]. In terms of geometry, rectangular coils have been found as a better option for WPT systems [30]. The obtained experimental results and analysis show that the ferrite bars and ferrite substrates employed in the coils remarkably enhance the mutual inductance. Nonetheless, ferrite bars tend to increase system loss too [11].

Conventional design schemes would lead to large coils for high efficiency, which may enjoy low-power density within coils. The product of coupling coefficient and quality factor of the coils has been considered as a design figure of merit [31]. Thus, high area-related power density has been considered in the optimization of WPT circular coils as well as system efficiency [31]. As a result, a power density of 147 kW/m^2 and an efficiency of 96.5% have been obtained [31].

Double D (DD) winding structure containing two adjacent square coils without overlap has been proposed in the literature [32]. It is proved that a higher coupling coefficient can be achieved as well as lower losses in the aluminum shield. However, the coupling coefficient of DD coils becomes zero for two specific values of lateral misalignment. An empirical formula is presented in [32] to correct the null coupling position of DD pads. The double D quadrature (DDQ) structure is obtained by adding another overlapping square coil to the DD structure. The DDQ arrangement adds the benefit of tolerance for lateral misalignment in the coils [33]. The bipolar

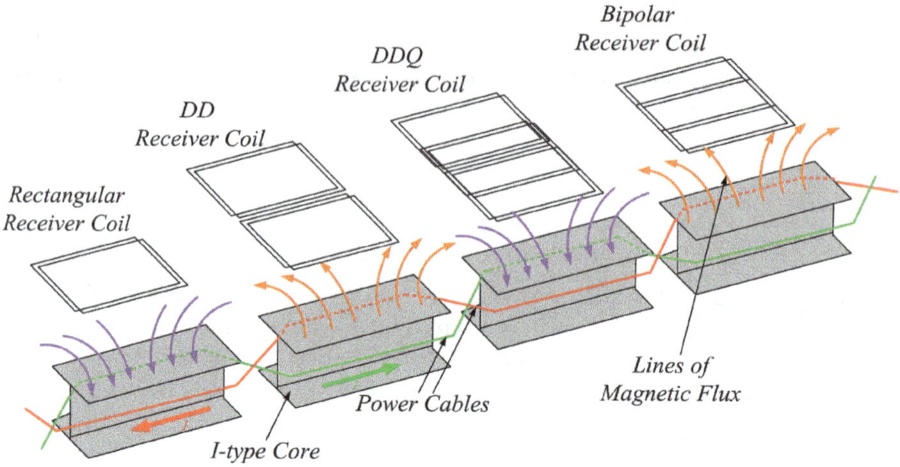

Figure 3.3 *The I-type transmitter rail and different possible receiver configurations including rectangular receiver, DD receiver, DDQ receiver, and bipolar receiver*

pad has been constructed by overlapping coils used in the DD structure [34]. It is observed that besides being lower in cost, the bipolar pads exhibit the same amount of tolerance to lateral displacements as DDQ configuration [34]. Pictorial depictions of rectangular, DD, DDQ, and bipolar receiver coils are shown in Figure 3.3.

The induced voltage in a single-phase receiver within an EV fluctuates due to variation in the coefficient of coupling as it moves along the segmented pads. In order to mitigate the effect of these fluctuations, researchers have proposed the dual-phase coil receiver in [35]. It is shown that not only the fluctuation decreases from 1.0 to 0.3 but also the average induced voltage doubled as compared with the case of employing a single-phase receiver [35].

This idea can also be extended to multi-phase receiver structures [36]. It is shown that the three-phase and four-phase receivers reduce the fluctuation factor from 1.0 to 0.13 and 0.08, respectively [36]. It is noteworthy that by increasing the number of phases cable cost increases and also the system power loss decrease in parallel connection of phases [36]. Meanwhile, in series connection the voltage level at the resonator decreases with the increase in the number of phases. Moreover, design guidelines for the multiphase receiver have been proposed considering the optimizations of the single-phase coil size and the distance between phases [36]. These optimization techniques reduce the cost, loss, and voltage level of the resonator.

Incorporating an extra coil and compensation capacitor (LC) branch short circuit, also called a repeater circuit on the transmitter side, improves the system performance. While the efficiency and transferred power can be improved simultaneously by using a repeater, it cannot be realized in a classical 2-coil system [37]. Here, the self-resonant frequencies of the transmitter circuit, receiver circuit, and repeater circuit, as well as the operating frequency, have been used as the optimization variables. Researchers have proved that their proposed optimization has improved both the efficiency and the transferred power against load variations [37].

Tri-spiral repeaters have been introduced in [38]. For optimizing this configuration, a novel figure of merit has been developed considering the efficiency and the transferred power along with receiver misalignment tolerance in 6 degrees of freedom [38]. While it is shown that system performance improves against misalignment, a slight performance reduction was observed in the alignment [38]. The striking fact is that the proposed figure of merit and design scheme can be modified according to different applications with respect to their possible misalignment degree of freedom [38]. In order to accentuate the DWC systems rather than SWCs, the proposed designs of transmitter structure are reviewed in this section. The design of the transmitter-side configuration is of great importance in the DWCs due to the following reasons.

a. Unlike the cable-based chargers and stationary wireless chargers (SWCs), the DWC structure is not limited to the specialized locations of cities. Spread along the urban streets or roads, the transmitter structure comprises the majority of DWC structure expenses.
b. Due to the discrete presence of EVs on the DWC structure, it is needed to excite the charging track when an EV enters it and turn it off when an EV leaves it. Otherwise, the efficiency declines due to the constant primary losses while the current always flows through the transmitter coil. Furthermore, the high-frequency current produces a high-frequency magnetic field with a considerable magnitude, which may harm the pedestrians and other creatures.
c. In practical cases, however, it is possible that several drivers want to charge their EVs using one charging lane. Therefore, a proper structure must be able to charge several EVs simultaneously.

Both long power rail and segmented short primary power pads have been proposed for dynamic WPT systems [11, 12, 39]. Advantages such as low electromagnetic field (EMF), low weight, and compact structure are achieved in segmented short pads, but the costs of power converters increase in this configuration due to the complexity of the control system. On the other hand, using long power rails reduces the number of power converters and simplifies the control system. But the EMF and energy waste increase in this configuration [11].

On the basis of components such as bipolar pads and unipolar ones, a complex coupling structure for DWC has been proposed in [40]. It is claimed that employing the proposed structure would result in stable mutual inductance and subsequent stable power transfer during the movement of EV [40]. In the proposed configuration, unipolar square coils have been used for both transmitter and receiver sides. Furthermore, in order to mitigate the mutual inductance between unipolar transmitter coils, bipolar coils have been added on the transmitter side [40]. The bipolar coils consist of two DD windings with perfect overlapping. Thus, each transmitter segment is a series circuit of a DD winding located in the middle of the previous

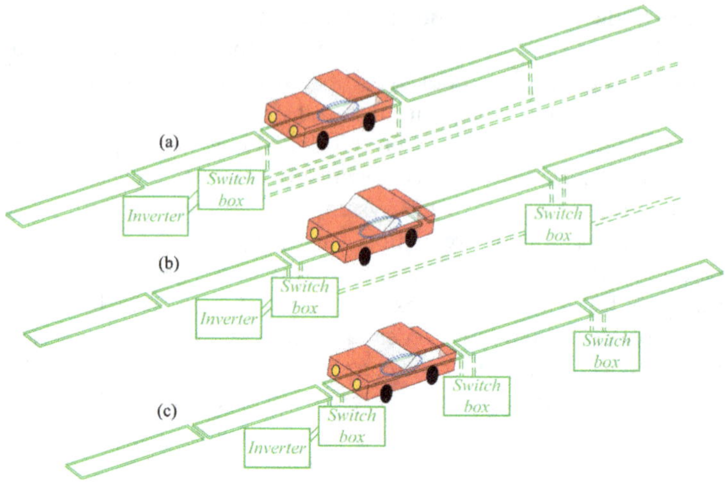

Figure 3.4 Different configuration of DWC: (a) with centralized switch box; (b) with distributed switch boxes; (c) with cross-segmented X-Rails

transmitter segment, a unipolar square coil located in its own transmitter segment, and a DD winding located in the middle of the next transmitter segment. For the sake of avoiding excessive electromagnetic interference, automatic control has been employed to energize the transmitter segments [40].

In a DWC structure depicted in Figure 3.4(a) single inverter is used in order to supply several transmitter coils [41]. It is noteworthy that a bundle of power supply cables and a centralized switch box are required to transmit the produced current to the coils and control their excitation status, respectively. Meanwhile, the striking fact about the DWC structure is that cable cost comprises 20% of the total construction cost [41]. Thus it is cost-effective to reduce the expenses of this sector.

The number of power cables and their total length can be reduced by employing the configuration demonstrated in Figure 3.4(b) [41]. In this configuration, several distributed switch boxes are substituted for one centralized switch box. Each of them can control the excitation status of two adjacent transmitter coils.

In spite of the fact that the total length of cables reduces significantly, using the common power supply cable for several coils increases the construction cost and cable conduction loss [41]. While these configurations could charge merely a single EV, the intuitive design proposed in [41], which is called Cross-Segmented Power Supply Rails (X-Rail), can support multiple EVs. Moreover, the required total cable length is just 50% of the distributed switch box configuration. Hence, this structure could be significantly more cost-effective compared with the previous ones [41]. The schematic view of the X-Rail is depicted in Figure 3.4(c). In terms of core and winding, U-type core monorail, W-type core dual rail, and I-type core power rail are proposed in order to realize the configuration. I-type core is depicted in Figure 3.3. Moreover, two pairs of power cables are used in the I-type X-Rail. Therefore, the activation and deactivation of the rail can be achieved by changing the direction of each pair using a special switch box. As the two adjacent cables in each pair conduct currents in the same direction, a considerable magnetic field is generated when the coil is activated. By contrast, the produced magnetic field from one pair would be canceled by the oppositely oriented magnetic field produced by another pair, if the cables within each pair conduct currents in opposite direction [41]. Therefore, by employing four sets of bidirectional switches the current direction of each cable can be controlled. Furthermore, a few capacitors and a coupling transformer are also used so as to cope with the compensation of variable inductance [41]. Even though the results have shown that the produced EMF decreases remarkably during the deactivated operation of coils by using the twisted cable and the deactivation method, the ICNIRP standard is violated at the point of intersection of cables. Nonetheless, the problem can be solved by using a copper net shield around the cables [41].

In the Korea Advanced Institute of Science and Technology several generations of online electric vehicles have been implemented [12, 39]. For the first generation, a DWC for a little golf car was constructed with a 45-m power track and a 1-cm air gap [12]. In this case, core has been used for both transmitter and receiver sides. Focusing on the improvement of the air gap, in the second-generation, designers increased it to 17 cm so as to comply with the Japanese and Korean road regulations [12]. By employing a 60-m U-type power supply rail and the I-type pick-up, it was observed that the effective area of the magnetic path increases. The W-type power supply rail and the flat pick-up with overlapped double coils were incorporated in the third generation in order to achieve high power and large lateral tolerance. Here, magnetic shield protection is not required due to a decrease in the leakage flux [12].

While the system efficiency had not been as high as the previous ones, the design of the overall system has been improved in the third plus generation [12]. By employing the W-type core for the transmitter side, the power transfer increased significantly due to the declined total magnetic resistance of the system. Moreover, the width of the transmitter rail was halved as compared to the second generation resulting in the decreased capital cost of the system [12]. Using I-type configuration in the fourth generation resulted in the decrease in the rail width significantly, consequently resulting in a 20% decrease in the overall cost as compared to the third generation [12]. The transmitter was manufactured in a modular configuration containing the rail power supply and capacitor banks making them robust against humidity and mechanical impacts. The transmitter track width was taken to be 4 cm in the fifth generation, which used the S-type configuration [12]. Besides having less cost of construction and deployment time is considered another virtue of this design [12]. The cost of construction in the sixth generation design can be decreased further by using coreless rails [39]. This is because the concrete layers used for protecting the core plates in previous generations are not required for the coreless structures [39].

The Endesa research team developed a DWC system for Roadway Powered Electrical Vehicles (RPEVs) in 2014 in Spain, Malaga. Incorporating 30 m U-type power tracks in the roadside system rather than multiple small pads, the configuration has continuous power transfer, low construction cost and time, as well as simple control characteristics. Furthermore, a control system is incorporated in order to guide the drivers for keeping the vehicle trajectory on the charging pad and improving the system alignment [39].

Another implemented system in Lathen, Germany, developed by the INTIS research has a 25-m power rail with a double U-type power supply rail, which has two U-type power rails in parallel. The system is able to transfer 200 kW power at an operating frequency of up to 35 kHz [39].

3.4.3 Types of compensation circuits

The WPT systems having large leakage inductance and operating at high frequencies produce a large value of reactive power until they are effectively compensated. Hence, compensation circuits have been proposed to mitigate the reactive power produced by such systems [42]. Compensation in WPT systems is a major decisive issue since it affects the operation in many ways. First of all, a complete compensation would result in the minimum apparent power derived from the power source for constant active power [9]. It would decrease the input current of the system if the input voltage remains constant. Higher currents in the circuits result in higher power loss in the wires consequently decrease their efficiency. Moreover, the realization of soft-switching depends on the phase difference between inverter voltage and current. Rephrase this sentence to make its meaning more clear and make it more connected with the previous sentence [43]. Another striking fact is that sensitivity of system operation to coils' misalignment, operating frequency variations, load power change, and resonant frequency on both the sides depends on the compensation circuit. Hence, the compensation circuits must be designed as per the desired operating criteria.

Different compensation circuits have been studied comprehensively in the literature [44–46]. Employing capacitors in series with transmitter and receiver windings, S-S compensation, has been regarded as a straightforward and suitable compensation configuration for wireless EV charging systems. The S-S topology can be employed in both applications of constant current and constant voltage WPT systems [44]. It is proved that S-S compensation increases system efficiency for high rates of load power [47]. Moreover, the zero phase angle between input voltage and input current waveforms, ZPA, can be achieved by setting the operating frequency of transmitter and receiver sides at a common value of resonant frequency [9]. On the other hand, the optimum frequency in this configuration is the resonant frequency of the receiver side [25]. Therefore, the optimum frequency is not dependent on the coupling coefficient or equivalent load resistance. Thus, the optimum frequency does not change with the variations of coupling coefficient and load resistance. This can be regarded as an extolled virtue for DWC since variations of coupling coefficient and load resistance are frequent during system operation. Moreover, the input impedance of DWC declines significantly when EV leaves the charging pad. This increases the current remarkably if it is not regulated by the controllers [11]. Furthermore, the load voltage increases by the increase of load resistance or misalignment, which may result in harming system components [11].

Using a series capacitor on the transmitter side and a parallel capacitor on the receiver side results in series-paralell (S-P) configuration. In this case, the receiver side operates as a current source, which makes it suitable for battery charging. On the other hand, employing the series compensation on the transmitter side results in a reduced voltage of a power source [48]. Unlike the previous structures, in this method the resonant frequency of the system is dependent on the coupling coefficient and load resistance. However, DWC must incorporate current limiting control in order to obviate overcurrent conditions when there is no receiver on the pad [11]. Moreover, the input *PF* is not suitable for low load-resistance values. It is therefore advised to use the S-S compensation topology when the load resistance is small and the S-P topology when the load resistance is large [49].

Paralell-Series (P-S) configuration occurs if the parallel capacitor on the transmitter side and series capacitor on the receiver side is used. Though this configuration exhibits almost the same efficiency profile as S-S configuration, the input *PF* cannot reach unity for low load resistances [11]. Another possible compensation configuration, Parallel-Paralell (P-P), comprises parallel capacitors with the transmitter and receiver coils. Meanwhile, it has not been employed in wireless EV chargers due to its deficiencies in high-power applications [11].

Among the other possible configurations obtained by combining series and parallel capacitors and inductors, using a series inductor with parallel capacitor and transfer coil (L-C-L structure) has gained considerable attention [50–53]. It can be implemented by adding an extra series inductor to the parallel LC network of the transmitter. While the receiver side can be compensated by using either series or parallel capacitor, the parallel one is proposed in the SAEj2954. The proposed system can be considered as a constant current source with no change under variations of equivalent load resistance and coils coupling coefficient [52]. Also, the transmitter coil current remains constant in this case [54].

In addition to the aforementioned configurations, a few more complex ones have also been proposed in the literature. In order to cope with the change in transmitter inductance caused by activation and deactivation of segments of X-Rail structure, researchers have proposed a configuration incorporating two capacitors and a transformer [41]. They have shown that the system resonance frequency can be made constant if the capacitors are selected according to their proposed design criterion.

3.4.4 Parameter design methods

Several issues like system efficiency and power transfer are considered as crucial parameters for designing the WPT systems [37, 38]. However, the choice of inductance and compensation capacitors is made from the limited practical range [47]. Moreover, economic considerations too are dominant in the design of WPT systems [12]. A general parameter design method for WPT systems has been proposed in the literature [47]. The value of both side inductance and capacitance varies in their limited range along with several iterations of the algorithm. For each inductance and capacitance, system efficiency is calculated for a range of operating frequencies [47]. The parameters will be recorded if the obtained efficiency is higher than its value in the previous iteration. It is declared that the proposed method can be applied in three different scenarios [47]. In the first case, the capacitors are kept constant, and different resonant frequencies are realized by changing the inductances. The resonant frequency is kept constant by changing both capacitance and inductance of each side in the second case so as to keep the product of L and C constant. Meanwhile, the resonant frequency is changed in addition to inductances and capacitances in the third case [47]. Due to the limitations of the design, each of the cases can be selected.

Another parameter design algorithm has been proposed in the literature [55]. System parameters can be obtained so as to optimize efficiency while minimizing the voltage across the resonant capacitors. The process uses the models of self-inductance and mutual inductance of the transmitter and receiver coils to calculate them for a given number of winding turns. The operating frequencies and the area of cross-section of the wire are also assessed. The number of winding turns increases in each iteration and the process terminates when the number of winding turns reaches their maximum. From the resultant possible cases, the one that provides the required power with the lowest copper quantity is selected [55].

A parameter design method based on the variable frequency phase shift control strategy and optimal operation frequency range was proposed in the literature [56]. Overall, this method determines multiple boundary conditions according to the requirements of the system. The required charging current and voltage, the minimum allowable value of the quality factor, and both sides resonant voltage and current are considered in this design method. The formulated limitations have been translated into boundary conditions for the transmitter and receiver inductances. Hence, by plotting the result of boundary conditions on the plane of transmitter and receiver coils inductance, an optimum area composed of pairs of transmitter and receiver coils inductance satisfying all of the mentioned boundary conditions can be obtained [56]. Subsequently, the value of compensation capacitors can be determined according to the selected inductances and desired resonant frequency.

3.4.5 Considerations for soft-switching of inverter

Recently, there is an idea where batteries on EVs are used as energy storage systems for utility grids [57]. For this, the bidirectional WPT (BWPT) technique is expected to help the connection between the batteries and the utility grid [58–60]. The reason is that the EV battery systems with BWPT can receive and transmit power without manual charging operation. Due to the high-frequency operation of systems, high-speed semiconductor switching devices such as Silicon Carbide Metal–Oxide–Semiconductor Field-Effect Transistors (SiC MOSFETs) are used for converters in BWPT. However, these switching devices cause many electromagnetic noises during high-speed switching [61]. From this background, the soft-switching technique is often applied to the WPT systems [60, 62, 63]. It is inevitable to achieve soft-switching in order to make the switching secure without any severe loss of power.

The circuit configuration of BWPT can be obtained by substituting the passive receiver side rectifier of Figure 3.1 with an H-bridge active converter [64]. Here, δ,

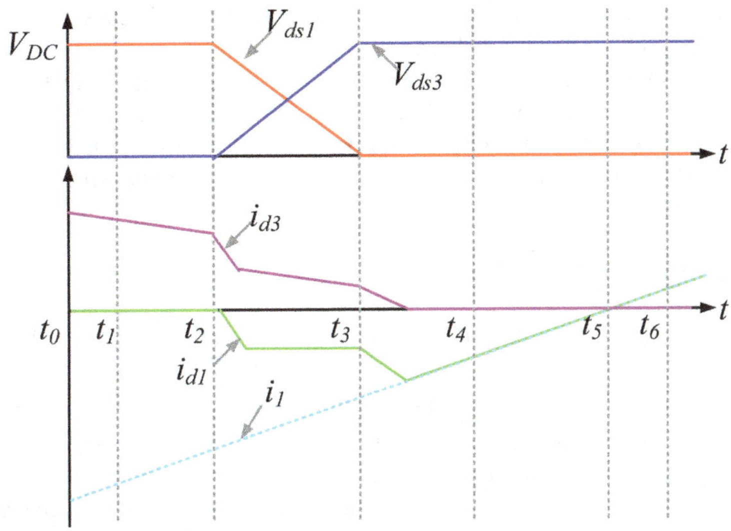

Figure 3.5 *Variations of DS voltages of switches* S_1 *and* S_3 *and currents* D_1 *and* D_3 *while achieving soft-switching operation in switches* S_1 *and* S_3

α_1, and α_2 are defined as the phase shift between the transmitter and receiver switching modulation, the phase angle between the legs in the transmitter side, and the phase angle between the legs in the receiver side, respectively. In BWPT system, the voltages and the currents in the resonant circuit can be easily controlled by δ, α_1, and α_2. Thus, soft-switching can be achieved according to the following requirements.

The soft-switching operation can be classified as "turn-on soft-switching" and "turn-off soft-switching." Here, Figure 3.5 and Table 3.1 show the transition of the switches S_1 and S_3 during the soft-switching operation. In Figure 3.5, S_1 and S_3 turn-on and -off with soft-switching, respectively. In order to achieve the turn-on soft-switching, it is necessary that the drain-source (D–S) voltage reaches zero before the turn-on transition. In other words, during the dead time before the turn-on transition of the switch, it is necessary that the electric charge in the D–S parasitic capacitor of the switch is completely discharged. In this case, the current from the resonant network works to discharge, which is that reactive current. Therefore, in order to apply "turn-on soft-switching" to the active bridges, the reactive current is essential. On the other hand, S_3 can achieve the turn-off soft-switching by the effect of the

Table 3.1 *Soft-switching operation*

Sub-period	$t_0 - t_1$	$t_1 - t_2$	$t_3 - t_4$	$t_4 - t_5$	$t_5 - t_6$	$t_6 - t_7$
Conducting elements	S_3	S_3	S_3 D_1 C_1: discharging, C_3: charging	D_1	S_1	S_1

D–S parasitic capacitor in the whole operation area. This is because the capacitor prevents the rapid rise of D–S voltage.

From Section 3.4.3, in order to apply soft-switching to the active bridges, it is necessary to discharge the electric charge in the D–S parasitic capacitor of the switch before the switch turns on. When it is assumed that the discharge is finished during the dead time before the switch turns on, the requirement can be expressed by (3.19) and (3.20).

$$-\pi + \frac{\theta_d}{2} \leq \theta_{i1} \leq -\frac{\theta_d}{2} - \alpha_1, \tag{3.19}$$

$$\delta + \frac{\theta_d}{2} \leq \theta_{i2} \leq \pi - \frac{\theta_d}{2} - \alpha_2 + \delta, \tag{3.20}$$

where θ_{i2} (rad) means the phase angle of $\bar{i}_T$, which is represented as $\theta_{i1} = -\frac{\pi}{2} + \delta - \frac{\alpha_1}{2} - \frac{\theta_d}{2}$. Similarly, θ_{i2} (rad) means the phase angle of $\theta_{i1} = -\frac{\pi}{2} + \delta - \frac{\alpha_1}{2} - \frac{\theta_d}{2}$, which is represented as $\theta_{i2} = \frac{\pi}{2} - \frac{\alpha_1}{2} + \frac{\theta_d}{2}$. The inequalities (3.19) and (3.20) represent the soft-switching requirement for the primary-side and secondary-side active bridge inverters.

However, the soft-switching operation affects the efficiency of the resonant circuit. The maximum efficiency of the resonant circuit can be achieved by optimizing the power factor of the resonant circuit and the power loss balance between the transmitter side and the receiver side [64]. In other words, during maximizing the resonant circuit efficiency, the reactive currents in the system are the lowest. However, from the above, soft-switching is achieved by utilizing the reactive currents in the resonant circuit. Therefore, the soft-switching operation and the maximization of the resonant circuit efficiency cannot be achieved simultaneously. In order to improve the efficiency during soft-switching operation, it is necessary to explore the maximum efficiency point under the soft-switching requirement [62, 63].

3.5 Control system for DWC

Controlling the operation of the system is a decisive requirement for DWCs regarding the system's intrinsic features and special requirements of its application. It will be shown that system efficiency will decline if the operating frequency is not tuned properly. Similarly, the resultant load power and load voltage depend also on several operational variables. If all system parameters are considered known and constant, system operation can be set so as to satisfy the requirements. Yet, in DWCs this assumption is rarely valid due to the following reasons:

a. The coupling coefficient of the transmitter and receiver coils depends on their relative positions. While the coupling coefficient can be considered in the design stage, practically there is no warranty that the coils are aligned so as to attain the considered coupling coefficient. Moreover, in DWC application the EV is moving along the transmitter coils. Thus, not only the coupling coefficient may not be equal to its considered value, but also it may vary during the motion of the EV. The variation of the coupling coefficient, which results in the variation of the mutual inductance, directly affects the load power and system efficiency. Therefore, it is necessary to acquire a control scheme to make the DWC system robust against variation of coupling coefficient.

b. Either the DWC system charges the batteries or feeds the motor directly. In both cases, the equivalent load resistance of the system does not remain constant. In case of battery charging the equivalent load, resistance is a function of the SoC of the batteries. On the other hand, if the motor drive is fed directly by the DWPT system, the equivalent load resistance depends on the EV's required load power. Since in different driving conditions the required load power varies, the equivalent load resistance varies during the operation.
c. A proper DWC infrastructure must be compatible with all types of EVs. According to the SAEj2954, each WPT system with a special power class must be interoperable with the systems with other classes. Definitely, the same regulation would be inflicted on DWCs. It behooves manufacturers to employ a control that enables the DWC to operate efficiently for different EV types. On the other hand, it is probable that the circuit parameters of an EV deviate from the defined standard values due to the aging effect or deficiencies in components. Thus the control of a suitable DWC system must be able to deal with deviated circuit parameters of the EVs so as to provide the required power with high efficiency for them. Hence it is necessary to incorporate a closed-loop control in the DWCs.

According to the aforementioned reasons, the suitable closed-loop control system must achieve the following goals:

a. The control system must maintain the desired level of efficiency during the situations of coil misalignments and variations in the power required for the load. Obviously, reaching the maximum possible efficiency is a virtue.
b. The control system must also be able to provide desired load power or load voltage. Particularly, the load power regulation and the load voltage regulation are needed in the battery charging process and feeding the motor drive directly.

Fortunately, the incorporated active power electronic converters in the WPT systems provide several controllable variables. Thus, many control methods can be applied to the system. For instance, the duty ratio and frequency of the output inverter voltage can be controlled so as to achieve maximum efficiency. While acquiring more active power converters in the system enables incorporating more control schemes, the fewer number of converters is considered as an upside due to the lower capital cost of the system and declined power loss in converters. Therefore, it is preferable to use fewer power converters, while achieving all control requirements.

Apart from that, a feasible control method must be fast enough. Even though, there is not an intense restriction for the settling time in the control of SWC, in the DWC the time period is remarkably important due to the movement of the EV. In other words, since the vehicles must be charged while their drivers prefer to drive

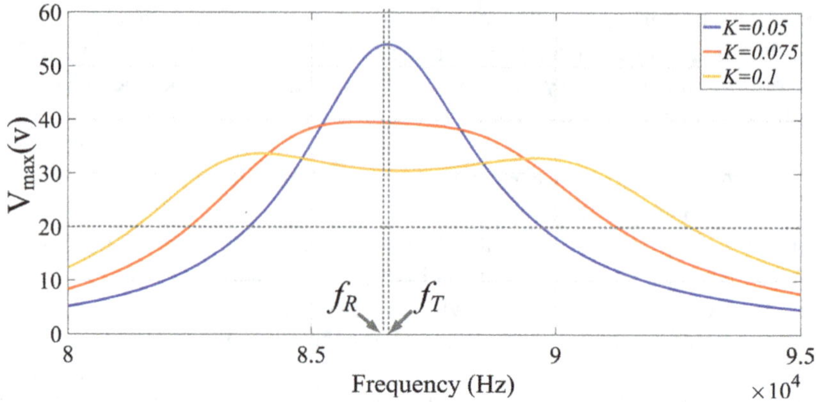

Figure 3.6 The maximum possible load voltage for different operating frequencies and coils coupling coefficients [9]

at a reasonable pace, there is a short time period available for each charging pad to transfer a considerable amount of power. Therefore, the allocation of settling time must be little compared to the whole period of charging on a specific charging pad.

3.5.1 Load voltage and power regulation

The load voltage of DWC system has been modeled in Section 3.3 according to (3.17) and (3.18). It is obvious that the load voltage is dependent on many variables of the systems including the primary DC link voltage, the duty ratio of the primary inverter, the operating frequency of the system, the coupling coefficient of coils, the duty ratio of the secondary DC-DC converter, and the equivalent load resistance. Hence, in spite of the fact that the load voltage may be affected by many system variables, many of them can be adjusted too to achieve regulated load voltage. The maximum possible load voltage of a typical DWC system is illustrated in Figure 3.6 according to (3.18). It is observed that the maximum load voltage and consequently, the load voltage itself is highly dependent on the variations of operating frequency and coupling coefficient. Therefore, a suitable closed-loop control system must be acquired for tuning load voltage on its reference value. Several methods of achieving this have been discussed in [65]. It has been observed in [66] that adjustments of the operating frequency in order to achieve the output voltage regulation would result in a decrease in the system efficiency. This solution, although being effective, may lead to deviations from the optimum operating frequency, causing a decline in system efficiency. Using a DC-DC secondary converter and controlling its duty ratio have been proposed in the literature in order to regulate the load voltage [65]. Nonetheless, the variation of its duty cycle changes the equivalent load resistance and declining efficiency, consequently. Moreover, employing an extra converter increases the system capital cost. Adding a DC-DC converter to the primary side in order to tune the DC link voltage before the inverter has the same disadvantage [67]. Researchers have proposed adjusting the inverter duty ratio in order to achieve the

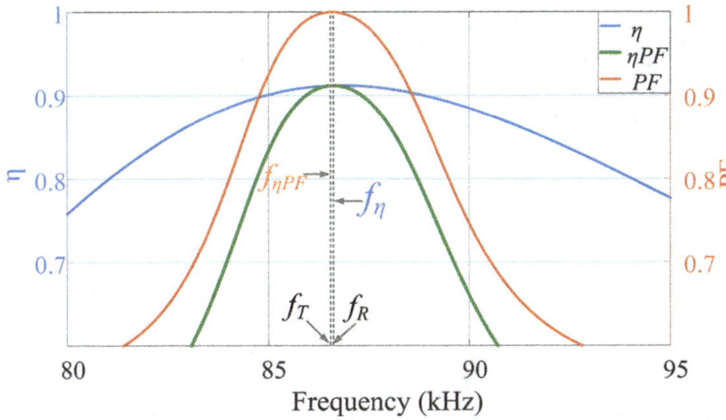

Figure 3.7 *System efficiency, power factor, and product of efficiency and power factor versus operating frequency [9]*

reference load voltage in [20]. The proposed method uses the samples of the load voltage sensor and compares it with the reference voltage. The obtained voltage error is applied as input to a partial integral controller in order to attain the proper duty cycle of the inverter. No extra converter has been exploited in this method while the system can maintain its optimum operating frequency. Furthermore, conditions for realizing soft-switching will be provided. Meanwhile, the total capacity of the system will not be used if the inverter duty cycle is set at small rates.

3.5.2 Tuning of operating frequency

WPT system operation is highly dependent on the operating frequency. Therefore, making the system operate at its optimum frequency would improve its performance. Some investigations regarding the tuning system operating frequency are reviewed in Sections 3.5.2.1 and 3.5.2.2.

3.5.2.1 Optimum frequency for maximum efficiency

The effect of operating frequency on system efficiency has been elaborated in the literature [9]. The efficiency of a typical WPT system versus operating frequency is shown in Figure 3.7. It is observed that there is an optimum frequency, denoted as f_η, in which that efficiency reaches its peak value. This frequency is so close to the common resonant frequency of the transmitter and receiver sides.

In the SS compensated WPT systems, by setting the operating frequency on the resonant frequency of the secondary side, the imaginary part of the denominator in (3.6) would be canceled. Therefore, it has been determined as a straightforward optimum frequency in the literature. If it is assumed that the on-board circuit resonance frequency is known, a constant frequency control system could be employed in order to provide the optimum operation. Meanwhile, the operating resonant frequency changes with the vertical spacing between the charging coils and due to the presence of proximal metallic objects [68]. Therefore, an automatic frequency-tuned wireless charging system is employed in order to achieve maximum power transfer efficiency [68].

An adaptive frequency tracking control method based on the fuzzy proportional-integral (PI) compound controller has been proposed in the literature so as to maintain the system operation on the resonant frequency [69]. The optimum parameters of the PI controller are updated by adding the incremental values to the old proportional and integral gains, respectively [69]. In order to obtain the incremental values for both of the proportional and integral parts, seven fuzzy subsets have been defined including negative big, negative medium, negative small, zero, positive big, positive medium, and positive small. Triangular membership functions and weighted average methods have been employed as fuzzification and defuzzification methods [69]. By using this method, the ZPA in the input of the system is achieved [69].

In order to achieve more accuracy, the optimum frequency has been sought in [70]. By solving (3.21), the optimum frequency can be obtained as (3.22):

$$\frac{\partial}{\partial f}\eta = 0. \tag{3.21}$$

$$f_\eta = \frac{1}{2\pi\sqrt{L_R C_R - \frac{1}{2}\left(C_R\left(r_R + R\right)\right)^2}} \tag{3.22}$$

It is observed that the term containing equivalent load resistance and receiver coil resistance has been neglected in obtaining the optimum frequency for maximum efficiency in the literature [9]. In DWC application in which load resistance may vary due to the variation of load power, the optimum frequency may change during the system operation. In the proposed control method in [70], the optimum frequency is calculated after each load resistance variation. Subsequently, the obtained optimum frequency can be applied to the system as the operating frequency.

A three-loop control scheme has been proposed in [56] in order to realize the variable frequency phase shift control strategy. The control method tunes the frequency within the optimal operating frequency range so as to provide the required charging current or voltage for the battery packs as well as satisfying zero-voltage switching (ZVS) conditions without extra DC-DC converters. Nonetheless, the ZVS operation cannot be realized for the resonant control methods in WPT systems if load voltage regulation is employed. Unlink methods considering the resonance frequency as the best operating frequency, the inverter conductive loss and switching loss can be reduced by using the optimum frequency mentioned in the literature [71].

In the outer loop, the load current and voltage are sampled and compared with their reference values. Two possible reference values of the transmitter current are obtained by the proportional integral derivative (PID) controller and the smaller value is selected. Subsequently, the transmitter current is sampled and compared with its desired value in the inner loop. The required inverter duty cycle is obtained by entering the error to the PID controller.

Realizing ZVS of the inverter, the third control loop uses samples of the inverter current and adjusts the inverter operating frequency [56]. A phase-detection method has been implemented by using Op-amps and D flip-flops. The phase-detection method has been elaborated comprehensively in the literature [43]. It calculates the phase difference between the rising edge of inverter voltage and zero points of the sinusoidal waveform of transmitter current [56]. The time delay of analog devices may affect the resultant phase difference. However, the possible delays of the system are considered in the literature [43]. The ZVS would be realized if the calculated phase difference is larger than zero. The calculated phase difference is compared with its desired value. The required inverter frequency is obtained by entering the error to the PID controller [56].

3.5.2.2 Optimum frequency for high efficiency and power transfer

The enhancement of system efficiency has been elaborated in Section 3.5.2.1 since it results in the optimum operation of the system. Further to the efficiency, the input power factor of the system can be considered as a decisive criterion as it determines how much of the power transfer capacity of the system is used during the operation. Operating in unity power factor means using the whole power transfer capacity of the system. In contrast, a small fraction of system capacity is employed in low power-factor operating points. In other words, high *PF* can be interpreted as minimizing the capital cost of the system, while high efficiency results in minimizing the system's operating cost. In Figure 3.7, the efficiency and power factor of a typical system are plotted versus operating frequency. It is seen that both efficiency and power factor achieve the peak value at the common resonant frequency of transmitter and receiver. Furthermore, the maximum possible *PF* reaches unity while the maximum efficiency is limited to a lower value.

In Figure 3.7, the resonant frequency of transmitter and receiver were considered identical. However, in practice, these may not be the same. As it was elaborated in this section, a suitable DWC system must be compatible with diverse EV types. There is no warranty that all the receiver circuits implemented on EVs, and all the DWC pads have been manufactured so as to have the same resonant frequency. On the other hand, the aging effect may degrade the circuit components so as to change the resonant frequency to some extent. Therefore, a practical DWC must be able to cope up with the receivers with diverse resonant frequency values. The inequality of the resonant frequency of transmitter and receiver sides has been investigated in the literature [9]. Figure 3.8(a) and (b) illustrates system efficiency and power factor while the resonant frequencies of transmitter and receiver sides are not the same. The inequality has been realized with a 6% change in the capacitance of the receiver side [9]. Hence, while the transmitter resonant frequency has been set at 86 571 Hz, the receiver side resonant frequency has reached 84 159 and 89 161 Hz in Figure 3.8(a) and (b), respectively.

As shown in Figure 3.8(a) and (b), maximum values of *PF* and efficiency do not take place at the same operating frequency. While the system efficiency remains close to its maximum as far as the operating frequency is set at the secondary

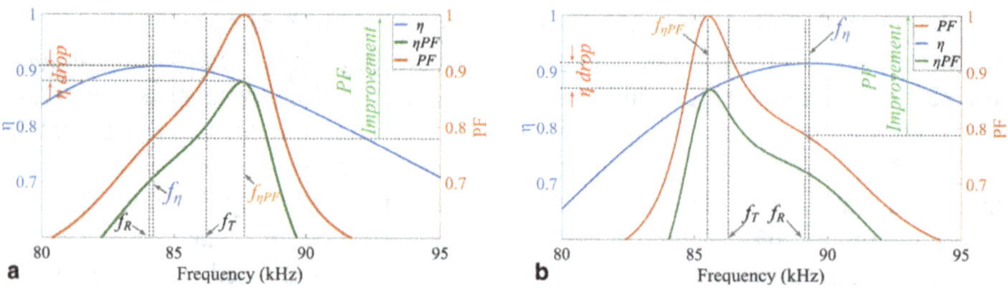

Figure 3.8 *Variation of system efficiency, power factor, and product of efficiency and power factor versus operating frequency while (a) the resonant frequency of secondary side is larger than primary one and (b) the resonant frequency of primary side is larger than secondary one [9]*

resonant frequency, the *PF* may deviate significantly from unity at this frequency [9]. Several solutions can be conceived for dealing with the issue. A novel albeit straightforward figure of merit has been introduced in [9] in order to maintain both system efficiency and power factor in high values. The product of system efficiency and power factor, $\eta.PF$ has been considered as an objective function in order to optimize system operation. The criterion can be obtained by multiplying (3.6) and (3.9) as follows [9].

$$\eta.PF = \frac{1}{U_1}\frac{P_L}{i_T} = c\frac{P_L}{i_T}. \tag{3.23}$$

It is seen that $\eta.PF$ not only incorporates both system efficiency and power factor but also represents the output power per input current while the input voltage is maintained constant. Thus maximizing $\eta.PF$ can be interpreted as maximizing the load power for constant input current derived from the inverter in case the input voltage is kept constant. However, the input voltage may be considered a control variable to regulate the load voltage. If so, it cannot be considered constant and (3.24) can be corrected as follows [9].

$$\eta.PF = \frac{P_L}{U_1 i_T} = \frac{P_L}{S_{in}}. \tag{3.24}$$

It is seen that the criteria represent output power per input apparent power of the system. Hence maximizing $\eta.PF$ results in maximum output power for specific input volt-ampere of the system. In other words, for a particular amount of load power, the minimum required input volt-ampere and a minimum rating of the primary power electronic switches are acquired. $\eta.PF$ can be written as a function of system impedances as follows, using (3.10), (3.2), and (3.3) [9].

$$\eta.PF = R\frac{\omega^2 M^2}{|z_1 z_2 + \omega^2 M^2||z_2|}. \tag{3.25}$$

The criterion has been plotted versus operating frequency for a typical WPT system in Figures 3.8 and 3.7 for the equal and unequal resonant frequency of two sides, respectively. It is noteworthy that the $\eta.PF$ gets maximum at a special

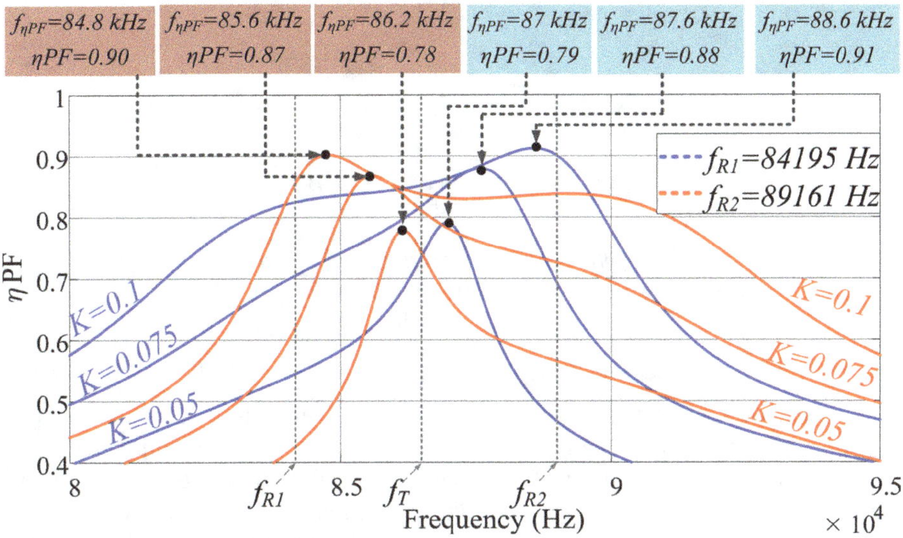

Figure 3.9 *Effect of coupling coefficient variation on* η.PF *and optimum frequency while resonant frequencies of the two sides are not equal [9]*

frequency denoted as $f_{\eta \cdot PF}$. The frequency is equal to the common resonant frequency of transmitter and receiver sides if they have the same resonant frequency. Otherwise, an optimum frequency is obtained, which is different from the resonant frequencies of either side. The $\eta.PF$ is plotted for different coupling coefficients in Figure 3.9.

It is notable that the optimum frequency changes with variation in the value of the coupling coefficient. Therefore, the figure of merit can be maximized by the frequency tracking method. Moreover, the optimum frequency opts to the transmitter side resonance frequency for loosely coupled systems [9]. This can be attributed to the dominance of *PF* in the $\eta.PF$, as shown in Figure 3.8(a) and (b). The other striking fact is that the maximum *PF* is unity regardless of the coupling coefficient of the system. However, the maximum efficiency declines for loosely coupled systems. Needless to say, by maximizing this figure of merit, the efficiency has been declined compared to its maximum value to some extent. Nonetheless, the *PF* improvement is significant enough to consider the proposed figure of merit as a proper choice [9].

The effectiveness of the proposed method in [9] has been verified by simulation and the experimental results, which are illustrated in Figure 3.10(a) and (b). As shown in Figure 3.10(a), after a few tracking steps, the optimum frequency is sought successfully.

As a result the $\eta.PF$ has reached 0.88 and 0.76 for coupling coefficients of 0.1 and 0.05, respectively [9].

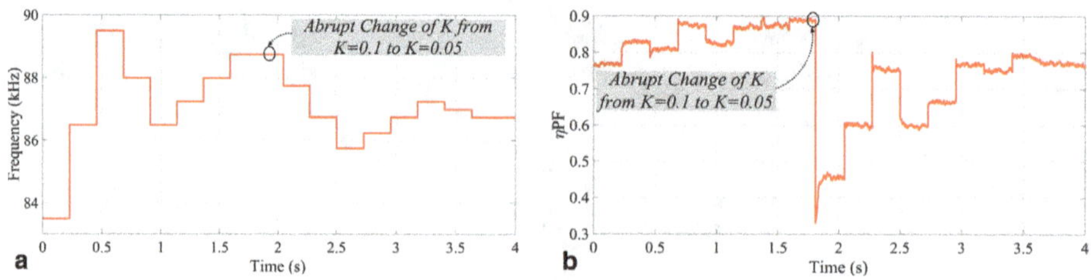

Figure 3.10 Results of optimum η.PF *tracking method: (a) variations of operating frequency and (b) variations of* η.PF *[9]*

3.5.3 Load impedance matching

The impact of frequency variations on system efficiency has been discussed in Section 3.5.2. Meanwhile, another decisive factor affecting system efficiency is the equivalent load resistance. Hence, it becomes essential to discuss the control methods for load resistance tuning used for impedance matching. Due to the variations of the WPT system load, the equivalent load resistance cannot be assumed constant for different load operating conditions. In DWCs, driving conditions including road slope, vehicle's speed, and acceleration may change the required load and the equivalent load resistance. System efficiency is plotted versus the equivalent load resistance in Figure 3.11 while the system is operating on the receiver side resonance

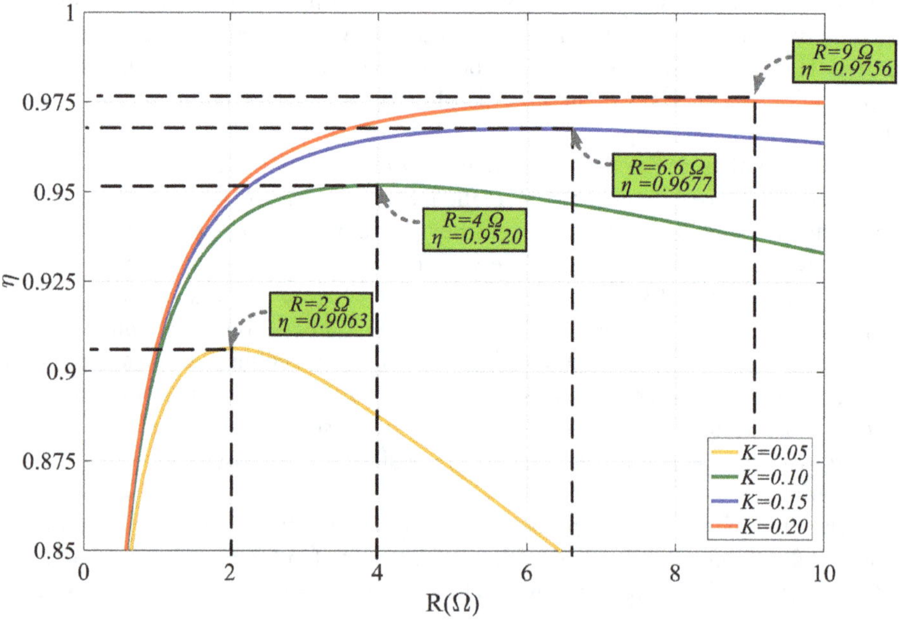

Figure 3.11 System efficiency versus the equivalent load resistance in different coupling coefficients

frequency. It is observed that the efficiency reaches its peak for a specific value of equivalent load resistance [42].

By solving (3.26), the optimum equivalent load resistance, R_o, can be obtained as (3.27) [42].

$$\frac{\partial}{\partial R}\eta = 0. \tag{3.26}$$

$$R_o = \omega M\sqrt{\frac{r_R}{r_T}}. \tag{3.27}$$

In order to realize the impedance matching, a DC-DC power converter can be employed. The equivalent load resistance seen from the input side of the converter can be tuned by adjusting the converter duty cycle [42]. Exploiting Boost and Buck-Boost converters for this purpose has been proposed in the literature [67, 72, 73]. Regarding the usual low values of equivalent load resistance in the DWC application, using a Buck converter can be more effective due to its ability to increase the equivalent load resistance seen from its input side. The resultant equivalent load resistance by employing a Buck converter is given by (3.28) [25].

$$R = \frac{8}{\pi^2 D_B^2} R_L, \tag{3.28}$$

where D_B is the Buck converter duty ratio. For this case, the optimum D_B, denoted as D_o, can be obtained by (3.29) [42].

$$D_o = \sqrt{\frac{8}{\pi^2}\frac{R_L}{\omega M}\sqrt{\frac{r_T}{r_R}}}. \tag{3.29}$$

By considering (3.27) and (3.29), it is observed that R_o and D_o are functions of the operating frequency (ω), the load resistance (R_L), the mutual inductance (M), the transmitter coil resistance (r_T), and the receiver (r_R) [42]. Among these parameters, the mutual inductance and the load resistance may vary during the operation of DWC. While the load resistance can be calculated online by using the measurements of load voltage and current which are usually available from the associated on board sensors, coping with the mutual inductance variations is not simple [25]. The effect of the coupling coefficient on the system efficiency and the consequent R_o is depicted in Figure 3.11. It is seen that the value of R_o changes with the variations of the coupling coefficient [74]. Thus, the main feature of an effective impedance matching scheme for DWCs is its adaptation with the variations of coupling coefficient. Therefore, two distinct approaches have been proposed for controller design: the tracking method and the coupling coefficient estimation method. These approaches are investigated in Sections 3.5.3.1 and 3.5.3.2.

3.5.3.1 Tracking method

In order to cope with the variations of K and R, the D_o can be determined by a searching algorithm. In the tracking scheme proposed in [70], different values of D_B were applied to the Buck converter and the corresponding efficiency was observed.

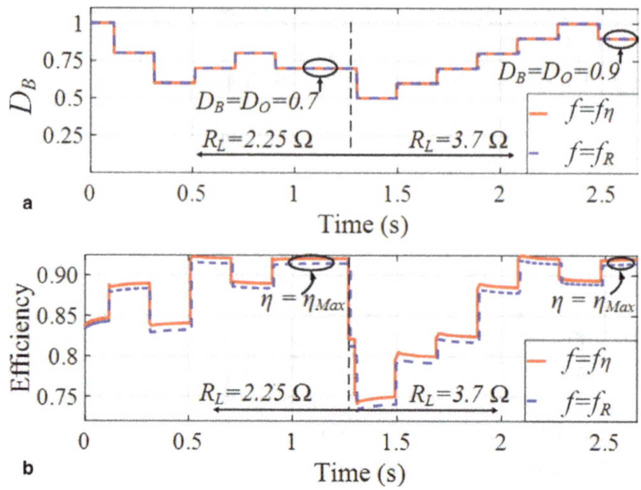

Figure 3.12 Results of optimum D_b method: (a) variations of applied D_b and (b) variations of system efficiency [70]

Generally, if the previously applied D_B leads to increase the efficiency, the same pitch and direction of change would be applied to D_B. Otherwise, the pitch and direction of change would be halved and reversed, respectively [70]. The performance of the control algorithm is depicted in Figure 3.12, where the results of the tracking method are illustrated [70]. It is observed that the control has reached the optimum operating point in less than 1 s. In order to assess the system's ability to find the new optimum operating point in case of variations of R_L, the load resistance was abruptly reduced from 2.25 to 3.7 Ω intentionally. It is seen that the tracking method started to search for the new *Do* again and achieved it in less than 1.5 s.

3.5.3.2 Estimation of mutual inductance

As it has been mentioned before, the optimum equivalent load resistance is a function of the mutual inductance of coils. Since variations of mutual inductance or coupling coefficient are possible in wireless chargers, proper control schemes must be employed to tune the equivalent load resistance on its optimum value so as to obtain the highest possible efficiency. While the previously mentioned method exploited the tracking method to achieve this goal, some other solutions based on estimation methods have been proposed in the literature.

The initial value of the coupling coefficient and the dynamic value of it have been studied separately in the literature [75]. The system initial coupling coefficient has been obtained as a function of circuit parameters, voltages, and currents. Meanwhile, the equation has two solutions based on one instance of the output voltage. Therefore, two instances of the output voltage can be applied and the four distinct values of the coupling coefficient are obtained. The initial coupling coefficient is determined by averaging the two closest obtained values [75]. On the other hand, another equation has been presented for obtaining the dynamic value of the coupling coefficient. According to the incremental model of the equation, the dynamic coupling coefficient can be obtained according to (3.30) [75].

$$K_2 = K_1 + \Delta D_i \frac{\mathrm{dF}(\mathrm{D_i})}{\mathrm{dD_i}}, \tag{3.30}$$

where K_1, K_2, and F are denoted as the initial coupling coefficient, the latest coupling coefficient, and the function relating the coupling coefficient to the inverter duty ratio. By each variation in the coupling coefficient, the K_1 must be updated in order to effectively track the optimum operating point [75]. The results demonstrate the proper estimation of the coupling coefficient of the system. Nonetheless, the accuracy of the results decreases when the difference between K_2 and K_1 is significant. It can be attributed to the fact that (3.30) is obtained based on the linearization method. Hence, by using this method accuracy may decrease when the abrupt change in the coupling coefficient occurs.

Based on the coupling circuit equations, another estimation method has been proposed in the literature. The measurements of both primary and secondary current are required in order to obtain accurate estimation. However, the proposed equation is simple and not dependent on many parameters of the system. Furthermore, it can cope with load power variation affecting the optimum equivalent load resistance.

Based on the voltage regulation method proposed in [70], an estimation method has been proposed in [25].

$$\omega M = \frac{V_{DC} D_B R \sin\left(\frac{\pi}{2} D_i\right) + \sqrt{D_B^2 V_{DC}^2 R^2 \sin^2\left(\frac{\pi}{2} D_i\right) - 4 V_L r_T (r_R + R)}}{2 V_L}. \tag{3.31}$$

Comparing with the previously mentioned method, the equation is more complex and dependent on the system parameters. However, the striking feature is that it merely needs load voltage and current sensors that are usually available on conventional chargers. Nevertheless, reduction of sensors has been realized at the expense of slower dynamics of the estimation because it is required to perform voltage regulation process, which may enforce the system to run the estimation process subsequently again. It is noteworthy that both of the methods are realized assuming that the system is operating on its common resonance of primary and secondary. Meanwhile, this assumption may not be valid in many practical cases. Thus, researchers proposed a general estimation method in [73], which not only estimates the coupling coefficient but also calculates the circuit parameters of each EV using the charging structure. Therefore, merely the primary circuit parameters are presumed known while the parameters of the EV circuit and coupling coefficient are assumed to be unknown. This is a more practical assumption due to the two following facts [73].

a. A practical DWC infrastructure must be capable of operating efficiently with diverse individual EVs that may have different resonant frequencies and coil resistances.
b. The circuit parameters of the transmitter side will be checked during frequent maintenance observations.

When an EV comes on the charging track, the controller applies two different operating frequencies during two 15 ms periods. Then the receiver-side inductance, capacitance, and resistance can be estimated by solving a set of non-linear equations [73]. At the same time, the mutual inductance would be estimated as well. Hence, the optimum frequency and the optimum equivalent load resistance would be determined. In spite of the fact that this method is highly adaptive to the variations of many parameters, it requires the phase angle measurement of primary and secondary currents, which may be practically difficult because of the high operating frequency of the system.

3.6 Future trends

While wireless communications have become prevalent now, the information signals used to be transferred by wires in the previous century. In order to have safer and more convenient usage of household and industrial devices, it would not be surprising if the thriving trend of WPT trend flourishes so as to omit many present cables transferring power to electrical devices. The study of future trends of WPT concept can be categorized regarding its diverse applications. Nonetheless, the EV charging application with a focus on the DWCs is considered in this section.

Due to the environmental issues, using renewable energy sources in the future smart grids is more probable. Thus, the effective integration of these distributed generation units with wireless EV charging infrastructure would be a decisive challenge. On the other hand, connected vehicles in the transportation systems of smart cities will be one of the decisive future implications of developed technology. Connected vehicles would be able to contact different components of smart cities including the traffic observer system, power grid infrastructure, and even other vehicles. Hence, all of these interactions must be studied so as to attain the most possible profit from the implemented infrastructure. The WPT from vehicle to grid (V2G) and vehicle to vehicle (V2V) is also discussed in this section.

3.6.1 Integration of WPT system and renewable energy systems

Wireless EV charging structures can be employed in either the urban areas or the roads between them. While in the urban areas the distribution power grid as the input power supply is often accessible, it may not be available along the whole roads. Therefore, using distributed generation units can be accounted for as an effective option to cope with this challenge. Yet, in order to produce maximum power, most of these sources need to work on a special operating point. In photovoltaic (PV) applications, for instance, the maximum power point tracker (MPPT) fulfills this task by adjusting the PV voltage at its optimum value. However, in many kinds of research, the MPPT requires an independent DC-DC converter. The researchers have proposed a multi-objective topology based on loosely coupled fly-back converters with a multi-stage switching algorithm in the literature [20], instead of the common resonant converters. The PV source with the suitable MPPT algorithm has been applied to the input, without needing an extra converter. Compared to the resonant converters, the proposed topology can achieve the same efficiency with lower operating frequency without needing the compensator circuits on both sides [20]. At the primary side, five switches are used to manage the power between the PV source, the backup battery source, and an energy recovery circuit. The MPPT is also guaranteed with these multi-functional switches without any extra converter.

3.6.2 Vehicle to grid connection

Grid to vehicle (G2V) charging tends to increase the grid power demand, which would require additional investment in power generation and distribution infrastructure [76]. As a fascinating concept, flowing electric power from the V2G would help reduce this additional cost and provide grid services like load leveling, peak shaving, frequency control, operate as a backup power supply, etc. [76–78]. The vehicles consisting of V2G technology are generally charged at the hours that electricity production is higher than grid demand and the price of charging is low. On the other hand, they can deliver power to the grid during the hours that grid demand hits a peak. During these hours, the costumer can sell the stored power in EV battery to the grid with higher price comparing the charging cost at low demand hours [79]. In addition, V2G can also supply the required reactive power for grid voltage support as well as fault blocking capabilities, increasing the reliability and efficiency of the energy system [77].

Like cable-based chargers, wireless chargers can also be employed in the V2G concept. A BWPT system that can be exploited for V2G is proposed in the literature [80]. The proposed circuit is almost similar to the system shown in Figure 3.1. Meanwhile, the secondary-side full-bridge rectifier is substituted by an H-bridge converter. Moreover, L-C-L compensation has been proposed for both the primary and secondary sides [80]. The obtained mathematical model has shown that both the amount and direction of power flow between EVs and charging structure can be controlled through either phase or/and magnitude modulation of voltages generated by converters of each system [80]. For any given voltages, the maximum power transfer is achieved when the phase angle of the pickup system is set on ±90 with respect to the transmitter voltage. A leading phase and lagging phase angle imply power transfer from the EV to the primary and from the primary to the EV, respectively. Thus, for any given primary and pickup voltages, both the amount and direction of power flow between the primary and the EV can be regulated by controlling the relative phase angle between voltages generated by primary and EV reversible rectifiers [80]. In order to regulate the track current and the output power at the desired value, two control loops are included in the primary and secondary sides, respectively. Each controller determines the duty ratio of its converter, while the relative phase angle is set on +90 or −90 depending on the desired power flow direction [80].

The dual-phase shift (DPS) control method has been also studied for a bidirectional wireless EV charger in the literature [81]. Considering losses in WPT system, researchers have shown that for each required load power, the power transfer efficiency reaches its maximum for a particular pair of duty ratios of primary and secondary converters. Meanwhile, the high stress imposed on system components during the transient state was indicated as the major deficiency of DPS control when the coupling coefficient is unknown. Therefore, in order to alleviate the transient overshoot, the optimal duty ratios have been broken into smaller steps that are applied consecutively with a short time interval between them. According to the obtained results, the proposed method is able to maximize the power transfer efficiency, while limiting the system stresses when the coupling coefficient is between 5% and 35% [81].

EV battery degradation and grid-side instability have been regarded as the two main issues for Wireless V2G in [76]. Regarding the first one, it has been claimed that if the cost of EV battery depreciation is considered, the profit per vehicle in V2G would significantly reduce. The effect would be more adverse in DWC due to the pulsating intrinsic feature of power [76]. On the other hand, the peak power encountered by the grid is significantly higher than the average charging power in both V2G and G2V conditions. This pulsation problem deteriorates in microgrids, where grid size could be small enough to trigger instability. In order to address these issues, an additional buffering unit containing a current smoothing choke inductor and a super capacitor has been employed in the primary so as to provide a grid-side energy buffer for both G2V and V2G charging [76]. Moreover, two controllers, named the track-current controller and the grid-current controller, are proposed in order to regulate the primary coil current and control charge/discharge of the buffering unit, respectively [76]. The acquired results indicate that the remarkable power difference while the charging period and standby period would be diminished by using the proposed method, while the required output power has been provided sufficiently [76].

3.6.3 V2V power transfer

The vehicle to vehicle WPT platform (V2V WPT) has been introduced as an innovative solution, which can address the challenges impeding the wide adoption of EVs [82]. This would provide a platform for vehicles in demand of electric power to receive energy from peer EVs, or a designated mobile charger, either in the stopped or in moving condition. The supplier of power can be compensated based on the level of the charge they provide [82].

Although the WPT in the V2V platform can lead to energy loss compared to the plug-in transfer, eliminating the need for detours to reach charging stations can create energy savings. As such, the net energy efficiency would depend on the efficiency of the non-contact power transfer technology as well as the density of charging stations in the network [82]. The solution also decreases the pressure on the electric grid during the day when the electricity demand is the highest since the supplier vehicles are typically charged during the night, thereby maximizing the use of electricity generation resources. Additionally, no extra grid enhancements are needed for the V2V WPT platform, since it utilizes its power source from other EVs. As such, the V2V WPT can be conceptualized as a platform for crowd-sourcing mobile, on-demand, and distributed energy storage units [82].

V2V WPT platform is introduced as an option that takes advantage of the sharing economy to promote a greener transportation system since it turns any EV into a mobile source of energy [82]. A mathematical program is proposed to optimize the logistics of the V2V WPT system. It routes supplier and recipient vehicles by placing them in spatiotemporal proximity of each other to facilitate the transfer of power between them [82]. A dynamic programming solution methodology has also been proposed to optimize vehicle routes on this expanded network [82]. Moreover, a series of numerical studies have been conducted, using commuter and long-distance trips from the state of Michigan, in order to quantify the degree to which the V2V WPT technology can reduce the dependency of EVs on charging infrastructure [82].

In order to cope with the limited accessibility to charging stations, researchers at the University of Sheffield have focused on V2V WPT too [83]. The proposed infrastructure would solve the issue of the limited number of plug-in stations as well as provide auxiliary charging between moving vehicles, thus reducing the need to locate a fixed charging station. The proposed system can be employed in two applicable scenarios [83].

a. Plug-in with wireless V2V charging system: the limited capacity of charging stations and necessary regular maintenance and service are the two main drawbacks of the plug-in charging stations. The combination of plug-in charging with a wireless V2V stationary charging system can help to solve the above issues. While the transmitter coil and receiver coil are embedded in the front and rear of the car, respectively, the first car that is connected to the grid via charging cable can charge wirelessly the car behind it.
b. Vehicle assistance wireless charging system when the vehicle's battery runs low, the driver can request the help of other EVs on the road to wirelessly charge the vehicle during motion. Using an application on a smartphone, the driver can find a near suitable vehicle that indicates available donor vehicle energy, pricing, distance, and other factors. This concept can reduce the anxiety range of EV consumers since they can get to the nearest charging station by using the available energy in another vehicle battery.

The angular offset due to the change in the location of the vehicle has been regarded as the main challenge of the wireless V2V charging technology [82]. Therefore, a mathematical model for the mutual inductance of multi-turn coils with angular offset is derived in [83]. As a conclusion, it is stated that if analog coils for transmitter and receiver are employed, the maximum mutual inductance increases by reducing the space between the two adjacent conductors. On the other hand, by using a larger transmitter coil than the receiver, the maximum mutual inductance is obtained by choosing the optimum space between the two adjacent conductors rather than the least space possible. Apart from that, increasing the number of turns will increase the mutual inductance remarkably [83].

3.6.4 Integration of WPT system and motor drive

Providing the required thrust, the electric motor is the most decisive component of EVs [27]. Thus, the way of feed it must be considered carefully in the electrical design of EVs [24]. While batteries or a combination of batteries with fuel cells and supercapacitors are considered the most prevalent EV energy sources, WPT systems can provide a new option. Rather than merely charging the EV battery, WPT systems are able to feed the motor drive system directly [9]. Thus, controlling vehicle operation can be carried out by on-board system as it is available in conventional systems. Meanwhile, driving EVs on-road systems only from the off-board primary side of WPT system may be a virtue in future smart cities. The striking fact is that the integrated configuration enables transportation systems to control the driving cycle of all vehicles and power-sharing protocols (V2V and V2G) merely with one decision center.

The operating frequency of WPT systems and electric motors differs significantly. In order to acquire considerable efficiency, the operating frequency of WPT systems must be set at several kHz. On the other hand, electric motors operate at frequencies much lower than 1 kHz. Thus, it is necessary to exploit a system that converts the high frequency of the received voltage on the receiver side to the low frequency that is needed in the electric motor. Nonetheless, the wireless charging systems and motor drive systems are considered distinctly separated in conventional designs. Generally, a transmitter power transfer circuit delivers AC voltage in the receiver circuit. Afterwards, the AC voltage is converted to DC voltage by a rectifier so as to preserve the energy in a capacitor tank or battery. The motor drive sub-system incorporates an inverter, which is controlled in a way that provides the required voltage amplitude and frequency for an AC motor. Therefore, such structures suffer from complexity and cost deficiency, in addition to reduced reliability and power transfer efficiency. Thus, eliminating a DC electrolyte capacitor and power conversion stage together with the switching converter enhances system simplicity and cost-effectiveness.

Researchers have proposed a new class of wireless secondary converter-less switched reluctance motor (SRM) drives in order to develop a unique WPT system directly driving the SRM [24]. The proposed motor drive needs only one transmitter to directly drive the phase windings of the SRM while the motor side does not involve any controller or converter. A controller at the transmitter side is responsible for both WPT and motor speed control. Meanwhile, the corresponding selective WPT and speed control are realized by the proposed compensated capacitor switcher and signal duty ratio control, respectively [24]. For the sake of simplicity, the position signal will be fed back to the controller at the transmitter side via Bluetooth communication.

Another investigation has focused on the integration of the WPT system and single-phase induction motor [84, 85]. The overall schematic view of the system is given in Figure 3.13. In the proposed system, the primary side inverter performs two functions, simultaneously. As the first fucntion, it provides high-frequency square-wave voltage to transfer power through the loosely magnetic-coupling, with

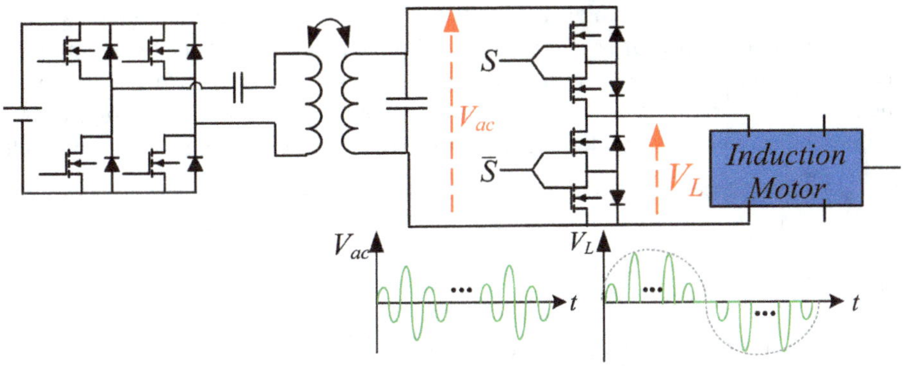

Figure 3.13 An integrated topology of WPT and motor drive and the key waveforms

acceptable efficiency. The high-frequency voltage is produced by a bipolar pulse width modulation (PWM) switching algorithm of the full-bridge inverter [84, 85]. As the second function, it drives the electric motor with a closed-loop control scheme that maintains the ratio of motor supply voltage magnitude over its frequency constant. A look-up table is applied to the PWM switching algorithm regarding the reference wave produced from the mentioned scheme. In the next step, the modulated high-frequency sinusoidal voltage in the primary is demodulated by the AC-AC chopper in the secondary side and, subsequently, supplies the motor with a low-frequency sinusoidal current [84, 85]. Incorporating two reversed series switches in the upper and lower sections, the AC-AC chopper topology and its symbolic input and output voltages denoted as V_{ac} and V_L are demonstrated in Figure 3.13.

Another simple and low-cost direct single-to-three-phase frequency converter based on the cycloconverter has been proposed in the literature in order to be used in IPT systems [86]. Reduction in switching frequency and elimination of the capacitor link are the main virtues of this method. Space vector modulation has been employed as a switching strategy due to its ability to improve the quality of output voltage by decreasing total harmonic distortion [86]. Providing the required frequency and amplitude from its single-phase input, the cycloconverter incorporates bidirectional switches without diodes that enable the converter to provide bidirectional power flow.

3.7 Conclusion

In order to cope with the challenges of EV batteries, WPT has been proposed as a novel idea. By wireless charging of EVs during motion, battery weight, capacity, and cost can be decreased. Furthermore, the improved accessibility to charging units provided by wireless chargers would result in a declined depth of discharge in EV batteries. Thus, their life span would be extended remarkably. In this chapter, several features of the system have been studied. The basic principles of wireless

EV chargers have been explained as well as the main challenges regarding system design and control. It is observed that, despite serious challenges, the intuitive schemes proposed in the literature would solve many decisive issues. Moreover, several possible future trends of these systems reviewed in the last section of the chapter can convey a suitable prospect to the readers considering further research in this field.

References

[1] Tie S.F., Tan C.W. 'A review of energy sources and energy management system in electric vehicles'. *Renewable and Sustainable Energy Reviews*. 2013, vol. 20(4), pp. 82–102.

[2] Castaings A., Lhomme W., Trigui R., Bouscayrol A. 'Comparison of energy management strategies of a battery/supercapacitors system for electric vehicle under real-time constraints'. *Applied Energy*. 2016, vol. 163, pp. 190–200.

[3] Bonges H.A., Lusk A.C. 'Addressing electric vehicle (EV) sales and range anxiety through parking layout, policy and regulation'. *Transportation Research Part A: Policy and Practice*. 2016, vol. 83(1), pp. 63–73.

[4] Sujitha N., Krithiga S. 'RES based EV battery charging system: a review'. *Renewable and Sustainable Energy Reviews*. 2017, vol. 75(3), pp. 978–88.

[5] Garole D.J., Hossain R., Garole V.J., Sahajwalla V., Nerkar J., Dubal D.P. 'Recycle, recover and repurpose strategy of spent Li-ion batteries and catalysts: current status and future opportunities'. *ChemSusChem*. 2020, vol. 13(12), pp. 3079–100.

[6] Hasanzadeh S., Vaez-Zadeh S. 'A review of contactless electrical power transfer: applications, challenges and future trends'. *Automatika*. 2015, vol. 56(3), pp. 367–78.

[7] Xie S., Hu X., Qi S., *et al.* 'Model predictive energy management for plug-in hybrid electric vehicles considering optimal battery depth of discharge'. *Energy*. 2019, vol. 173(4), pp. 667–78.

[8] Musavi F., Eberle W. 'Overview of wireless power transfer technologies for electric vehicle battery charging'. *IET Power Electronics*. 2014, vol. 7(1), pp. 60–6.

[9] Zakerian A., Vaez-Zadeh S., Babaki A. 'A dynamic WPT system with high efficiency and high power factor for electric vehicles'. *IEEE Transactions on Power Electronics*. 2020, vol. 35(7), pp. 6732–40.

[10] Grazian F., Shi W., Dong J., van Duijsen P., Soeiro T.B., Bauer P. 'Survey on standards and regulations for wireless charging of electric vehicles'. 2019 AEIT International Conference of Electrical and Electronic Technologies for Automotive (AEIT AUTOMOTIVE); Turin, Italy, IEEE, Jul; 2019. pp. 1–5.

[11] Patil D., McDonough M.K., Miller J.M., Fahimi B., Balsara P.T. 'Wireless power transfer for vehicular applications: overview and challenges'. *IEEE Transactions on Transportation Electrification*. 2017, vol. 4(1), pp. 3–37.

[12] Choi S.Y., Gu B.W., Jeong S.Y., Rim C.T. 'Advances in wireless power transfer systems for roadway-powered electric vehicles'. *IEEE Journal of Emerging and Selected Topics in Power Electronics*. 2014, vol. 3(1), pp. 18–36.

[13] García-Vázquez C.A., Llorens-Iborra F., Fernández-Ramírez L.M., Sánchez-Sainz H., Jurado F. 'Comparative study of dynamic wireless charging of electric vehicles in motorway, highway and urban stretches'. *Energy*. 2017, vol. 137, pp. 42–57.

[14] Vilathgamuwa D., Sampath J. 'Wireless power transfer (WPT) for electric vehicles (EVS)—Present and future trends' in Rajakaruna S., Shahnia F., Ghosh A. (eds.). *Plug in Electric Vehicles in Smart Grids*. Springer; 2015. pp. 33–60.

[15] Knecht O., Bosshard R., Kolar J.W. 'High-efficiency transcutaneous energy transfer for implantable mechanical heart support systems'. *IEEE Transactions on Power Electronics*. 2015, vol. 30(11), pp. 6221–36.

[16] Knecht O., Bosshard R., Kolar J.W., Starck C.T. 'Optimization of transcutaneous energy transfer coils for high power medical applications'. IEEE 15th Workshop on Control and Modeling for Power Electronics (COMPEL); US, IEEE, Jun; 2014. pp. 1–10.

[17] Knecht O., Kolar J.W. 'Performance evaluation of series-compensated ipt systems for transcutaneous energy transfer'. *IEEE Transactions on Power Electronics*. 2018, vol. 34(1), pp. 438–51.

[18] Chabalko M.J., Sample A.P. 'Three-dimensional charging via multimode resonant cavity enabled wireless power transfer'. *IEEE Transactions on Power Electronics*. 2015, vol. 30(11), pp. 6163–73.

[19] Ahn W., Jung S., Lee W. 'Design of coupled resonators for wireless power transfer to mobile devices using magnetic field shaping'. IEEE International Symposium on Electromagnetic Compatibility. IEEE; Pittsburg, Pennsylvania, US, Aug; 2012. pp. 772–6.

[20] Babaki A., Vaez-Zadeh S., Moghaddam M.F., Zakerian A. 'A novel multi-objective topology for in-motion WPT systems with an input DG source'. 10th International Power Electronics, Drive Systems and Technologies Conference (PEDSTC). IEEE; Shiraz, Iran, Feb; 2019. pp. 787–92.

[21] Chung Y.D., Lee C.Y., Kim D.W., Kang H., Park Y.G., Yoon Y.S. 'Conceptual design and operating characteristics of multi-resonance antennas in the wireless power charging system for superconducting MAGLEV train'. *IEEE Transactions on Applied Superconductivity*. 2017, vol. 27(4), pp. 1–5.

[22] Chung Y.D., Lee C.Y., Lee W.S., Park E.Y. 'Operating characteristics for different resonance frequency ranges of wireless power charging system in superconducting MAGLEV train'. *IEEE Transactions on Applied Superconductivity*. 2019, vol. 29(5), pp. 1–5.

[23] Huang L., Hu A.P., Swain A.K., Su Y. 'Z-impedance compensation for wireless power transfer based on electric field'. *IEEE Transactions on Power Electronics*. 2016, vol. 31(11), pp. 7556–63.

[24] Jiang C., Chau K.T., Liu C., Han W. 'Design and analysis of wireless switched reluctance motor drives'. *IEEE Transactions on Industrial Electronics*. 2018, vol. 66(1), pp. 245–54.

[25] Zakerian A., Vaez-Zadeh S., Babaki A. 'Efficiency maximization control and voltage regulation for dynamic wireless EV charging systems with mutual induction estimation'. IECON 2019-45th Annual Conference of the IEEE Industrial Electronics Society. IEEE; Lisbon, Portugal, Oct; 2019. pp. 4298–303.

[26] Kalra G.R., Riar B.S., Thrimawithana D.J. 'An integrated boost active bridge based secondary inductive power transfer converter'. *IEEE Transactions on Power Electronics*. 2020, vol. 35(12), pp. 12716–27.

[27] Su J., Gao R., Husain I. 'Model predictive control based field-weakening strategy for traction EV used induction motor'. *IEEE Transactions on Industry Applications*. 2017, vol. 54(3), pp. 2295–305.

[28] Iwamoto T., Arima T., Uno T., Wake K., Fujii K., Watanabe S. 'Measurement of electromagnetic field in the vicinity of wireless power transfer system for evaluation of human-body exposure'. 2014 International Symposium on Electromagnetic Compatibility. IEEE; Tokyo, Japan, May; 2014. pp. 529–32.

[29] Budhia M., Covic G.A., Boys J.T. 'Design and optimization of circular magnetic structures for lumped inductive power transfer systems'. *IEEE Transactions on Power Electronics*. 2011, vol. 26(11), pp. 3096–108.

[30] Luo Z., Wei X. 'Analysis of square and circular planar spiral coils in wireless power transfer system for electric vehicles'. *IEEE Transactions on Industrial Electronics*. 2017, vol. 65(1), pp. 331–41.

[31] Bosshard R., Kolar J.W., Mühlethaler J., Stevanović I., Wunsch B., Canales F. 'Modeling and η-α-Pareto optimization of inductive power transfer coils for electric vehicles'. *IEEE Journal of Emerging and Selected Topics in Power Electronics*. 2014, vol. 3(1), pp. 50–64.

[32] Ke G., Chen Q., Xu L., Wong S.C., Chi K.T. 'A model for coupling under coil misalignment for DD pads and circular pads of WPT system'. *2016 IEEE Energy Conversion Congress and Exposition (ECCE). IEEE*; Milwaukee, US, Sep; 2016. pp. 1–6.

[33] Rasekh N., Kavianpour J., Mirsalim M. 'A novel integration method for a bipolar receiver pad using LCC compensation topology for wireless power transfer'. *IEEE Transactions on Vehicular Technology*. 2018, vol. 67(8), pp. 7419–28.

[34] Zaheer A., Covic G.A., Kacprzak D. 'A bipolar pad in a 10-kHz 300-W distributed IPT system for AGV applications'. *IEEE Transactions on Industrial Electronics*. 2013, vol. 61(7), pp. 3288–301.

[35] Wang Z., Cui S., Han S., *et al.* 'A novel magnetic coupling mechanism for dynamic wireless charging system for electric vehicles'. *IEEE Transactions on Vehicular Technology*. 2017, vol. 67(1), pp. 124–33.

[36] Cui S., Wang Z., Han S., Zhu C., Chan C.C. 'Analysis and design of multiphase receiver with reduction of output fluctuation for EV dynamic wireless

charging system'. *IEEE Transactions on Power Electronics*. 2018, vol. 34(5), pp. 4112–24.
[37] Jayathurathnage P.K.S., Alphones A., Vilathgamuwa D.M. 'Optimization of a wireless power transfer system with a repeater against load variations'. *IEEE Transactions on Industrial Electronics*. 2017, vol. 64(10), pp. 7800–9.
[38] Sampath J.P.K., Alphones A., Vilathgamuwa D.M. 'Figure of merit for the optimization of wireless power transfer system against misalignment tolerance'. *IEEE Transactions on Power Electronics*. 2016, vol. 32(6), pp. 4359–69.
[39] CC M., Buja G., Choi S.Y., Rim C.T. 'Modern advances in wireless power transfer systems for roadway powered electric vehicles'. *IEEE Transactions on Industrial Electronics*. 2016, vol. 63(10), pp. 6533–45.
[40] Li X., Hu J., Wang H., Dai X., Sun Y. 'A new coupling structure and position detection method for segmented control dynamic wireless power transfer systems'. *IEEE Transactions on Power Electronics*. 2020, vol. 35(7), pp. 6741–5.
[41] Choi S., Huh J., Lee W.Y., Lee S.W., Rim C.T. 'New cross-segmented power supply rails for roadway-powered electric vehicles'. *IEEE Transactions on Power Electronics*. 2013, vol. 28(12), pp. 5832–41.
[42] Zakerian A., Vaez-Zadeh S., Babaki A., Moghaddam M.F. 'Efficiency optimization of a dynamic wireless EV charging system using coupling coefficient estimation'. 10th International Power Electronics, Drive Systems and Technologies Conference (PEDSTC); Shiraz, Iran, IEEE, Feb; 2019. pp. 629–34.
[43] Jiang Y., Wang L., Wang Y., Liu J., Li X., Ning G. 'Analysis, design, and implementation of accurate ZVS angle control for EV battery charging in wireless high-power transfer'. *IEEE Transactions on Industrial Electronics*. 2018, vol. 66(5), pp. 4075–85.
[44] Zhang W., Mi C.C. 'Compensation topologies of high-power wireless power transfer systems'. *IEEE Transactions on Vehicular Technology*. 2015, vol. 65(6), pp. 4768–78.
[45] Wang Y., Yao Y., Liu X., Xu D., Cai L. 'An LC/S compensation topology and coil design technique for wireless power transfer'. *IEEE Transactions on Power Electronics*. 2017, vol. 33(3), pp. 2007–25.
[46] Abou Houran M., Yang X., Chen W. 'Magnetically coupled resonance WPT: review of compensation topologies, resonator structures with misalignment, and EMI diagnostics'. *Electronics*. 2018, vol. 7(11), p. 296.
[47] Hasanzadeh S., Vaez-Zadeh S. 'Efficiency analysis of contactless electrical power transmission systems'. *Energy Conversion and Management*. 2013, vol. 65(15–17), pp. 487–96.
[48] Lee S.H., Lorenz R.D. 'A design methodology for multi-kW, large airgap, MHz frequency, wireless power transfer systems'. IEEE Energy Conversion Congress and Exposition; Arizona, US, IEEE, September; 2011. pp. 3503–10.
[49] Fang C., Song J., Lin L., Wang Y. 'Practical considerations of series-series and series-parallel compensation topologies in wireless power transfer system

application'. IEEE PELS Workshop on Emerging Technologies: Wireless Power Transfer (WoW); Chongqing, China, IEEE, May; 2017. pp. 255–9.

[50] Su Y., Tang C., Wu S., Sun Y. 'Research of LCL resonant inverter in wireless power transfer system'. International Conference on Power System Technology; Chongqing, China, IEEE; 2006. pp. 1–6.

[51] Hao H., Covic G.A., Boys J.T. 'An approximate dynamic model of LCL-T based inductive power transfer power supplies'. *IEEE Transactions on Power Electronics*. 2013, vol. 29(10), pp. 5554–67.

[52] Liu C., Ge S., Guo Y., Li H., Cai G. 'Double-LCL resonant compensation network for electric vehicles wireless power transfer: experimental study and analysis'. *IET Power Electronics*. 2016, vol. 9(11), pp. 2262–70.

[53] Xia C., Chen R., Liu Y., Chen G., Wu X. 'LCL/LCC resonant topology of WPT system for constant current, stable frequency and high-quality power transmission'. 2016 IEEE PELS Workshop on Emerging Technologies: Wireless Power Transfer (WoW); Knoxville, US, Oct 2016; 2016. pp. 110–3.

[54] Li W., Zhao H., Deng J., Li S., Mi C.C. 'Comparison study on SS and double-sided LCC compensation topologies for EV/PHEV wireless chargers'. *IEEE Transactions on Vehicular Technology*. 2015, vol. 65(6), pp. 4429–39.

[55] Villa J.L., Sallán J., Llombart A., Sanz J.F. 'Design of a high frequency inductively coupled power transfer system for electric vehicle battery charge'. *Applied Energy*. 2009, vol. 86(3), pp. 355–63.

[56] Jiang Y., Wang L., Wang Y., Liu J., Wu M., Ning G. 'Analysis, design, and implementation of WPT system for EV's battery charging based on optimal operation frequency range'. *IEEE Transactions on Power Electronics*. 2018, vol. 34(7), pp. 6890–905.

[57] Yan Q., Zhang B., Kezunovic M. 'Optimized operational cost reduction for an EV charging station integrated with battery energy storage and PV generation'. *IEEE Transactions on Smart Grid.* 2018, vol. 10(2), pp. 2096–106.

[58] Neath M.J., Swain A.K., Madawala U.K., Thrimawithana D.J. 'An optimal PID controller for a bidirectional inductive power transfer system using multiobjective genetic algorithm'. *IEEE Transactions on Power Electronics*. 2013, vol. 29(3), pp. 1523–31.

[59] Diekhans T., De Doncker R.W. 'A dual-side controlled inductive power transfer system optimized for large coupling factor variations and partial load'. *IEEE Transactions on Power Electronics*. 2015, vol. 30(11), pp. 6320–8.

[60] Samanta S., Rathore A.K. 'A new inductive power transfer topology using direct AC–AC converter with active source current waveshaping'. *IEEE Transactions on Power Electronics*. 2017, vol. 33(7), pp. 5565–77.

[61] Balcells J., Bogonez-Franco P. 'Effect of driver to gate coupling circuits on EMI produced by SiC MOSFETS'. 2013 International Symposium on Electromagnetic Compatibility; Colorado, US, IEEE; 2013. pp. 209–14.

[62] Ota R., Sudarmo Nugroho D., Hoshi N. 'A consideration on maximum efficiency of resonant circuit of inductive power transfer system with soft-switching operation'. *World Electric Vehicle Journal*. 2019, vol. 10(3), p. 54.

[63] Ota R., Thrimawithana D.J., Madawala U.K., Covic G.A. 'Boundary of soft-switching for efficient operation of bi-directional IPT systems'. IEEE PELS Workshop on Emerging Technologies: Wireless Power Transfer (WoW); Seoul, Korea, IEEE, Nov; 2020. pp. 164–9.
[64] Ota R., Kakomura K., Hoshi N., Sato D. 'Maximum efficiency control scheme and design method for resonant circuit of bi-directional inductive power transfer system'. 2017 19th European Conference on Power Electronics and Applications (EPE'17 ECCE Europe); Warsaw, Poland, IEEE, Sept; 2017. pp. 1.
[65] Li H., Li J., Wang K., Chen W., Yang X. 'A maximum efficiency point tracking control scheme for wireless power transfer systems using magnetic resonant coupling'. *IEEE Transactions on Power Electronics*. 2014, vol. 30(7), pp. 3998–4008.
[66] Ahn D., Hong S. 'Wireless power transmission with self-regulated output voltage for biomedical implant'. *IEEE Transactions on Industrial Electronics*. 2013, vol. 61(5), pp. 2225–35.
[67] Fu M., Ma C., Zhu X. 'A cascaded boost–buck converter for high-efficiency wireless power transfer systems'. *IEEE Transactions on Industrial Informatics*. 2013, vol. 10(3), pp. 1972–80.
[68] Kar D.P., Nayak P.P., Bhuyan S., Panda S.K. 'Automatic frequency tuning wireless charging system for enhancement of efficiency'. *Electronics Letters*. 2014, vol. 50(24), pp. 1868–70.
[69] Zheng Z., Wang N., Ahmed S. 'Adaptive frequency tracking control with fuzzy Pi compound controller for magnetically coupled resonant wireless power transfer'. *International Journal of Fuzzy Systems*. 2020, vol. 1, pp. 2199–3211.
[70] Babaki A., Vaez-Zadeh S., Zakerian A. 'Performance optimization of dynamic wireless EV charger under varying driving conditions without resonant information'. *IEEE Transactions on Vehicular Technology*. 2019, vol. 68(11), pp. 10429–38.
[71] Babaki A., Vaez-Zadeh S., Zakerian A., Covic G.A. 'Variable-frequency retuned WPT system for power transfer and efficiency improvement in dynamic EV charging with fixed voltage characteristic'. *IEEE Transactions on Energy Conversion*. 2021, pp. 1558–9.
[72] Chinthavali M.S., Onar O.C., Miller J.M., Tang L. 'Single-phase active boost rectifier with power factor correction for wireless power transfer applications'. 2013 IEEE Energy Conversion Congress and Exposition; Denver, US, Sep; 2013. pp. 3258–65.
[73] Fu M., Yin H., Zhu X., Ma C. 'Analysis and tracking of optimal load in wireless power transfer systems'. *IEEE Transactions on Power Electronics*. 2014, vol. 30(7), pp. 3952–63.
[74] Zakerian A., Vaez-Zadeh S., Babaki A. 'Maximum efficiency control of a wireless EV charger with on-line parameter calculation'. 2019 IEEE PELS Workshop on Emerging Technologies: Wireless Power Transfer (WoW); London, UK, IEEE, June 2019; 2019. pp. 85–8.

[75] Dai X., Li X., Li Y., Hu A.P. 'Maximum efficiency tracking for wireless power transfer systems with dynamic coupling coefficient estimation'. *IEEE Transactions on Power Electronics*. 2017, vol. 33(6), pp. 5005–15.

[76] Azad A., Pantic Z. 'Bidirectional grid-side power management in DWPT systems for EV charging applications'. 2019 IEEE Energy Conversion Congress and Exposition (ECCE); Baltimore, US, IEEE, Sep 2019; 2019. pp. 557–62.

[77] Semsar S., Soong T., Lehn P.W. 'On-board single-phase integrated electric vehicle charger with V2G functionality'. *IEEE Transactions on Power Electronics*. 2020, vol. 35(11), pp. 12072–84.

[78] Noel L., Zarazua de Rubens G., Kester J., Sovacool B.K. 'The technical challenges to V2G'. *Vehicle-to-Grid. Energy, Climate and the Environment*; Cham: Palgrave Macmillan; 2019. pp. 65–89.

[79] Vadi., Bayindir., Colak., Hossain. 'A review on communication standards and charging topologies of V2G and V2H operation strategies'. *Energies*. 2019, vol. 12(19), p. 3748.

[80] Madawala U.K., Thrimawithana D.J. 'A bidirectional inductive power interface for electric vehicles in V2G systems'. *IEEE Transactions on Industrial Electronics*. 2011, vol. 58(10), pp. 4789–96.

[81] Moghaddam M.F., Vaez-Zadeh S., Zakerian A., Babaki A. 'Design and analysis of a modified dual phase shift control method for a wireless EV charger considering coupling uncertainty'. 10th International Power Electronics, Drive Systems and Technologies Conference (PEDSTC); Shiraz, Iran, IEEE, Feb; 2019. pp. 635–40.

[82] Abdolmaleki M., Masoud N., Yin Y. 'Vehicle-to-vehicle wireless power transfer: paving the way toward an electrified transportation system'. *Transportation Research Part C: Emerging Technologies*. 2019, vol. 103(1), pp. 261–80.

[83] Mou X., Zhao R., Gladwin D.T. 'Vehicle to vehicle charging (V2V) bases on wireless power transfer technology'. IECON 2018-44th Annual Conference of the IEEE Industrial Electronics Society; Washington, DC, US, IEEE, Oct; 2018. pp. 4862–7.

[84] Babaki A., Vaez-Zadeh S., Zakerian A., Jafari Natanzi A. 'Analysis and control of wireless motor drives with a single inverter in primary side'. *IEEE Transactions on Energy Conversion*. 2020, vol. 36(2) 930–9.

[85] Babaki A., Vaez-Zadeh S., Jahanpour-Dehkordi M., Zakerian A. 'Wireless motor drives with a single inverter in primary side of power transfer systems'. 2019 IEEE PELS Workshop on Emerging Technologies: Wireless Power Transfer (WoW); London, UK, June; 2019. pp. 125–8.

[86] Sahraneshin S., Ameri M.H., Varjani A.Y. 'A single to three-phase AC/AC cycloconverter for inductive power transfer'. 4th Annual International Power Electronics, Drive Systems and Technologies Conference; Tehran, Iran, IEEE, Feb; 2013. pp. 389–93.

Chapter 4

Considerations on dynamic inductive charging: optimizing the energy transfer at a high efficiency and experimental implementation

Ioannis Karakitsios[1] *and Nikos Hatziargyriou*[1]

4.1 Introduction

Dynamic inductive charging of electric vehicles (EVs) can be proven quite effective in resolving problems related to the high cost of the EV battery, while also addressing the limited range capabilities of EVs, which is one of the main issues impeding their wide market adoption. More specifically, dynamic inductive charging allows EV drivers to wirelessly charge while moving on the road eliminating the need to interrupt their trips to stop at a charging station and charge their vehicle. Moreover, studies indicate that the ability of EVs to charge at various spots during their trips allows a considerable decrease in the EVs' battery capacity [1] and, hence, a significant decrease in the battery cost.

Dynamic inductive charging is quite similar to static inductive charging; energy is transferred between a primary pad installed in the charging station and a secondary pad placed under the vehicle. However, certain considerations shall be taken into account when studying a dynamic inductive charging system. In this respect, the variations in the system's inductance values while the EV moves over the station demand a system analysis different than the one in the case of static inductive charging, where the EV is parked over the station. Additionally the bifurcation phenomenon, indicating that the efficiency of the system may be maximized at frequencies different than the ones maximizing the transferred power, shall be carefully examined in the case of dynamic inductive charging. Moreover, the selection of the compensation capacitors' values shall be examined according to the variations of the self-inductance values of the magnetic coupler's coils which are noted due to the movement of the vehicle over the station.

In order to achieve a high power transfer and efficiency in dynamic inductive charging not only do the compensation capacitors need to be appropriately selected

[1]School of electrical and computer engineering, National Technical University of Athens, Greece

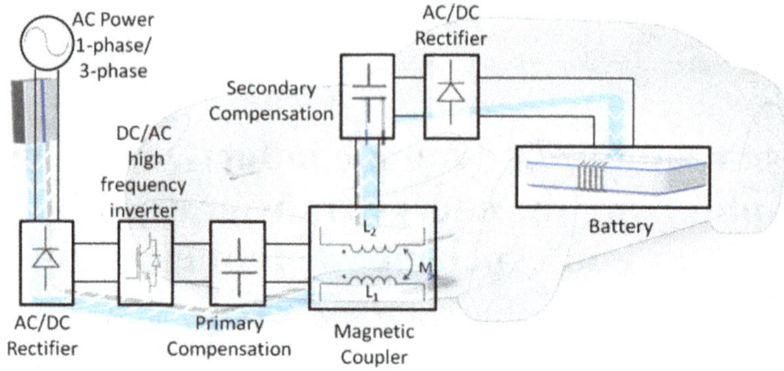

Figure 4.1 General concept of an inductive power charging system

but the optimal values of all the system's parameters shall also be optimally defined. In this respect, an optimization method shall be defined in order to determine the optimal values for the operational frequency, the values of the compensation capacitors as well as the load resistance in a dynamic inductive charging system. Taking into account the requirements for the implementation of this optimization method, a control scheme is also examined, which does not require any additional detection mechanisms to identify the EV's position, while eliminating communication requirements between the station and the vehicle.

The optimization method is applied in the case of a 60 cm diameter circular pad, and the suggested control scheme is validated by simulating the operation of the system over the whole EV movement range. The effective operation of the proposed control scheme is also validated by an experimental setup, comprising a circular magnetic coupler as well as the inverter, implementing the suggested control scheme.

4.2 Differences among static and dynamic inductive charging

4.2.1 Analysis of a dynamic inductive charging system

The general concept of inductive charging is presented in Figure 4.1. Initially, the AC grid power is rectified to DC, and a DC/AC high-frequency inverter ensures that the required high-frequency current is supplied to the primary coil. This high-frequency current induces an AC voltage of the same frequency in the secondary side, which is finally rectified to DC and is supplied to the EV's battery. A network of capacitors may be employed in the primary and secondary sides in order to increase the transferred power and the system's efficiency [2].

The equivalent circuit of an Inductive Power Transfer (IPT) system is depicted in Figure 4.2. L_{1_i} and L_{2_i} indicate the values of self-inductance in the primary and secondary coils, respectively, while M_i denotes the mutual inductance. Index i indicates the misalignment x_i between the primary and the secondary stations at which the values

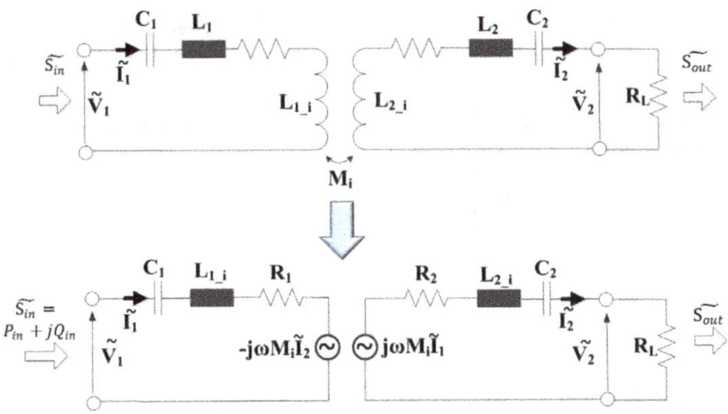

Figure 4.2 Equivalent circuit of the IPT system in static inductive charging

of inductances $\{L_{1\_i}, L_{2\_i}, M_i\}$ are considered. In the case of static inductive charging, the values of L_{1_i}, L_{2_i}, and M_i can be considered constant since the vehicle remains parked above the charging station at a certain misalignment, x_i, between the primary and the secondary sides as depicted in Figure 4.3. It shall also be noted that positioning mechanisms may also be considered to facilitate the proper position of the EV over the station and to ensure that the secondary pad is aligned at the best possible way above the primary pad, allowing sufficient amount of energy to be transferred [3, 4].

In order to increase the efficiency and the power transfer capability of the inductive charging system, compensation capacitors may be used in the system's design. C_1 and C_2 in Figure 4.2 indicate the compensation capacitors in the primary and secondary sides, respectively. Many compensation methods have been proposed [5] according to the connection of the capacitors in the system (in parallel, in series, etc.) Among them, the Series-Series (SS), which is depicted in Figure 4.2, is often used both in static [6] and dynamic inductive charging applications [7]. $\omega = 2\pi f$ represents the angular speed at the operational electrical frequency, f, of the system, while R_1 and R_2 are the coil resistances of the primary and secondary coils, respectively. The resistor R_L indicates the equivalent battery load [8], which is expected to be connected in the secondary side after an inverter and a filter [5].

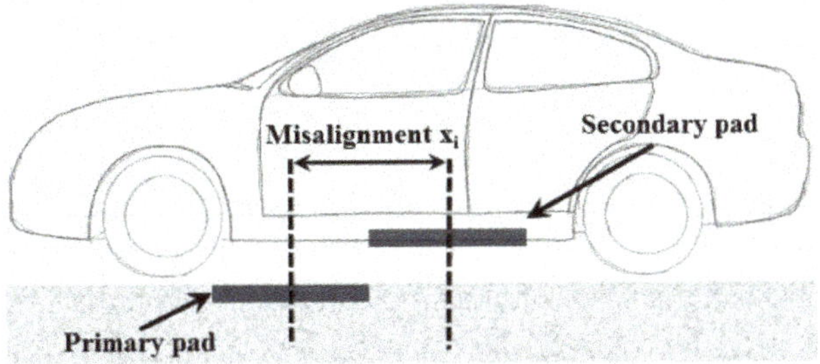

Figure 4.3 Misalignment x_i between the primary and the secondary pads

Due to the filtering effect of the compensation network, a first-order analysis can be performed [7, 9]. In this respect, nearly sinusoidal currents can be considered for the analysis of the system.

The voltage values V_1 and V_2 at the primary and secondary sides in Figure 4.2 can be calculated as

$$\widetilde{V_1} = \left(j\omega L_{1\_i} - \frac{j}{\omega C_1} + R_1\right)\widetilde{I_1} - j\omega M\widetilde{I_2} \tag{4.1}$$

$$\widetilde{V_2} = R_L * \widetilde{I_2} \tag{4.2}$$

Moreover, applying Kirchhoff's law in the secondary side, the current I_2 will be equal to

$$\widetilde{I_2} = \frac{j\omega M_i}{R_2 + j\omega L_{2\_i} - \frac{j}{\omega C_2} + R_L}\widetilde{I_1} \tag{4.3}$$

The output apparent power transferred to the load resistance R_L can be calculated as

$$\widetilde{S_{out\_i}} = R_L \left|\widetilde{I_2}\right|^2 \tag{4.4}$$

and the active power transferred to the load resistance is equal to

$$P_{out\_i} = \text{Re}\left\{\widetilde{S_{out\_i}}\right\} = R_L \left|\widetilde{I_2}\right|^2 \tag{4.5}$$

Substituting (4.3) in (4.5)

$$P_{out\_i} = \left|\widetilde{I_1}\right|^2 \frac{\omega^2 M_i^2 R_L}{\left(\omega L_{2\_i} - \frac{1}{\omega C_2}\right)^2 + (R_2 + R_L)^2} \tag{4.6}$$

The input apparent power to the circuit depicted in Figure 4.2 can be calculated as

$$\widetilde{S_{in\_i}} = \widetilde{V_1} * \widetilde{I_1}^* \tag{4.7}$$

Therefore, according to (4.1), (4.3), and (4.7), S_{in} is calculated as

$$\widetilde{S_{in\_i}} = \left\{\left[R_1 + \frac{\omega^2 M_i^2 (R_L + R_2)}{\left(\omega L_{2\_i} - \frac{1}{\omega C_2}\right)^2 + (R_L + R_2)^2}\right] + j\left[\omega L_{1\_i} - \frac{1}{\omega C_1} - \frac{\left(\omega L_{2\_i} - \frac{1}{\omega C_2}\right)\omega^2 M_i^2}{\left(\omega L_{2\_i} - \frac{1}{\omega C_2}\right)^2 + (R_L + R_2)^2}\right]\right\} \left|\widetilde{I_1}\right|^2 \tag{4.8}$$

The efficiency of the system will be equal to

$$\text{eff} = \frac{\left|\widetilde{S_{out}}\right|}{\left|\widetilde{S_{in}}\right|} \tag{4.9}$$

In the case of static inductive charging, the capacitors C_1 and C_2 can be calculated according to (4.10) and (4.11) in order to maximize the output power and achieve maximum efficiency, as explained in [2, 7, 10].

$$C_1 = \frac{1}{\omega^2 L_{2\_i}} \tag{4.10}$$

$$C_2 = \frac{1}{\omega^2 L_{1\_i}} \tag{4.11}$$

However, in the case of dynamic inductive charging, the misalignment x_i is expected to vary while the EV moves over the station. For instance, when considering a constant EV speed υ, as in (4.12), the misalignment will vary according to the EV speed and is equal to $x_i = \upsilon t_i$ (i.e. If the EV enters the station at a time equal to

$t_0 = 0$, then t_i indicates a misalignment equal to x_i.)

$$\upsilon = \frac{x_i}{t_i} \tag{4.12}$$

Therefore, the values of L_{1_i}, L_{2_i}, and M_i will not remain constant during the charging process [7] but will rather vary according to x_i. Thus, variations are expected both in the efficiency of the system in (4.9) and in the output power in (4.6) while the EV moves over the station.

4.2.2 Bifurcation in dynamic inductive charging

One of the main aspects concerning the design of an IPT system is the investigation of the bifurcation phenomenon [5]. In particular, the bifurcation phenomenon indicates that the efficiency in an IPT system may be maximized in more than one frequency value. On the other hand, the power transferred to the vehicle may only be maximized at one of those values. Therefore, the bifurcation phenomenon indicates that frequencies may occur where the efficiency of the system is maximized, yet the output power may receive almost zero values at those frequencies. It is, therefore, crucial to take into account the bifurcation phenomenon when designing an IPT system in order to assure that the system operates with high efficiency, while sufficient amounts of power are also transferred to the vehicle.

Studies on dynamic inductive charging [11, 12] usually investigate the bifurcation phenomenon with an approach similar to static inductive charging. In particular, such studies are limited to studying the operational frequency of the system to investigate the maximization of the efficiency and the output power. However, since the system's parameters vary according to the misalignment among the primary and secondary coils, the EV movement shall be expected to affect both the efficiency of the system and the power transferred to the vehicle.

For instance, when considering the inductive charger depicted in Figure 4.4[a], in the case of static inductive charging, the values for the self-inductance and the mutual inductance can be assumed constant through the whole duration of the charging process.

[a]The coupler in Figure 4.4 has been designed in the Ansys simulation software according to the specifications in [13] and comprises two identical pads in the primary and secondary sides, each comprising the aluminum shielding, the coils, and the ferrite bars. The system is considered to operate with a constant primary current of 23 A.

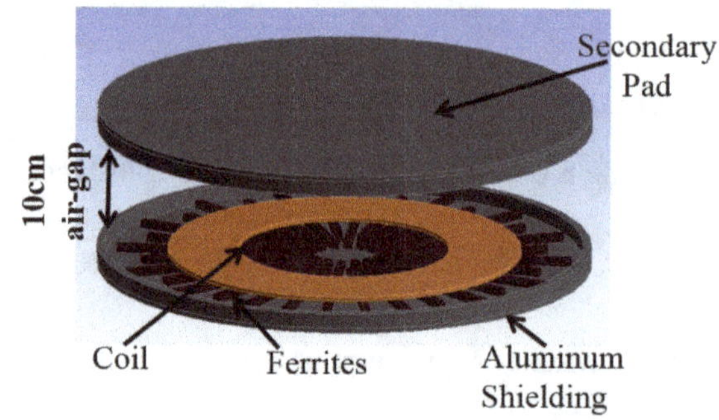

Figure 4.4 Inductive coupler, 60 cm in diameter, designed in Ansys

At the position where the primary is perfectly aligned with the secondary, the misalignment x_p as depicted in Figure 4.3, becomes zero and the inductance values for the magnetic coupler in Figure 4.4 are $L_1 = 301$ μH, $L_2 = 198$ μH, and $M = 109$ μH. The IPT system is considered to operate at a frequency of 30 kHz and transfers power to a resistance load equal to 5 Ω [10]. In order to achieve the maximum possible output power in this position, the value for the compensation capacitor in the secondary side shall be calculated according to (4.11) [10]. Similarly, in order to achieve maximum efficiency, the value of C_2 shall be calculated according to (4.10). For the examined case, the value of C_2 shall be equal to 136 nF, while the value of C_1 shall be equal to 135 nF.

The efficiency and the power transferred to the EV, calculated according to (4.9) and (4.10), are depicted in Figure 4.5. It is evident that both the efficiency and the output power are maximized in the frequency of 30 kHz. However, the efficiency can achieve its maximum values at additional frequencies of 26 kHz and 36 kHz, indicating the bifurcation phenomenon. Although the operation of the system at 30 kHz is expected to maximize both the efficiency and the output power, the operation at either 26 kHz or 36 kHz is expected to result

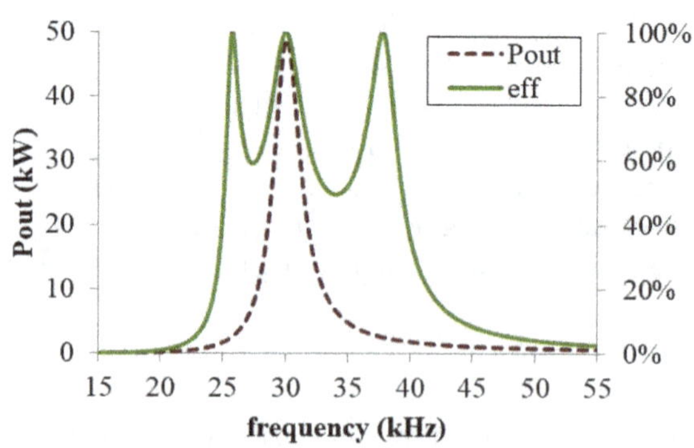

Figure 4.5 Output power and efficiency when the secondary is exactly above the primary [10]

in very small values for the output power. In particular, for both of these cases, the output power is equal to around 5% of the maximum possible value.

In the case of dynamic inductive charging, the misalignment between the primary and the secondary pads does not remain constant while the EV moves over the station. More specifically, the misalignment x_i (in Figure 4.3) will achieve a value equal to $x_i = -60$ cm when the EV is about to enter the station, be equal to zero when the vehicle is exactly above the station, and receive a value equal to +60 cm when the EV exits the station. The variation of the inductance values according to the different misalignment positions is indicated in Figure 4.6.

Similar to static inductive charging, the bifurcation phenomenon shall be evident when considering the variation of the output power and the efficiency according to the operational frequency. In fact, when considering the misalignment point when the EV is exactly above the station ($x_i = 0$ cm), the output power and the efficiency will vary according to Figure 4.5.

However, both the efficiency of the system and the output power are expected to vary while the EV moves over the station. In particular, when considering the magnetic coupler in Figure 4.4 operating at a frequency of 30 kHz, the output power and the efficiency, calculated as in (4.9) and (4.10), are depicted in Figure 4.7.

It is evident that the efficiency gets very high values for a movement area with misalignment points from −25 cm to +25 cm. In particular, the efficiency is maximized at the misalignment points −20 cm, −5 cm, +5 cm, and +20 cm. However, the output power is only maximized when the EV is exactly above the station, where the value of x is equal to 0 cm. Additionally, when considering the misalignment points $x_i = -20$ cm and $x_i = +20$ cm, although the efficiency is equal to 100%, the output power is less than 4% of the maximum value. Thus a phenomenon similar to bifurcation is also noted in dynamic inductive charging. In this case, however, such a phenomenon not only depends on the operational frequency but is also linked with the movement of the vehicle over the station [10].

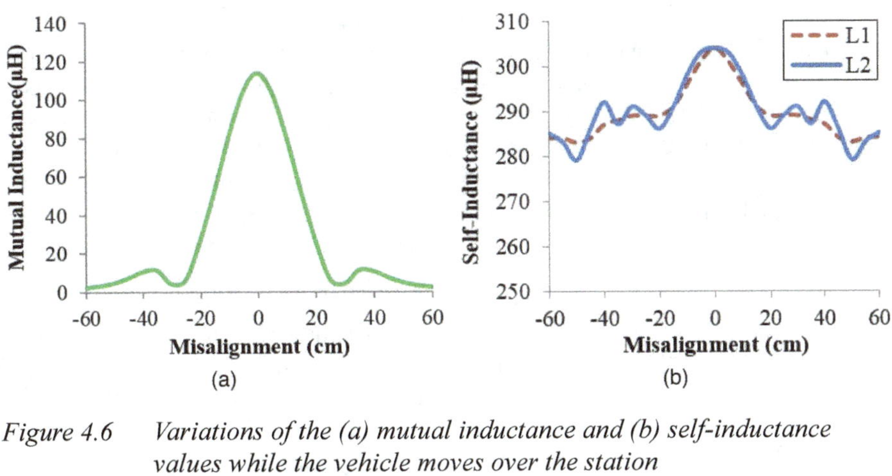

Figure 4.6 Variations of the (a) mutual inductance and (b) self-inductance values while the vehicle moves over the station

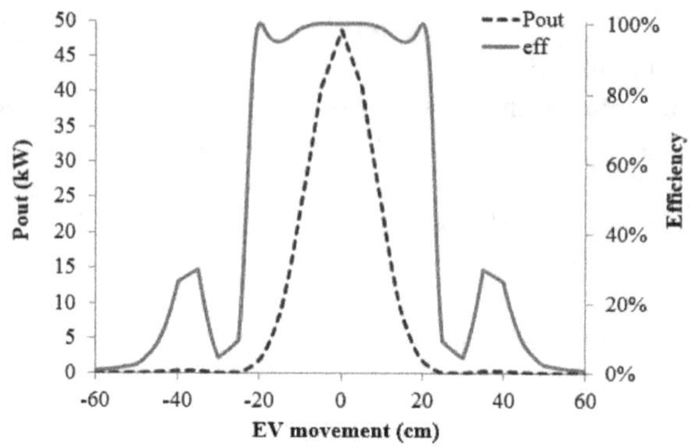

Figure 4.7 Output power and efficiency while the EV moves over the station [10]

It is therefore quite significant to take into account these variations in the efficiency and the output power when designing a dynamic IPT system in order to ensure that high amounts of power are transferred to the vehicle's battery, while also achieving high values of efficiency [10].

4.2.3 Self-inductance variations in dynamic inductive charging

One of the main parameters that shall be carefully selected in the case of inductive charging is the values of the compensation capacitors C_1 and C_2 in Figure 4.2 in order to ensure that a sufficient amount of power is transferred to the vehicle, while achieving increased system efficiency. When considering the EV to be parked over the station in static inductive charging applications, the values of the compensation capacitors can easily be selected according to (4.10) and (4.11) in order to maximize the system's efficiency and the power transferred to the vehicle [10].

In the general case of static inductive charging, if the capacitor values are selected according to (4.10) and (4.11) at a position $x_i = x_C$, where the self-inductance are equal to L_{1_C} and L_{2_C}, then the efficiency and the transferred power will be maximized for this particular position [2]. A similar approach is usually applied in the case of dynamic inductive charging in literature [14–16]. It shall be mentioned, however, that in the case of dynamic inductive charging, the misalignment x_i constantly varies while the EV moves over the station. Therefore, in any position x_i, while the EV moves over the station, the self-inductance values shall be given by (4.13) and (4.14) [2]:

$$L_{1\_i} = L_{1\_C} + k_1 L_{1\_C} \tag{4.13}$$

$$L_{2\_i} = L_{2\_C} + k_2 L_{2\_C} \tag{4.14}$$

where k_1 and k_2 indicate the variation of the self-inductance values relative to L_{1_C} and L_{2_C}. Apparently, if $x_i = x_C$, k_1 and k_2 are equal to zero.

The use of Litz wire containing multiple strands of wires of smaller diameter in the system design can significantly reduce the AC copper losses at the high frequency of an Inductive Power Transfer system. In order to simplify the analysis in this section, nearly zero values are considered for R_1 and R_2 [2, 10]. Thus, substituting (4.13) and (4.14) in (4.6) and taking into account the value for the compensation capacitor C_2 in (4.11), the output power will be equal to [2]:

$$P_{out\_i} = \left|\widetilde{I_1}\right|^2 \frac{\omega^2 M_i^2 R_L}{\left(\omega k_2 L_{2\_C}\right)^2 + R_L^2} \tag{4.15}$$

When considering a misalignment x_i, the output power may receive a maximum value, P_{out_max}, in case $k_2 = 0$ (i.e. for the EV position where $x_i = x_C$). According to (4.15), the value of this maximum output power will be equal to [2]:

$$P_{out\_c} = P_{out\_max} = \left|\widetilde{I_1}\right|^2 \frac{\omega^2 M_i^2}{R_L} \tag{4.16}$$

According to (4.15) and (4.16), when considering misalignment positions where $x_i \neq x_C$, it can be observed that [2]:

$$P_{out\_i} = \frac{1}{1+\left(\frac{\omega k_2 L_{2\_i}}{R_L}\right)^2} P_{out\_max} \tag{4.17}$$

Thus, selecting a compensation capacitor C_2 according to (4.11) for a misalignment $x_i = x_C$ will result in the maximum possible output power in this particular position. However, in any other position $x_i \neq x_C$, the maximum achievable power will be reduced by a percentage, which, according to (4.17), will be equal to [2]:

$$\text{Percentage}_{\text{reduction}}\ (\%) = \frac{1}{1+\left(\frac{\omega k_2 L_{2\_i}}{R_L}\right)^2} 100 \tag{4.18}$$

Variations shall also be expected in the system's efficiency when considering positions where $x_i \neq x_C$. In particular, substituting (4.13) and (4.14) in (4.9), while taking into account (4.6) and (4.8), the efficiency in a dynamic IPT system can be rewritten as [2]:

$$\text{eff}_i = \frac{\dfrac{\omega^2 M_i^2 R_L}{\left(\omega k_2 L_{2\_C}\right)^2 + R_L^2}}{\sqrt{\left[\dfrac{\omega^2 M_i^2 R_L}{\left(\omega k_2 L_{2\_C}\right)^2 + R_L^2}\right]^2 + \left[\omega k_1 L_{1\_C} - \dfrac{\left(\omega k_2 L_{2\_C}\right)\omega^2 M_i^2}{\left(\omega k_2 L_{2\_C}\right)^2 + R_L^2}\right]^2}} \tag{4.19}$$

At the EV position, where $x_i = x_C$ and $k_1 = k_2 = 0$, the eff_i in (4.19) is equal to one. However, while the EV moves over the station and when $x_i \neq x_C$, k_1 and k_2 will not be equal to zero; thus variations shall be expected in the system's efficiency [2].

✓ ***Selecting C_1 and C_2 for* x_C = 0 cm**

Both the efficiency and the output power are expected to vary, according to (4.17) and (4.19), while the EV moves over the station. For instance, when considering that the magnetic coupler in Figure 4.4 operates at 80 kHz providing power to a 5 Ω equivalent load, the compensation capacitors can be selected according to (4.10) and (4.11), for the case where the EV is exactly above the station (x_C = 0 cm). In particular, taking into account the values of L_1 and L_2 at this specific position the compensation capacitors can be calculated as $C_1 = C_2$ = 13 nF. At this position (the

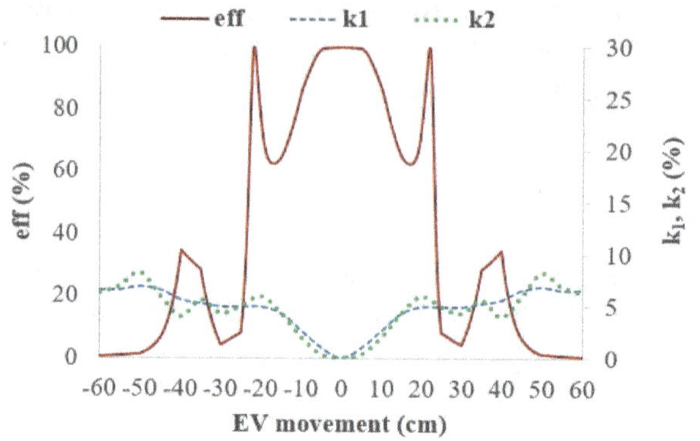

Figure 4.8 Efficiency when the EV moves over the station in the case that the compensation capacitors are selected for x_c *= 0 cm [2]*

misalignment is equal to 0 cm), the efficiency of the system is equal to 100% as evident in Figure 4.8.

In the case of dynamic inductive charging, additional positions shall be examined denoting the movement of the vehicle over the station. In particular, when examining other movement positions than the position x_C = 0 cm (i.e. the misalignment point at which the compensation capacitors have been selected), the efficiency of the system is expected to get the values denoted by (4.19) due to the variations in the values of the self-inductance L_1 and L_2. For instance, when x_i = −30 cm, the self-inductances are only reduced by k_1 = 4.9% and k_2 = 4.3% (k_1 and k_2 as denoted in (4.13)–(4.14)) compared to the position where x_i = 0 cm, as depicted in Figure 4.8. However, the efficiency is considerably reduced by 94% at this position (x_i = −30 cm), as evident in Figure 4.8.

It shall also be noted that in the case of dynamic inductive charging, the efficiency values in *all* potential moving positions shall be taken into account. More specifically, when considering all the moving positions above the station (from x_i = −60 cm up to x_i = −60 cm), the *average efficiency* in Figure 4.8 is equal to 39%.

✓ *Selecting C_1 and C_2 for* x_C = 15 cm

Figure 4.8 depicts the results for the efficiency when selecting the compensation capacitors for the position where the EV is exactly above the station (i.e. x_c = 0 cm). However, additional misalignment positions may be selected for calculating the compensation capacitors. For instance, when selecting C_1 and C_2 for x_c = 15 cm, the efficiency values for all the misalignment points are depicted in Figure 4.9. At this specific position, the self-inductance values L_1 and L_2 are only reduced by 4.3% compared to the previous case (x_c = 0 cm and the results in Figure 4.8). However, even such a small reduction in the self-inductance values can lead to significant variations in the system's efficiency

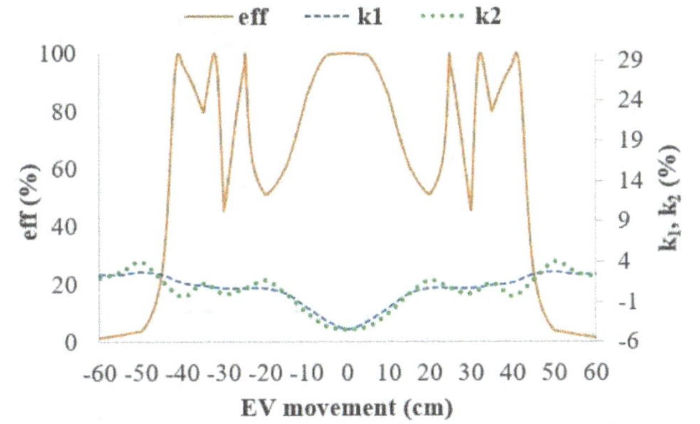

Figure 4.9 *Efficiency when the EV moves over the station in the case that the compensation capacitors are selected for $x_c \neq 0$ cm [2]*

as evident in Figure 4.9. In fact, the *average efficiency* in this case is considerably increased to 59.8%.

✓ *Variations in the output power*

Similar variations shall be expected in the power transferred to the vehicle. More specifically, according to (4.17), the output power is expected to vary according to the variations of k_2 in the self-inductance L_2. Thus, the proper selection of the compensation capacitor C_2 (according to (4.11)) is quite crucial in this case. For instance, when selecting the compensation capacitor C_2 considering the value of L_2 for the case where the EV is exactly above the station ($x_c = 0$ cm), the results for the output power are depicted in Figure 4.10.

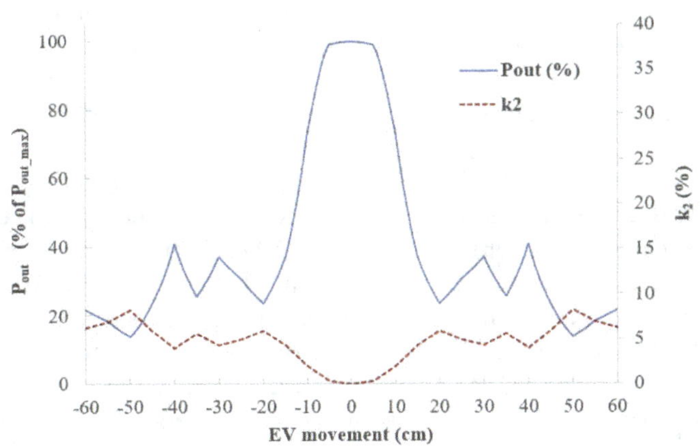

Figure 4.10 *Output power as a percentage of the maximum possible output power for each movement position, as defined in (4.17). The compensation capacitor C_2 has been selected for $x_c = 0$ cm [2].*

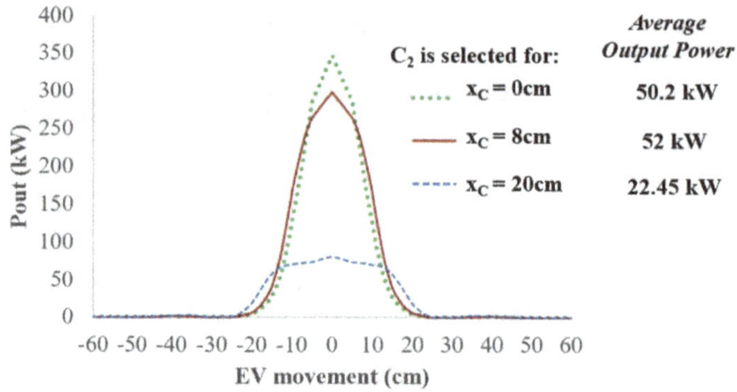

Figure 4.11 *Output power when the EV moves over the station when selecting C_2 for various positions of x_c [2]*

It is evident in Figure 4.10 that the selection of C_2 for $x_c = 0$ cm can allow the maximum power (P_{out_max}) to be received in the position where the EV is exactly above the station ($x_i = 0$ cm). However, a reduction is noted in the output power for all the other movement positions. For instance, when considering a misalignment position equal to $x_i = 20$ cm, the inductance L_2 is reduced by only 6% compared to the case where the EV is exactly above the station, yet the output power is significantly reduced, as evident in Figure 4.10. More specifically, according to (4.16), the maximum possible output power in this position can be achieved only when selecting C_2 for $x_c = 20$ cm. However, the selection of C_2 for the position $x_c = 0$ cm results in a much smaller value for the output power. In particular, the output power, as depicted in Figure 4.10, is equal to around 23% of the maximum possible one.

Apparently, an increase in the output power for the misalignment position $x_i =$ 20 cm could be achieved by selecting C_2 for $x_C = 20$ cm. However, this value for C_2 would result in a reduction in the output power for any other misalignment position, as evident in Figure 4.11. In this respect, the *average output power* when considering all the misalignment points is reduced from 50.2 kW (for $x_c = 0$ cm) to 22.45 kW (for $x_c = 20$ cm).

However, a different selection for the compensation capacitor C_2 could result in increased values for the system's output power. For instance, when selecting the compensation capacitor C_2 for the position $x_c = 8$ cm (Figure 4.11), an increase can be achieved in the *average output power*, which is now equal to 52 kW.

It is, therefore, evident that the selection of the compensation capacitors is linked with the position of the EV over the station and the relevant values of the self-inductance at these positions. Although the variations noted in the self-inductance values are not very significant in dynamic inductive charging, they can, however, introduce quite considerable variations in the output power and the efficiency when considering the selection of the capacitors C_1 and C_2.

The appropriate selection of C_1 and C_2 may be implemented according to [2], taking into account the maximization of both the average output power and the system's efficiency. However, in the case of dynamic inductive charging, additional parameters shall be taken into account, such as the proper selection of the system's frequency, resulting in a more complicated problem. The optimization of the relevant parameters in such a problem can be implemented according to the way described in [7], which is presented in detail in the next section.

4.3 Optimizing the power transfer and the efficiency in dynamic inductive charging

It is evident in (4.6) and (4.9) that both the efficiency and the output power depend on the compensation capacitors C_1 and C_2, the angular speed $\omega = 2\pi f$, and the equivalent load resistance R_L. An optimization problem shall, therefore, be defined in order to extract the optimal values for these variables and achieve high values for the efficiency and the output power during the whole EV range [7].

Concerning dynamic inductive charging, unlike static inductive charging, it is crucial to define not only the output power but also the *energy* transferred to the battery of the vehicle. According to [7], when considering that the EV enters the station at a misalignment x_0 and leaves the station at a misalignment x_N, the energy transferred to the battery can be calculated as:

$$E_{transferred} = \frac{\Delta x}{\upsilon}\left[\left(\sum_{i=0}^{N} P_{out\_i}\right) - \frac{P_{out\_0}+P_{out\_N}}{2}\right] \tag{4.20}$$

The values of the output power $P_{out\_1}, P_{out\_2}, \ldots, P_{out\_N}$ are calculated considering the inductance values $\{L_{1\_i}, L_{2\_i}, M_i\}$ at the misalignment points $x_1, x_2, \ldots, x_N$. Δx in (4.20) denotes the distance between the misalignment points $x_1 - x_0 = x_2 - x_1 = \ldots = x_N - x_{N-1} = \Delta x$, while υ is the EV speed.

Both the efficiency of the system in (4.9) and the energy transferred to the vehicle in (4.20) are affected by the angular speed $\omega = 2\pi f$, where f is the system's frequency. For a minimum and maximum frequency f_{min} and f_{max}, respectively, the constraints for the angular speed can be calculated as $\omega_{min} = 2\pi f_{min}$ and $\omega_{max} = 2\pi f_{max}$ [7].

As already mentioned in Section 4.2.3, the selection of the capacitors' values, C_1 and C_2, affects both the efficiency and the output power. As it was mentioned in the relevant section, the selection of C_1 and C_2 according to (4.10) and (4.11), when considering the relevant self-inductance values, can maximize the output power and the efficiency of the system at a specific position x_i with self-inductance values L_{1_i} and L_{2_i}. Taking into account all the possible misalignment points x_i in Figure 4.3 during the EV movement, a minimum and a maximum value can be defined for L_1 and L_2; hence the relevant values for the compensation capacitors can be defined as

$$C_{1,min} = \frac{1}{\omega_{max}{}^2 L_{1\_max}} \tag{4.21}$$

$$C_{1,max} = \frac{1}{\omega_{min}^2 L_{1\_min}} \tag{4.22}$$

$$C_{2,min} = \frac{1}{\omega_{max}^2 L_{2\_max}} \tag{4.23}$$

$$C_{2,max} = \frac{1}{\omega_{min}^2 L_{2\_min}} \tag{4.24}$$

The following optimization problem is, therefore, defined, according to [7]:

Objective Function	**Constraints**
Maximize the energy transferred to the EV: $E_{transfered} = \max$ (4.25)	Achieve an increased efficiency eff_{target} for all misalignment points x_i: $eff_i \geq eff_{target}$ (4.26)
	Physical limits for the variables:
	$\omega_{min} \leq \omega \leq \omega_{max}$ (4.27)
	$C_{1,min} \leq C_1 \leq C_{1,max}$ (4.28)
	$C_{2,min} \leq C_2 \leq C_{2,max}$ (4.29)
	$0 \leq R_L \leq R_{L,max}$ (4.30)

It has been observed in [7] that the optimization problem is infeasible when considering high values for the desirable efficiency eff_{target} in (4.26), in case the whole EV movement range is taken into account. However, the problem defined in (4.25)–(4.30) can be solved in a smaller section of the EV movement, or as it is called in [7], a High-Efficiency Section (HES). In this section, high amounts of energy can be transferred to the vehicle, while also achieving high values of efficiency.

In most IPT systems, the maximum power transferred to the vehicle is achieved when the EV is exactly above the station [4, 7, 13], as depicted in Figure 4.12(b). On the other hand, the power that is transferred to the vehicle when it enters (Figure 4.12(a) or leaves (Figure 4.12(c) the station is very close to zero. Thus, when considering misalignment points close to zero (Figure 4.12(b)) a HES can be defined, while two Low-Efficiency Sections can be defined considering the case where the EV enters or leaves the station, as depicted in Figure 4.12(a) and (c), respectively [7].

4.4 Control system in dynamic inductive charging

The maximum amount of energy can be transferred in the HES (where a high efficiency is also achieved), therefore the IPT system shall mainly operate in this specific section (compared to the Low-Efficiency-Sections). Operation only in the HES could be achieved by a detection mechanism (e.g. sensors) that would allow the system to be aware of the EV's position over the station. However, such a solution would result in an additional cost for the extra hardware required to be incorporated into the system.

Another option, which would allow operation only in the HES, is transferring information about the output power from the secondary to the primary side in order to allow calculations on the efficiency. However, such an option would result in

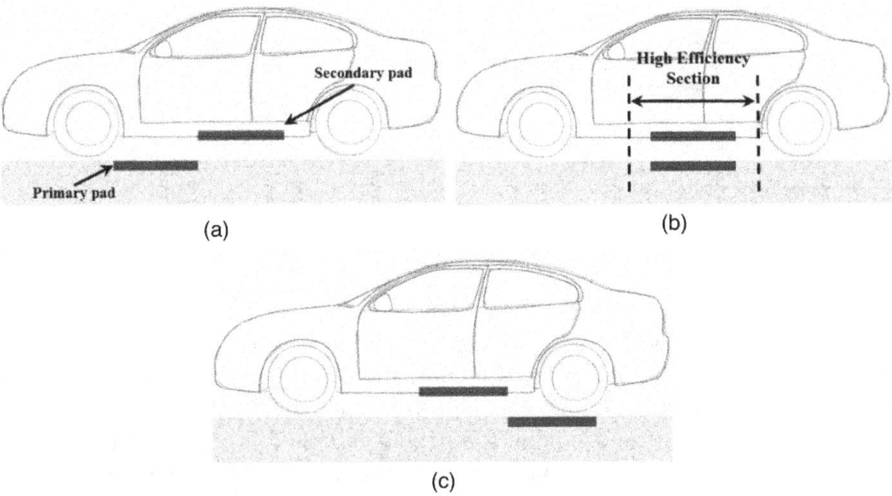

Figure 4.12 *(a) The EV enters the area of the station at a Low-Efficiency Section, (b) the EV is exactly above the station within the HES and (c) the EV leaves the station*

limited speeds for the vehicle in dynamic inductive charging due to delays in transferring the necessary information signals [7].

In order to avoid any additional hardware and/or any transfer of information between the primary and the secondary sides, the solution proposed in [7] can be incorporated into the design of an IPT system. In particular, specific values for the input power ($P_{in,enter}$ and $P_{in,leave}$) can indicate when the EV enters or leaves the HES [7].

In order, however, for the system to be aware of the input power, the operation in the first Low-Efficiency Section (as the EV enters the station) is necessary. In order to reduce the power losses in this section, a control method has been proposed in [7]. In particular, this control method allows variations in the primary current in order to reduce the power losses in the Low-Efficiency Section and achieve a maximum power transfer at the HES.

4.4.1 Primary side control

According to (4.6) and (4.8), it is evident that both the output power and the input power are proportional to the square of the input current I_1. Therefore, operation with a low primary current at the Low-Efficiency Section (as the EV enters the station) results in small amounts of power in the input side of the magnetic coupler, and, hence, small power losses. On the other hand, a high primary current at the HES allows the maximum power to be transferred to the vehicle. It is important to note that although the primary current affects both the input and the output power, it does not affect the efficiency of the system [7].

The control topology that can be applied in the system is depicted in Figure 4.13. The primary side control is responsible to vary the primary current, I_1, according to

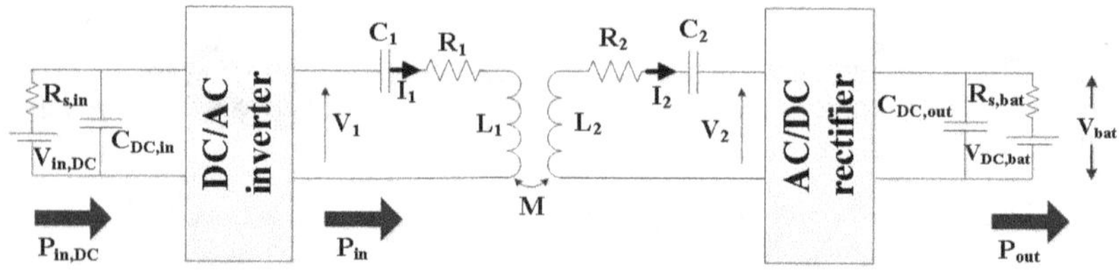

Figure 4.13 Circuit topology of the IPT primary and secondary sides

the position of the EV over the station, taking into account the variations in the input power P_{in}. A simplified description of the control procedure is depicted in Figure 4.14. In particular, the DC/AC high-frequency inverter in Figure 4.13 operates using a phase shift modulation scheme, as described in [6]. According to the values of the measured input power, the applied control is able to set a specific value for the primary current.

More specifically, the system will start operating at a low current, $I_{1,low}$, while the EV enters the station. The input power P_{in} is constantly monitored by the primary side control. As soon as the input primary gets a value greater than $P_{in,enter}$ indicating that the EV enters the HES, the primary current I_1 will receive its maximum value $I_{1,max}$, in order for the IPT system to transfer the maximum power at this specific section. As soon as the input power gets lower than a certain value, $P_{in,leave}$, indicating that the EV leaves the HES, power is no longer provided to the vehicle and the charging process ends [7].

4.4.2 Secondary side control

The optimization problem defined in (4.25)–(4.30) outputs the optimal values for the frequency, the compensation capacitors, and the load resistance. In order to ensure that the system operates with these optimal values while the EV passes over the station, it is important that the load resistance R_L remains constant (at the optimal value) during the movement of the vehicle over the station. A specific value for the load resistance in the secondary side can be achieved by controlling the voltage value V_2 in the secondary side (Figure 4.13).

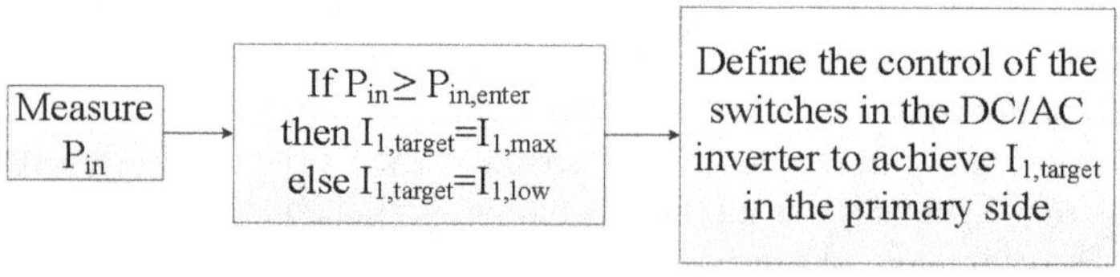

Figure 4.14 Description of the applied control scheme in the primary side

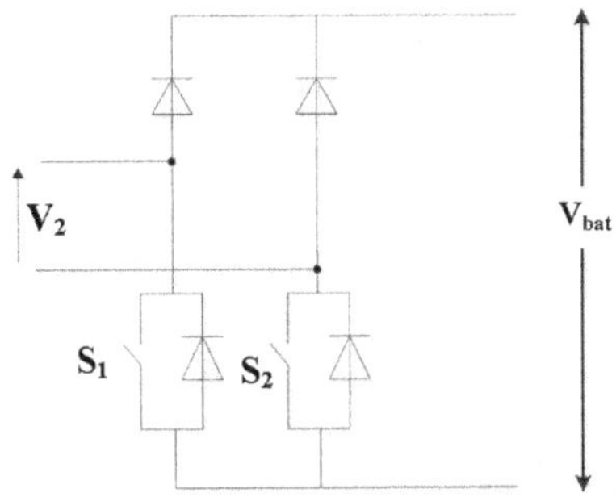

Figure 4.15 AC/DC rectifier in the secondary side [6, 7]

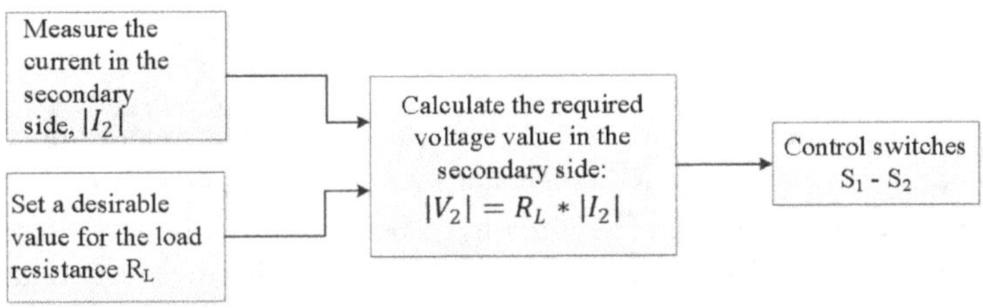

Figure 4.16 Control of the load resistance in the secondary side

In particular, the voltage V_2 can be controlled by introducing the two switches S_1 and S_2 in the secondary-side rectifier [6, 7], as depicted in Figure 4.15. A simplified description of the control procedure is provided in (Figure 4.16).[b] In particular, the control pulses in the switches S_1 and S_2 can be specifically adjusted in order to achieve a specific value for the fundamental output voltage $|V_2|$. Since $|V_2| = R_L * |I_2|$, the measurement of $|I_2|$ can allow a specific voltage $|V_2|$ to be achieved in the secondary side, hence ensuring a desirable value for the load resistance, R_L.

4.5 Application of the optimization problem and the control system in a circular magnetic coupler

The optimization problem described in (4.2) is applied for the circular magnetic coupler depicted in Figure 4.4, according to the specifications in [7]. The air gap between the primary and the secondary pads is considered equal to 15 cm and the inductance values are depicted in Figure 4.17. The EV is considered to be exactly

[b]A more detailed description concerning the control applied in the secondary side can be found in [7].

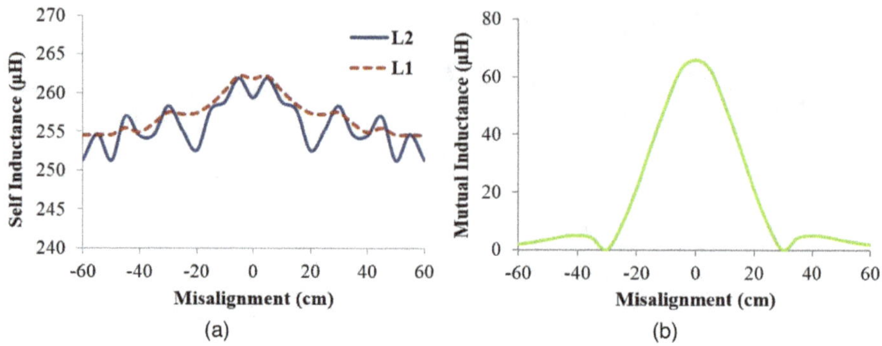

Figure 4.17 (a) Self-inductance and (b) mutual inductance values for the examined circular magnetic coupler, while the EV moves over the station

above the station when the misalignment between the primary and the secondary pads is equal to zero, while the EV enters and leaves the station with a misalignment equal to −60 cm and +60 cm. The coils have been simulated taking into account commercially available Litz wires.[c] The maximum value for the current, taking into account the design characteristics of the wires, is set to $I_{max} = 25$ A.

4.5.1 Application of the optimization problem

The optimization problem is applied in a real-case scenario of dynamic inductive charging. More specifically, the EV is considered to move over the station with a speed equal to 50 km/h. The system operates at a constant frequency that shall be defined according to the solution of the optimization problem. The upper and lower limits of the frequency in (4.27) are defined equal to 20 kHz and 90 kHz, respectively, taking into account the potential limitations in power electronics employed in the system or the characteristics of the magnetic coupler [13], as well as the maximum frequency defined in relevant standards for inductive charging of EVs [17]. The minimum and maximum values for the capacitance in (4.28)–(4.29) are calculated according to the minimum and maximum values of the frequency ($f_{min} = 20$ kHz and $f_{max} = 90$ kHz). The battery pack in [18] is taken into account for the examined system, indicating a maximum load resistance in (4.30) equal to $R_{L,max} = 9.72\ \Omega$ [7].

The optimization problem, as defined in Section 4.3, requires the maximization of the transferred energy in (4.25), taking into account constraints (4.26)–(4.30). The problem is solved for many cases regarding the length of the HES, as well as many different values for the desired efficiency eff_{target}. The optimization problem has been

[c]https://www.newenglandwire.com/

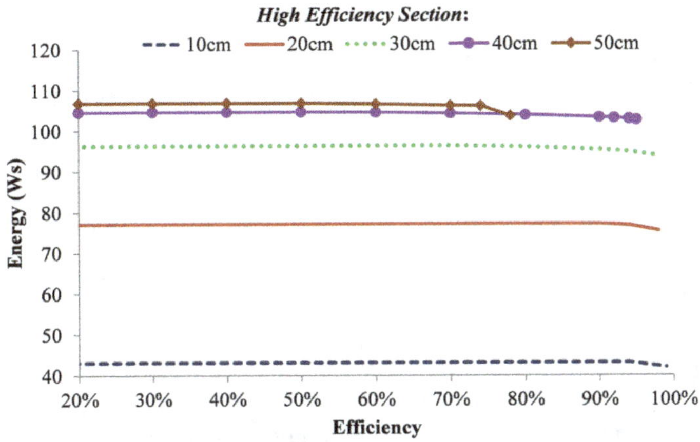

Figure 4.18 Solutions of the optimization problem: maximum energy transferred for different values of eff_{target} [7]

solved by employing the GAMS software and the LINDOGLOBAL solver, and the results are depicted in Figure 4.18.

Each line in Figure 4.18 indicates the results for a different length of HES. For instance, the red line indicates the results for a HES equal to 20 cm. Each point in the red line indicates the maximum energy transferred to the EV for the relevant value of the eff_{target} indicated in the diagram. It is evident that higher amounts of energy can be achieved when considering larger HESs. For instance, a significant increase can be noted in the transferred energy when considering a HES equal to 20 cm compared to the very limited HES of 10 cm. HESs of up to 50 cm are considered in Figure 4.18, since even larger HESs would lead to a non-feasible optimization problem. For instance, when considering a HES of 60 cm, the maximum allowable eff_{target} that would lead to a feasible optimization problem is less than 0.05% [7].

It shall be noted that for each line in the diagram (i.e. for each case of HES), the transferred energy remains almost constant until a value of efficiency, E_{limit}, before being reduced to lower values. This specific point in the diagram indicates the maximum value for the efficiency, eff_{limit}, that can achieve the maximum transferred energy, E_{limit}, in each case of HES [7]. Requesting a larger value for eff_{target} will result in a reduction in the transferred energy, as evident from each line in the diagram in Figure 4.18. The maximum efficiency point in each line indicates the maximum possible efficiency, eff_{feas}, that will lead to a feasible optimization problem [7].

For instance, for a HES equal to 40 cm in Figure 4.18, the maximum energy, E_{limit}, that can be achieved is 104.7 Ws. The maximum efficiency that can lead to transferring this amount of energy to the EV is eff_{limit} = 50%. Larger values for the eff_{target} in the constraint (4.26) will lead to a reduction in the transferred energy, as evident in Figure 4.18. The maximum efficiency eff_{target} that leads to a feasible

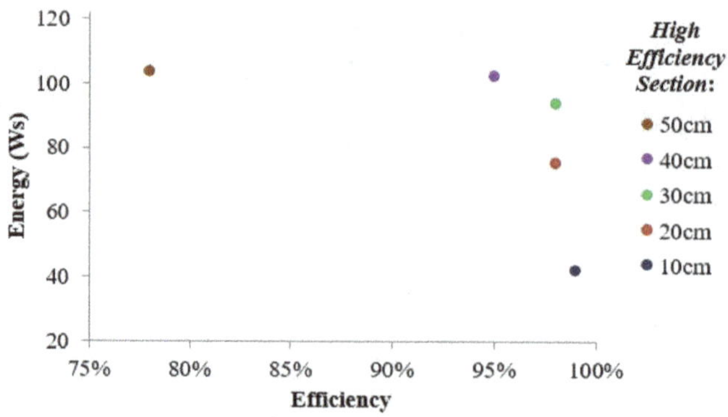

Figure 4.19 Values for eff_{feas} and E_{feas} for each HES [7]. Each color in the relevant index indicates a different HES, as depicted in Figure 4.18.

optimization problem in this case is equal to $eff_{feas} = 95\%$ with a relevant transferred energy $E_{feas} = 102.62$ Ws [7].

It shall be noted that the comparison of the values for E_{limit} and E_{feas} for the case of a HES equal to 20 cm indicates a reduction of 2% in the transferred energy. However, a significant increment is noted in the achieved efficiency. Similarly, for each examined case of HES in Figure 4.18, the E_{feas} is reduced by around 3% compared to E_{limit} while a great increment is noted in the efficiency [7]. In this respect, the values for eff_{feas} and E_{feas} can be considered indicative for each HES, considering the transferred energy and efficiency that can be achieved [7]. The relevant sets $\{eff_{feas}, E_{feas}\}$ for each case of HES are depicted in Figure 4.19.

A trade-off can be observed in Figure 4.19 between the energy that can be transferred to the EV's battery and the efficiency that can be achieved. In particular, high amounts of energy can be achieved when considering a HES equal to 50 cm, at the expense of a quite small efficiency. On the other hand, for a HES of 10 cm the efficiency is significantly increased; however, only small amounts of energy can be transferred to the vehicle's battery. Concerning the choice of $\{eff_{feas}, E_{feas}\}$, it shall be noted that for a dynamic inductive charging application, it is quite important to achieve high amounts of energy while the EV passes above the station with quite a high speed. It is evident in Figure 4.19 that high amounts of energy can be achieved for a HES of 50 cm or 40 cm. In both cases, the transferred energy is almost the same; however, for the case of a HES of 40 cm the efficiency is greatly increased. Thus, a HES of 40 cm is the choice for the examined application.

For the specific HES of 40 cm, the optimization problem outputs the optimal values for the operational frequency, the compensation capacitors, and the load resistance:

$f = 24.1$ kHz
$C_1 = 168$ nF
$C_2 = 159$ nF
$R_L = 9.72\ \Omega$

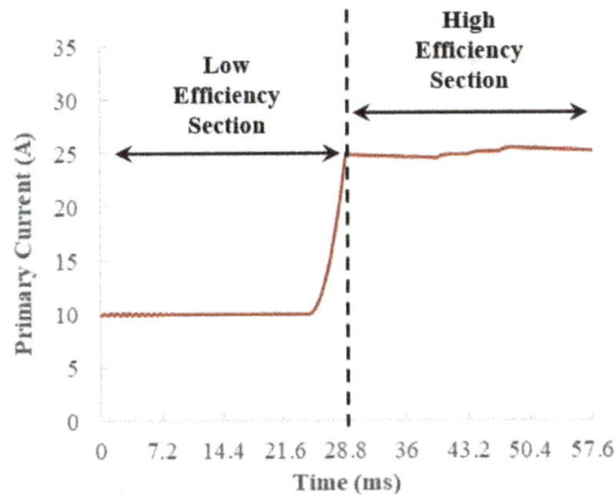

Figure 4.20 *Variation of the primary current, I1, while the EV moves over the station*

4.5.2 Simulation of the applied control

When the EV enters the station, the system is expected to operate with a low current in order to reduce the losses in the Low-Efficiency Section (Figure 4.12(a)). This current is set equal to $I_{low} = 10$ A. When the EV enters the HES (Figure 4.12(b)) at a misalignment of −20 cm, the current shall be increased to $I_{max} = 25$ A in order to transfer the maximum energy to the EV's battery. When the EV leaves the HES with a misalignment equal to +20 cm, power shall no longer be provided to the vehicle and the charging process shall end.

The modification of the current values shall be performed by the control scheme in the primary side. The secondary control scheme is responsible to maintain the load resistance at $R_L = 9.72\ \Omega$, which is the value indicated by the solution of the optimization problem. Both the primary and the secondary control schemes, as well as the whole IPT system (as depicted in Figure 4.13), have been simulated in MATLAB®-SIMULINK.

The variations in the primary current are indicated in Figure 4.20. It is evident that the primary control scheme can successfully vary the current's value from $I_{low} = 10$ A to $I_{max} = 25$ A, according to the efficiency section that the EV moves in (High- or Low-Efficiency Section). In particular, when the control scheme in the primary side measures the value of the input power $P_{in,enter}$ that indicates the transition from the Low-Efficiency Section to the HES ($P_{in,enter} = 98$ W in Figure 4.21), it increases the primary current to its maximum value. Additionally, the primary control scheme turns off the power provided to the station as soon as it detects the value of the input power ($P_{in,leave} = 618$ W in Figure 4.21) that indicates that the EV leaves the HES [7].

The control scheme can, therefore, efficiently employ variations in the primary current in order to reduce the power losses in the Low-Efficiency Section, while transferring the maximum possible energy in the HES.

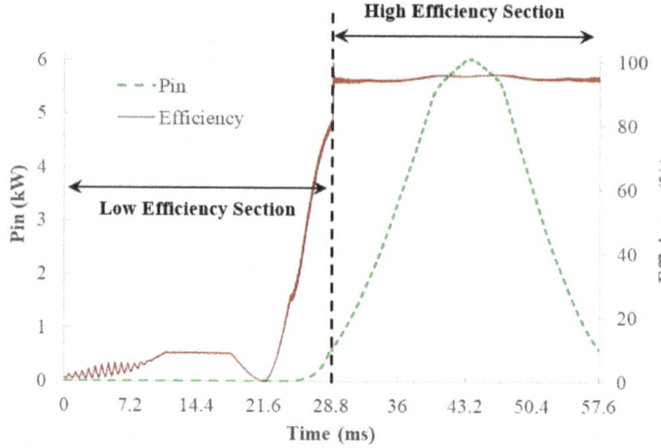

Figure 4.21 *Input-output power and efficiency while the EV moves over the station*

4.6 Experimental validation of the proposed optimization and control scheme

4.6.1 Implementation of the magnetic coupler

The circular magnetic coupler with ferrite bars in Figure 4.22 has been designed according to the optimal characteristics for the design of such couplers (comprising the optimal position of the ferrite bars, the characteristics, and the position of the Litz wires, etc.), which are provided in [13]. The designed system allows the movement of the secondary pad in order to evaluate the movement of the EV over the station in dynamic inductive charging conditions.

The 24.6 cm diameter primary and the secondary pads in Figure 4.22 are identically designed, each comprising 5 ferrite bars (type RDZEŃ I93/28/16 F-887). Each coil comprises 22 turns of Litz wire, which is constructed of 180 strands with 0.1 mm diameter. The maximum current in the primary and secondary coil is set equal to $I_{1,max} = I_{2,max} = 3$ A. The resistance values of the two coils are measured equal to R_1

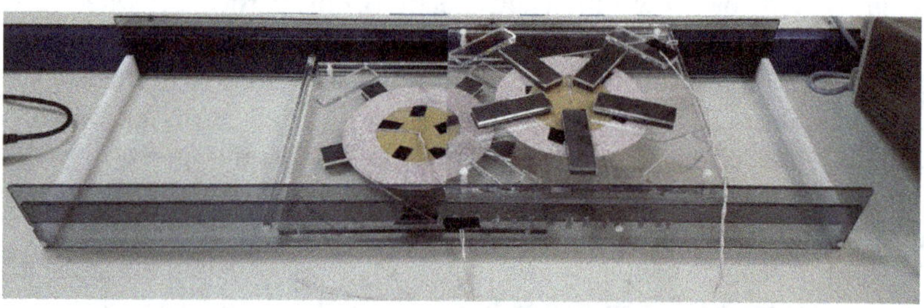

Figure 4.22 *Implementation of the primary and the secondary pads*

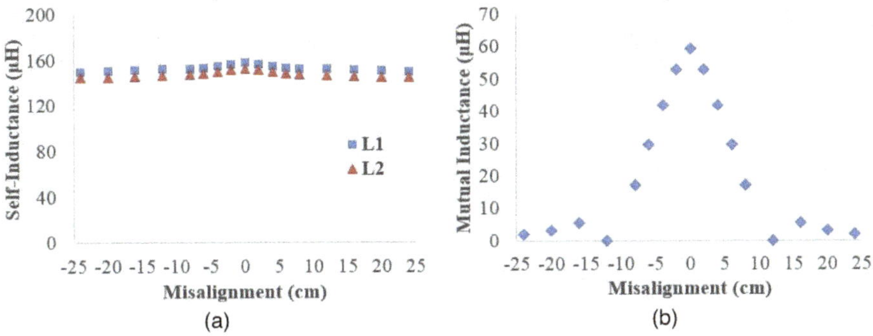

Figure 4.23 *(a) Measured self-inductance values and (b) measured mutual inductance values of the implemented magnetic coupler*

Table 4.1 *Minimum and maximum values of the optimization problem's variables*

Variable	Minimum value	Maximum value
Frequency, f (kHz)	20	50
Primary compensation capacitor, C_1 (nF)	64.2	422
Secondary compensation capacitor, C_2 (nF)	66.5	437.8
Load resistance, R_L(Ω)	0	8.65

= R_2 = 0.16 Ω. The measured inductance values for different misalignment positions are depicted in Figure 4.23.

4.6.2 Application of the proposed optimization method in the implemented magnetic coupler

The optimization problem defined according to (4.25)–(4.30) in Section 4.2 is applied in the designed magnetic coupler depicted in Figure 4.22 in order to identify the optimal values required for the compensation capacitors in the primary and secondary sides, the operational frequency, and the load resistance. The minimum and maximum values for the variables of the optimization problem are provided in Table 4.1. In particular, the minimum and maximum frequencies are considered equal to 20 kHz and 50 kHz, respectively. The minimum and maximum values for the primary and secondary capacitors are calculated according to (4.21)–(4.24). Finally, the maximum value of the load resistance is defined taking into account the

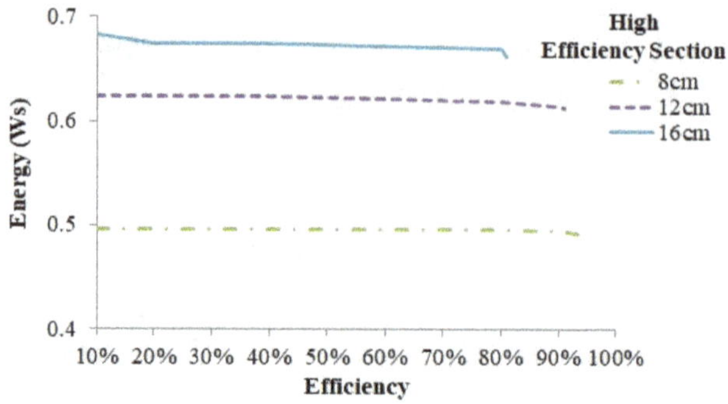

Figure 4.24 Maximum transferred energy for the implemented magnetic coupler, when considering different cases of HESs and different values for the eff_{target}

data for commercially available batteries[d] that can be used for applications similar to the one examined.

The optimization problem is solved for many cases of HESs as well as different values of the eff_{target} in (4.26). The optimization problem has been solved with the GAMS software, utilizing the LINDOGLOBAL solver. The results concerning the maximum energy that can be transferred to the EV's battery in each case are depicted in Figure 4.24. Each line in the diagram depicts the results for a specific HES.

It is evident that when a specific HES is taken into account, the transferred energy remains almost constant up until a certain value of efficiency (eff_{limit}). This value of energy indicates the maximum transferred energy that can be achieved for the examined HES. After the set E_{limit} and eff_{limit}, the transferred energy is slightly reduced up to a value E_{feas} indicating the maximum efficiency, eff_{feas}, which results in a feasible optimization problem.

It is evident in Figure 4.24 that E_{feas} is slightly reduced compared to E_{limit}. In particular, for every examined HES in Figure 4.24, E_{feas} is reduced by less than 3.1% compared to E_{limit}. However, the increment noted in the relevant efficiency values is quite significant. Thus, the high efficiency that is noted when taking into account the set of values $\{eff_{feas}, E_{feas}\}$, as well as the relevant small reduction in the transferred energy, indicates that these values $\{eff_{feas}, E_{feas}\}$ can be considered indicative of every HES. Those values are depicted in Figure 4.25.

It can be observed in Figure 4.25 that there is a trade-off in the energy that can be transferred to the vehicle and the efficiency that can be achieved. However, due to the requirements of dynamic inductive charging, high amounts of energy are

[d]https://www.dhgate.com

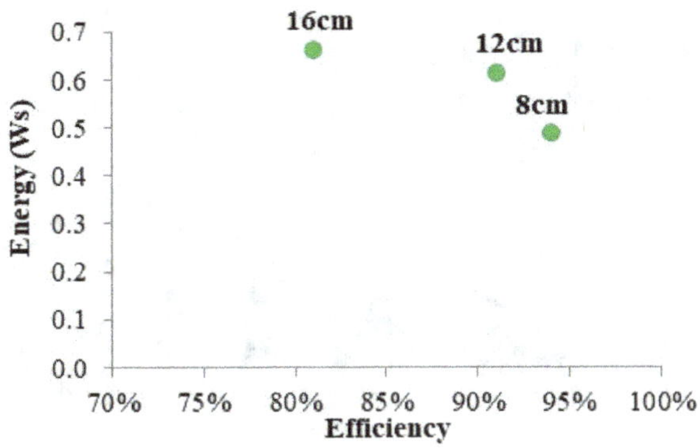

Figure 4.25 *Values of {eff_{feas}, E_{feas}} for each HES for the implemented magnetic coupler. The relevant HES is indicated near each point in the diagram.*

necessary, which can be achieved for HESs of 12 cm and 16 cm. Similar values are noted regarding the transferred energy for those two cases; however, the efficiency is increased from 81% to 91% when considering a HES equal to 12 cm. Thus, a HES equal to 12 cm is the preferable choice for the examined application. For this specific HES, the optimal values for the operational frequency, the compensation capacitors, and the load resistance (which are the output from the optimization problem's solution) are indicated in Table 4.2.

4.6.3 Implementation of the inverter and the control system

The experimental setup, comprising the DC/AC inverter, the compensation capacitors, and the magnetic coupler is depicted in Figure 4.26. The type-H DC/AC inverter comprises the IRFI4228PbF Power MOSFETs, which have low conduction losses. The driving of the switches is implemented with the Silicon Labs, Si8274, half-bridge, anti-shoot-through driver. The inverter capacitor bank ($C_{DC,in}$ in Figure 4.13) comprises 8 ECA-1JM Panasonic capacitors of 2 200 μF connected in parallel. A Texas Instrument C2000 Experimenter kit, which comprises the F28335 Delfino Control Card, is used to send the appropriate pulses to the inverter. Kemet

Table 4.2 *Optimal values of the optimization problem's variables, when considering a HES equal to 12 cm*

Variable	**Optimal value**
Frequency, f	27.91 kHz
Primary compensation capacitor, C_1	0.23 μF
Secondary compensation capacitor, C_2	0.27 μF
Load resistance, R_L	8.65 Ω

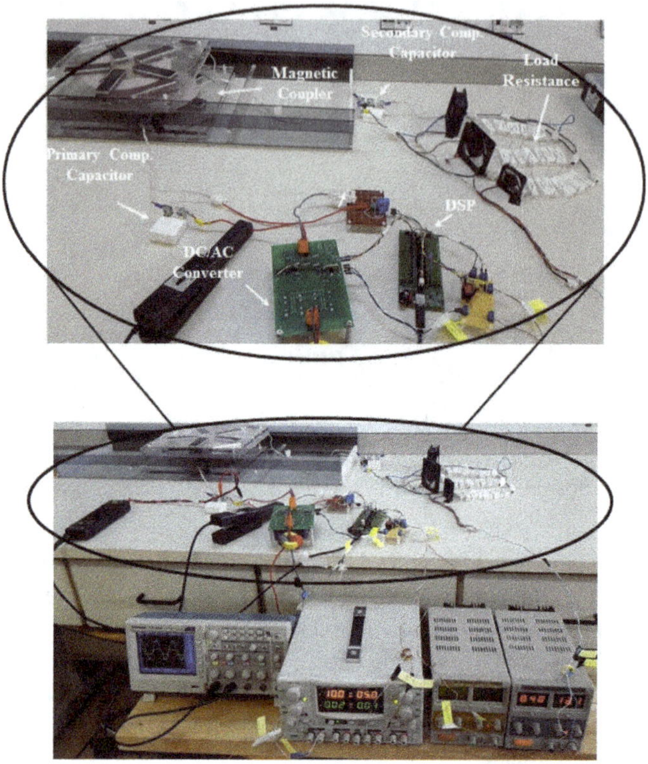

Figure 4.26 Implementation of the experimental setup

C4BSPBX Film Capacitors are used as compensation capacitors in the primary and secondary sides, which are suitable for the high voltages and currents noted in the required application. A programmable Regatron, TopCon Quadro, is used in the DC power supply, while a current transducer LTSR 6-NP is used for the measurement of the primary current in the output of the primary converter. A 9.6 Ω load resistance is placed in the secondary side, which consists of ceramic resistances suitable for the power requirements in the secondary side. Table 4.3 indicates a summary of the parameters and the components used in the experimental setup.

In order to reduce the number of calculations in the DSP, the transition between the Low- and High-Efficiency Sections is realized by observing the value of the switches' conduction angle, unlike observing the value of the input power (as described in 4.4.1). More specifically, the output voltage V_1 in Figure 4.13 (and accordingly the value of the primary current) can be controlled by modifying the conduction angle of the inverter's switching components. This angle is calculated by the control system in the primary side.

In particular, Figure 4.27 depicts the conduction angle that can achieve the desirable value for the primary current in each movement position. It shall be noted that the values for the angle in Figure 4.27 can be calculated before the EV enters the

Table 4.3 Parameters and values of the components used in the experimental setup

Component/Parameter	**Value/Type**
Frequency, f	27.9 kHz
Primary comp. capacitor, C_1	0.22 μF (C4BSWBX3220ZBFJ)
Secondary comp. capacitor, C_2	0.28 μF (C4BSPBX3680ZBFJ and C4BSWBX3470ZBFJ)
Load resistance	9.6 Ω
DC link capacitors	17.6 mF (array of ECA-1JM222)
MOSFET	IRFI4228PbF
DC power supply	Regatron, TopCon Quadro (30 ΩV)
DSP	F28335 Delfino Control Card
Litz wires	180 strands of 0.1 mm diameter
Ferrite bars	RDZEŃ I93/28/16 F-887, MnZn
Current transducer (measuring the value of the primary current)	LTSR 6-NP

station and before the initiation of the charging process. It is evident in Figure 4.27 that an angle equal to α_{enter} = 0.2 rad indicates that the EV enters the HES, while the EV leaves the HES with an angle α_{leave} = 0.6 rad.

In this respect, the overview of the implemented control scheme is depicted in Figure 4.28. In particular, the control scheme constantly measures the primary current, I_{in}, and compares it with the desired value $I_{1,target}$. $I_{1\ target}$ shall be equal to $I_{1,max}$ = 3 A when the EV moves in the HES. When the EV moves in the Low-Efficiency Sections (when it enters or leaves the station) the primary current is set to $I_{1,low}$ = 1 A in order to reduce the power losses in those sections.

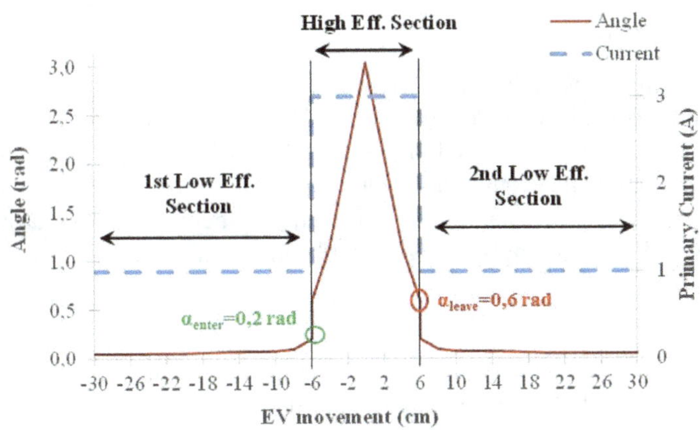

Figure 4.27 *Angle α that can achieve the desirable value of the primary current while the EV moves over the station*

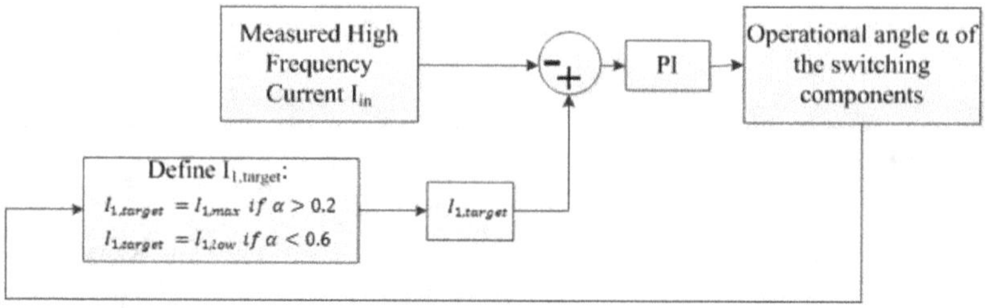

Figure 4.28 Implementation of the control scheme

In order to implement more realistic conditions for an inductive charging station operating in dynamic charging conditions, a slightly different operation than the one described in Section 4.4.1 is considered. More specifically, when the EV leaves the HES, the operation of the system is not terminated, yet the primary current is again set to $I_{1,low}$ as depicted in Figure 4.27. Such an operation will allow the station to be ready for another vehicle to pass as soon as the first EV leaves the station.

The primary current during the whole movement of the secondary pad over the primary pad is depicted in Figure 4.29, while a closer look at the primary current when the EV moves in the High- or Low-Efficiency Section is depicted in Figure 4.30. The charging process begins with a primary current equal to $I_{low} = 1$ A in the first Low-Efficiency section. The control scheme successfully manages to increase the primary current to $I_{max} = 3$ A as soon as the EV enters the HES, while the primary current is again set to $I_{low} = 1$ A when the EV enters the second Low-Efficiency Section.

The experimental efficiency between the DC input in the primary side ($P_{in\_DC\_exp\_i}$) and the load resistance ($P_{out\_exp\_i}$) for a specific misalignment position between the primary and the secondary pad can be defined as [7]

$$eff_{exp\_i} = \frac{P_{out\_exp\_i}}{P_{in\_DC\_exp\_i}} \tag{4.31}$$

The experimental efficiency for different cases of misalignment, according to the experimental results, is indicated in Figure 4.31. It is evident in this figure that high values for the efficiency are noted in the HES. In particular, the minimum value of the efficiency in the HES is measured equal to around 85%.

The theoretically calculated efficiency, eff_{theor_i}, between the input and output power, P_{in} and P_{out} accordingly in Figure 4.13 can be defined as [7]

$$eff_{theor\_\iota} = \frac{P_{out\_theor\_\iota}}{P_{in\_theor\_\iota}} \tag{4.32}$$

The theoretically calculated efficiency is also depicted in Figure 4.31. It shall be mentioned that the calculation of eff_{theor} does not take into account the DC/AC inverter losses in the primary side or the losses in the magnetic coupler. However, both the inverter losses and the losses in the magnetic coupler are considered when calculating eff_{exp}. In this respect, as evident in Figure 4.31, low losses can be observed in the

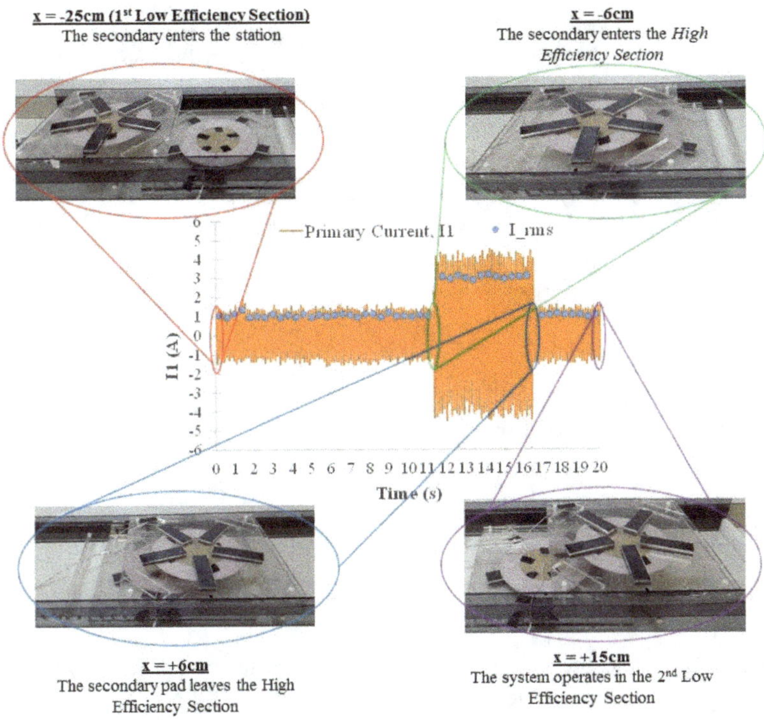

Figure 4.29 *Variations in the primary current from I_{low} = 1 A to I_{max} = 3 A with the application of the proposed control, while the secondary pad moves over the primary pad*

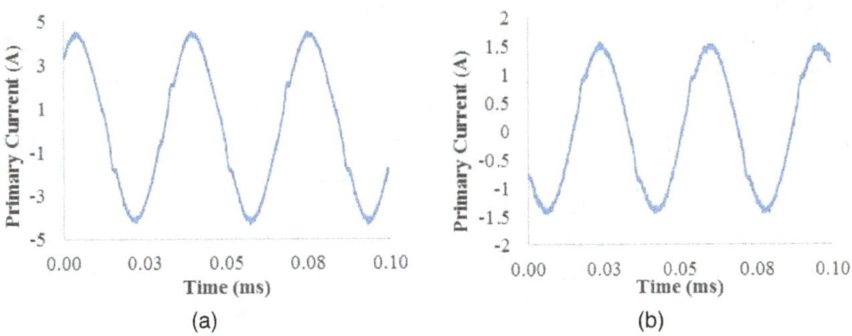

Figure 4.30 *Primary current when the EV moves in the (a) HES with a misalignment between the primary and the secondary equal to 0 cm and (b) Low-Efficiency Section with a misalignment equal to 10 cm*

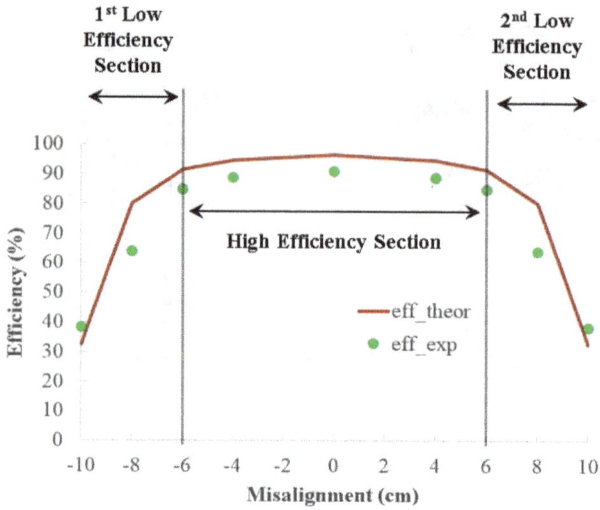

Figure 4.31 Experimental and theoretical efficiency for different points of misalignment

magnetic coupler and the DC/AC inverter [7], since in the HES, the experimental efficiency eff_{exp} is reduced to less than 7% compared to eff_{theor}.

4.7 Conclusions

Dynamic inductive charging of EVs can significantly resolve the issues related to the limited travel range of EVs allowing the vehicle to receive energy while moving on the road. Moreover, the ability to charge on multiple positions during a trip may allow a reduction in the EV's battery capacity and, accordingly, a lower battery cost.

The operation of a dynamic inductive charging system is similar to that of static inductive charging, where wireless power transfer can be achieved between the charging station and the vehicle, which is parked above the station. However, some significant issues shall be considered when the EV receives energy while passing over the station. In particular, the analysis of a dynamic inductive charging system requires all the misalignment positions among the primary pad (in the station's side) and the secondary pad (placed in the EV) to be taken into account.

Moreover, specific considerations shall be evaluated when studying the bifurcation phenomenon. In particular, in the case of dynamic inductive charging, not only does the operational frequency but also the EV movement over the station affect the system's efficiency and transferred power. In this respect, the system's efficiency may be maximized at misalignments points (among the primary and secondary pads) different than the ones maximizing the power transferred to the EV. In this respect, an analysis similar to the one presented in this chapter concerning the bifurcation phenomenon shall be considered in order to effectively evaluate the design of a dynamic inductive power transfer system.

Additionally, the results in Section 4.1.3 indicate that even small variations in the self-inductance values may result in quite significant variations in the system's efficiency and the power transferred to the vehicle. Thus, such variations in the self-inductance values shall be taken into account when selecting the compensation capacitors in the case of dynamic inductive charging in order to ensure high values of transferred power and efficiency while the EV moves over the station.

In the case of a dynamic inductive charging system, the efficiency and the transferred power are also affected by the battery's load resistance in addition to the system's operational frequency and the appropriate selection of the compensation capacitors. In this respect, an optimization problem shall be defined in order to indicate the optimal variables (frequency, compensation capacitors, and load resistance) that allow the maximization of the output power at high system efficiency. The optimization method has been applied in a circular magnetic coupler indicating the need for a compromise between the energy transferred to the EV battery and the system's efficiency. However, the appropriate selection of the HES (the movement section where high efficiency values can be achieved) allows the transfer of significant amounts of energy while also achieving a significantly increased efficiency.

A control scheme has also been examined in order to apply the required optimization method, while reducing the system's power losses. In particular, the control scheme allows the system to be aware of the EV's movement position that indicates the entrance to, and exit of, the HES. The control scheme eliminates any need for signal transfer between the primary and the secondary side, while not requiring any additional hardware (sensors, etc.).

Specific considerations shall also be taken into account in order to reduce the system's losses when the vehicle moves in Low-Efficiency Sections. More specifically, variations in the primary current between a low and a maximum value allow variations in the input power to the system. In this respect, a low input power (at the low-value current) in the Low-Efficiency Section can ensure low losses, while a high input power (at the maximum-value current) at the HES can ensure that the maximum energy is transferred to the EV's battery.

The simulation results of a dynamic inductive charging system indicate that the proposed control scheme can effectively be applied in a 60 cm diameter magnetic coupler. In particular, the control of the current in the primary side of the magnetic coupler allows the effective operation of the system both in the High- and Low-Efficiency Sections, while achieving high values for both the energy transferred to the EV and the system's efficiency.

The optimization method has also been applied in an experimentally implemented circular magnetic coupler. The results indicate that high-efficiency values, as well as increased values for the transferred energy, can be achieved when appropriately selecting the HES. Taking into account the output optimal values of the optimization method (i.e. the compensation capacitors, the equivalent load resistance, and the system's frequency), a dynamic inductive power transfer system has been experimentally implemented. In particular, the system comprises the magnetic coupler as well as the relevant DC/AC inverter that has been specifically developed for this application. The experimental results indicate that the control scheme can effectively be applied when the EV moves over the primary, allowing the successful detection of the vehicle's movement between the High- and Low-Efficiency Sections. Additionally, high-efficiency values can be achieved in the HES, while low losses are noted both in the inverter and the designed magnetic coupler.

References

[1] Lorico A., Taiber J., Yanni T. 'Effect of inductive power technology systems on battery-electric vehicle design'. IECON 2011 – 37th Annual Conference on IEEE Industrial Electronics Society; Nov. 2011.

[2] Karakitsios I., Hatziargyriou ND. 'Self-inductance variations in dynamic inductive charging of EVs'. *Mediterranean Conference on Power Generation, Transmission, Distribution and Energy Conversion*. 2018.

[3] Hudecek J., Küfen J., Langen O., Dankert J., Eckstein L. 'A system for precise positioning of vehicles aiming at increased inductive charging efficiency'. MedPower; Athens; 2014.

[4] Karakitsios I., Karfopoulos E., Madjarov N., *et al.*. 'An integrated approach for dynamic charging of electric vehicles by wireless power transfer – lessons learned from real-life implementation'. *SAE International Journal of Alternative Powertrains*. 2017, vol. 6(1), pp. 15–24.

[5] Wang C.-S., Covic G.A., Stielau O.H. 'Power transfer capability and bifurcation phenomena of loosely coupled inductive power transfer systems'. *IEEE Transactions on Industrial Electronics*. 2004, vol. 51(1), pp. 148–57.

[6] Diekhans T., De Doncker R.W., Rik W., Doncker D. 'A dual-side controlled inductive power transfer system optimized for large coupling factor variations and partial load'. *IEEE Transactions on Power Electronics*. 2015, vol. 30(11) 6320–8.

[7] Karakitsios I., Palaiogiannis F., Markou A., Hatziargyriou N.D. 'Optimizing the energy transfer, with a high system efficiency in dynamic inductive charging of EVs'. *IEEE Transactions on Vehicular Technology*. 2018, vol. 67(6), pp. 4728–42.

[8] Buja G., Bertoluzzo M., Dashora H.K. 'Lumped track layout design for dynamic wireless charging of electric vehicles'. *IEEE Transactions on Industrial Electronics*. Oct. 2016, vol. 63(10), pp. 6631–40.

[9] Zhao J., Cai T., Duan S., Feng H., Chen C., Zhang X. 'A general design method of primary compensation network for dynamic WPT system maintaining stable transmission power'. *IEEE Trans Power Electronics*. 2016, vol. 31(12), pp. 8343–58.

[10] Karakitsios I., Hatziargyriou N. 'Bifurcation in dynamic inductive charging of electric vehicles'. 2017 IEEE Manchester PowerTech; 2017. pp. 1–6.

[11] Kamineni A., Covic G.A., Boys J.T. 'Self-tuning power supply for inductive charging'. *IEEE Transactions on Power Electronics*. 2016.

[12] Lee K., Pantic Z., Lukic S.M. 'Reflexive field containment in dynamic inductive power transfer systems'. *IEEE Transactions on Power Electronics*. 2014, vol. 29(9), pp. 4592–602.

[13] Budhia M., Covic G.A., Boys J.T. 'Design and optimization of circular magnetic structures for lumped inductive power transfer systems'. *IEEE Transactions on Power Electronics*. 2011, vol. 26(11), pp. 3096–108.

[14] Gati E., Kampitsis G., Manias S. 'Variable frequency controller for inductive power transfer in dynamic conditions'. *IEEE Transactions on Power Electronics*. 2017, vol. 32(2), pp. 1684–96.

[15] Zaheer A., Neath M., Beh H.Z.Z., Covic G.A. 'A dynamic EV charging system for slow moving traffic applications'. *IEEE Transactions on Transport. Electrification*. 2016, pp. 1–18.

[16] Lu F., Zhang H., Hofmann H., Mi C.C. 'A dynamic charging system with reduced output power pulsation for electric vehicles'. *IEEE Transactions on Industrial Electronics*. 2016, vol. 63(10) 6580–90.

[17] 'Wireless Power Transfer for Light-Duty Plug-in/Electric Vehicles and Alignment Methodology'. *SAE standard J2954*

[18] Marcel M., Knakal T., Stifflemire T., Lock B. 'Testing of nanophosphate™ prismatic battery cells in the XM1124 hybrid electric HMMWV'. NDIA Ground Vehicle Systems Engineering and Technology Symposium, Power and Mobility (P&M) Mini Symposium; 2011.

Chapter 5

Converter classification, analysis, and control issues with EV

Anish Ahmad[1] and R K Singh[2]

This chapter presents a converter classification, its analysis, and control issues with electric vehicles (EVs). DC-DC converter and DC-AC inverter are the key power electronic interfaces used in EV applications. The input of the converters is given by the energy storage systems such as battery or fuel cells. The available voltage of the energy storage system is very less generally varying from 12 V to 96 V. To drive the EV motors, the required DC link voltage ratings are varying up to 400 V. So, the DC-DC converter and inverter are required that can be capable of providing higher voltage gain. In this chapter, the characteristics of converters and inverters used for EV applications and their control are discussed. The converters and inverters can be divided into different topologies such as cascaded, switched-inductors, switched-capacitors, coupled inductors, soft switching techniques, etc. The advantages, disadvantages, comparisons, and its key features are discussed. The converters are discussed with steady-state analysis, small-signal modeling, and passive components design, for example, a bidirectional quadratic boost converter analysis with simulation and experimental results are presented in the chapter. The EV drive is also an important issue with the converters and inverters. Therefore, the AC and DC motor drives' control issues are also discussed.

5.1 Introduction

Power electronic converters (DC-DC converter and DC-AC inverter) are the key converters used in the hybrid electric vehicle (HEV) or electric vehicle (EV) application. The DC input to these power converters is generally supplied from the battery storage, fuel cell, supercapacitors, etc. Drive systems such as four-wheel drive, two-wheel drive, rear-wheel drive, and all-wheel drive (in the wheel and near the wheel) are also commonly used drive system in battery-operated EVs and HEVs

[1]Department of Electrical Engineering, Tezpur University, India
[2]Department of Electrical Engineering, IIT (BHU) Varanasi, India

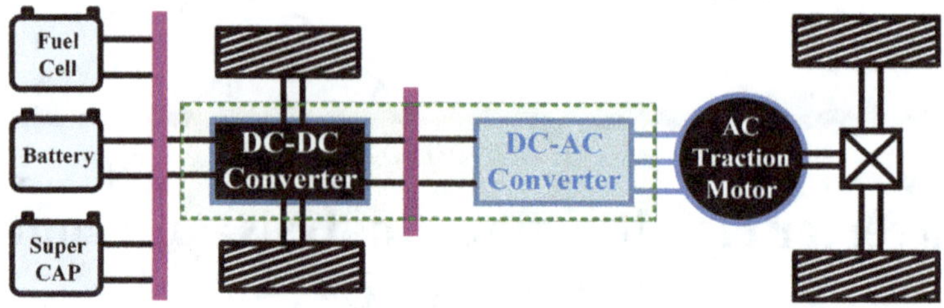

Figure 5.1 Typical representation of HEV/EV AC drive system

[1, 2]. The typical configuration of two-wheel HEV/EV drives system is shown in Figure 5.1.

The common inputs of the converter system are fuel cell, supercapacitors, and battery. Moreover, the battery is the most commonly used energy storage system in the HEV/EV system. The DC input is fed to the DC-DC converter followed by DC-AC inverter in most of the system. The choice of converter and inverter system depends upon the type of traction drives. The two drive systems commonly found in available literature are (1) DC drives and (2) AC drives. Figure 5.1 is a clear representation of AC drives system with two-stage converters. Obviously for DC motor drive system DC-AC inverter is not required because here both the input as well as required driving power are DC. Indeed, to meet voltage requirements suitable DC-DC converter is only required for the DC motor drives. DC motor drive representation is shown in Figure 5.2. However, for AC motor drive system either single-stage DC-AC inverter is used or in most of the cases two-stage converter, i.e., DC-DC converter followed by the inverter bridge (1-phase or 3-phase) is used.

The required power rating of the converters depends on the type of drive system. For main power conversion, bidirectional converter is required mandatorily to charge the energy storage system during regenerative or breaking operation of the traction motor [3–5].

The classical two-stage cascaded boost converter and VSI for AC drive system is shown in Figure 5.3. However, for DC drives system only boost converter is used and output from this converter is supplied to the DC motor. Since the motor wheel

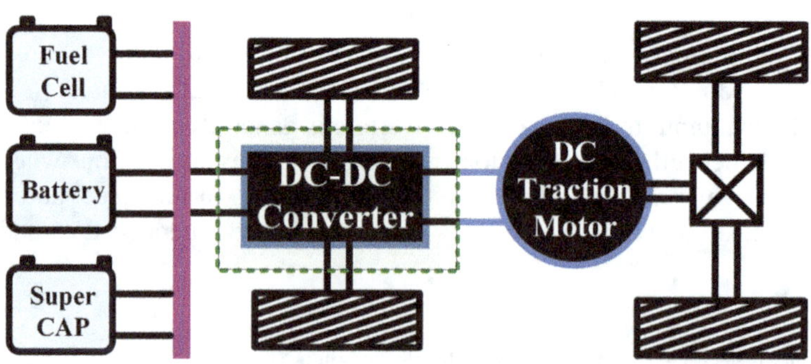

Figure 5.2 Typical representation of HEV/EV DC drive system

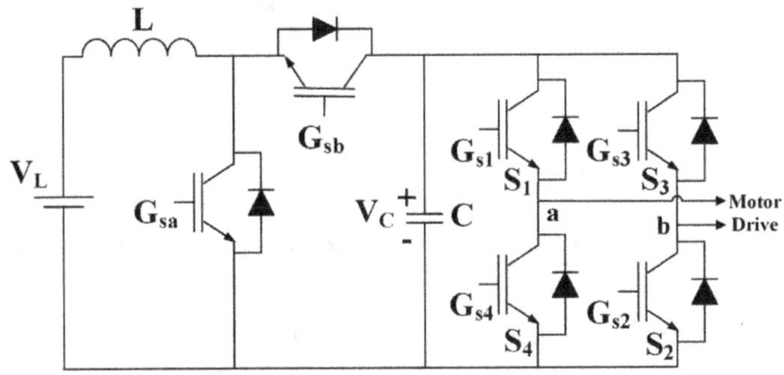

Figure 5.3 *Conventional two-stage cascaded boost and VSI*

needs to operate during motoring as well as in breaking modes; so, the converter needs to be bidirectional [6]. Therefore, the switches used here too should be bidirectional switches as depicted in Figure 5.3.

This conventional bidirectional converter has a limitation that it will not operate at extremely high duty cycle (for boost mode of operation) or extremely low duty cycle (for buck mode of operation). In boost mode, operating at extreme duty (duty cycle D_s = 0.8–0.85) leads to the maximum gain of 5–6.6. This activity causes severe reverse recovery of the diodes, increases electromagnetic interference (EMI), increases losses, and decreases efficiency. Moreover, the conventional buck mode also suffers from certain drawbacks [7]. The output voltage in buck mode cannot go below the internal reference voltage and internal circuit operation may limit the minimum turn-on time and maximum turn-on duty cycle [8]. The conventional bidirectional converter has issues with voltage conversion, current/voltage rating, losses, efficiency, etc. To overcome these drawbacks of conventional bidirectional converters, several different configurations of converters for HEV/EV applications are reported in the literature [9, 10].

In this chapter, first, a brief review of power converters used for EVs is given in Section 5.2. The quadratic bidirectional converter is described in Section 5.3. Section 5.4 explains the design criteria of the converter, which includes operation, steady-state behavior, passive components design, and small-signal analysis. Section 5.5 deals with the simulation and experimental verification of the bidirectional converter. Drives control issues are discussed in Section 5.6, and finally, conclusions are drawn in Section 5.7.

5.2 State of art of power converters used for EV application

The power converters used for the HEV/EV applications are DC-DC converter and DC-AC converter. The converters used in HEV/EV applications can be classified into isolated and non-isolated converters. But for both of the circuit configuration almost all types use same topologies. The classification of different topologies of converters is depicted in Figure 5.4. The DC-DC converter is broadly classified

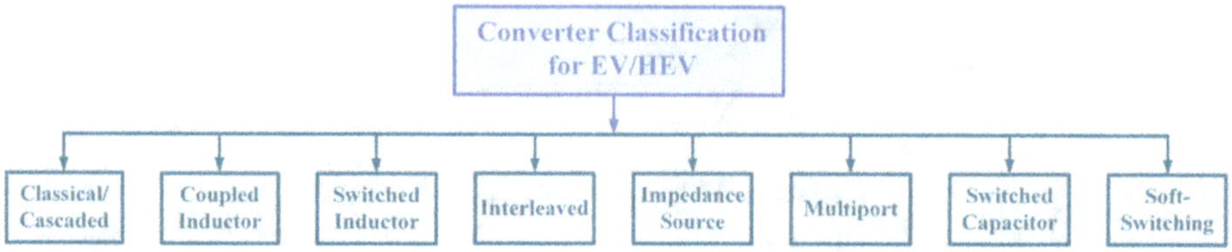

Figure 5.4 Power converters classification for HEV/EV applications

as conventional or classical bidirectional converter, transformer-based or coupled inductor, switched-inductors, interleaved, impedance source converters, soft switching, multilevel, and switched-capacitors [11, 12]. Each power converter topology has its advantages and drawbacks.

The conventional converters are the most common converters used for small- or low-power drives HEV/EV systems. Conventional boost, buck-boost, Cûk, and charge-pump are commonly used non-isolated converters. Half-Bridge/Full-Bridge, Fly-back, Forward, and Push-Pull converters are frequently used isolated converters [13]. However, depending upon the power rating some time these basic converters suffer from high power losses, efficiency, and higher cost of passive components. A cascaded converter system is also popular in the literature, which consists of multiple sub-modules of single converter. The modules may be used in series or in parallel, or in both [14, 15]. The main advantages of this type of converter are its modularity, simplicity, and low-voltage stress on output devices. However, the main drawback of this type of topologies is that it has a higher number of switches, diodes, and capacitors.

The coupled inductors or transformer-based topologies are also very common in the converter. Coupled inductors are used to eliminate the right-half-plane zero in boost mode of operation [16]. Transformer-based topologies are used to achieve higher voltage gain at very low duty cycle (for boost mode of operation) [17–21]. However, transformer-based topologies suffer from higher leakage inductance and have less efficiency.

The interleaved topologies are also popular due to the use of low-current-rated switching devices in high-power applications [22–24]. The limitations of the conventional interleaved boost converter are as follows: (1) the output diodes and switches suffer from high-current ripples, (2) the switch voltage stress is equal to the output voltage, which is large in high-voltage applications, and (3) high switch-voltage stress and the output diode reverse-recovery losses are large due to the hard switching operations that increase the losses. Nowadays, combinations of switched techniques as well as coupled techniques are introduced for the interleaved converters [24–27].

The switching used in the above-mentioned topologies is either hard switching or soft switching. The major advantage of the hard-switched converters is easy to design as well as controller design is easy, due to these attractive features the hard-switched converters are commonly used. However, hard switching has few limitations such as (1) switching losses, (2) device stress, (3) EMI due to higher dv/dt

and d*i*/d*t*, and (4) energy losses in stray inductances and capacitances. However, to minimize the above limitations while designing the converter a few possible solutions are as follows: (1) using snubbers to reduce d*v*/d*t* and d*i*/d*t*, (2) utilizing small/compact circuits to reduce stray inductances, (3) use high-speed gate driver to fast turn-on and turn-off the device, (4) or by using soft switching techniques to achieve surety of zero voltage switching (ZVS) or/and zero current switching (ZCS) [28]. Literature also finds high-gain converter with ZVS and/or ZCS by means of additional passive components [29, 30]. However, the possible drawbacks of soft switching are as follows: (1) components count increase, (2) higher power density, and (3) higher cost.

Multiport converter is an emerging topology used for high-power processing density and compact sizing [31, 32]. Multioutput converter reduces the number of converter requirements for simultaneous uses of operation such as main drives requirements as well as power required for auxiliaries. However, the multiport converter suffers from efficiency problems in addition to switching harmonics [33].

Multiport converter and single-port (single output) converters use different topologies for voltage boosting. The commonly used voltage boosting topologies in HEV/EV applications are voltage multiplier, switched-capacitor, switched-inductors, and voltage lift techniques [34–38]. These topologies may be used with the combination of other topologies. These voltage boosting techniques at the same time increase the passive components as well as diode and switches. The drawback of voltage multiplier techniques is that it has high voltage stress on components, also, they need several high rating cells to cater high-voltage rating requirements [39]. The switched-capacitor technique suffers from the high inrush current at the start-up and they are also sensitive towards equivalent series resistance (ESR) of capacitors. However, the switched-inductor techniques have increased the number of components count.

These power converters may be either voltage source converter (VSC) or current source converter (CSC). The conventional VSC and CSC have limitations such as (1) these converters cannot be buck-boost converters; they will be either buck or boost converters, (2) they cannot be open circuited or short circuited, (3) the main circuit is not interchangeable, and (4) they are susceptible to EMI noise [40]. Impedance source converter eliminates the disadvantages of classical voltage-fed and current-fed converters. The Z-source converters are one of the most promising converter topologies which overcome the disadvantages of classical VSC and CSC. The advantages of Z-source converter are single conversion stage, buck-boost capability, and increased reliability against EMI. Nowadays different topologies are utilized in the impedance source converter to improve the performance and efficiency of the converter [41–44]. Concise differences of different converter topologies are given in Table 5.1.

Based on the literature review and challenges associated with the HEV/EV power converters, their desired characteristics are identified and listed as follows:

Table 5.1 Power converter topologies' differences

Converter topologies	Advantages	Disadvantage
Cascaded/multistage	• High-power capability • Efficient and reliable	• Relatively more number of passive components • Bulky and efficiency may deteriorate
Coupled inductor/ transformer	• Variable voltage ability • High efficiency if soft switching	• Leakage inductance problem • Bulky and require precise design
Switched-inductor	• Ability of high gain • Easy to implement	• Increase passive components • Not recommended for high power
Soft switching	• Zero switching losses • High power density	• Complex design analysis • Requires additional passive components
Hard switching	• Easy design • Less passive components	• High switching losses • Low efficiency
Interleaved	• Less inductor current stress • Increased reliability	• Suitable for high-power applications • Fast switching transition required
Switched-capacitor	• Small size and lightweight • Fast dynamic response	• Inrush current • Sensitive to ESR
Multiport	• More than two simultaneous outputs • Lower cost due to components saving	• Lesser efficiency • Cross regulation problem
Impedance source	• Buck-boost capability • Single stage	• High switching current due to short circuit

1. The drive converter should be bidirectional in nature.
2. The converter should have less weight and volume. As electric power train has the limitation of space and weight, it directly affects the speed and mileage.
3. The continuous input current should be drawn from the input DC battery power supply, which is suitable for energy storage system. It also reduces the input current ripple and input filter requirement.
4. The converter should not have higher power losses and consequently lower efficiency.
5. The converter should operate for wide voltage variation. Also, it should have a wide range of duty cycle operations.
6. The converter should have a common ground to supply the system and should have low EMI.

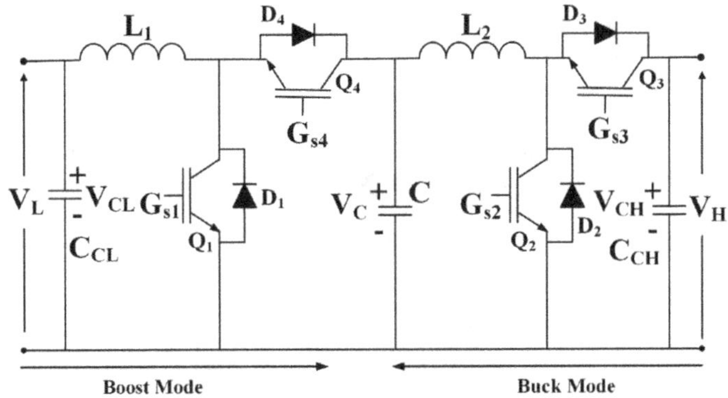

Figure 5.5 Quadratic boost converter

5.3 Quadratic converters

Conventional boost, buck, and buck-boost converter are commonly used for bidirectional conversion. The beauty of the classical converter is that if the diode is replaced with the bidirectional switch then the converter operates in both (source to load as well as load to source) directions. However, operating on higher voltage conversion ratio, the conventional bidirectional converter suffers from high stress on switches, increase in losses, and decrease in efficiency.

The drawback of the basic bidirectional converter can be eliminated with the modified converter or cascaded boost-based converter. The quadratic boost converters are basic converters that are formed by cascading two boost converters. For achieving the bidirectional conductivity, the diodes are replaced with the bidirectional switches [45]. The bidirectional quadratic boost converter is shown in Figure 5.5. From Figure 5.5 it is clear that in one direction low-voltage input is converted to high voltage (boost mode of operation) and from the high-voltage side, it is converted to low voltage (i.e., buck mode of operation).

The DC voltage conversion ratio of this converter has quadratic dependence on the duty cycle, thereby making the name quadratic converters most suitable for them. Therefore, by the use of this converter high conversion ratio is achieved. Due to its quadratic dependence on duty ratio, for achieving 10 times voltage boost in the input voltage, it is required to operate these converters at duty $D_s = 0.68$. Similarly, for 10 times voltage buck of the input voltage, the same converter has to be operated with duty = 0.32. The advantages of this quadratic converter are (1) continuous input current, (2) fast transient response, and (3) high voltage conversion ratio. Therefore, this converter is well suited for applications such as HEV/EV, UPS, and microgrid.

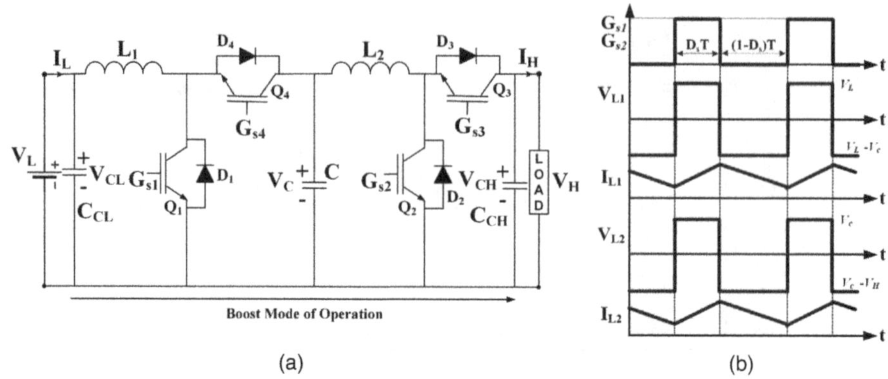

Figure 5.6 *Boost mode: (a) boost mode of operation and (b) operating waveform*

5.4 Design example of converter for HEV/EV

In this section, power converter design is discussed for HEV/EV applications. Basically, the design of quadratic bidirectional converter for HEV/EV is considered here. The complete design procedure can be divided into several operations such as steady-state analysis, passive components design, and small-signal modeling. Operations along with the working principle of bidirectional converters are discussed in the subsequent sections.

5.4.1 Working principle of bidirectional converter

The bidirectional quadratic converter works in both quadratic boost mode of operation as well as quadratic buck mode of operations. The bidirectional converter topology works in two modes. One is shoot-through mode (D_sT) and another is non-shoot-through mode ($(1\text{-}D_s)T$). Shoot-through and non-shoot-through modes are also known as ON mode and OFF mode, respectively, where D_s is duty cycle of the switch, T=1/ f_s; f_s is a switching frequency. The boost mode of operation is shown in Figure 5.6(a) and the operating waveform is given in Figure 5.6(b). In this operation, V_L is an input supply whereas load is connected on the high-voltage (V_H) side. Quadratic boost operation will be achieved when the switches S_1 and S_2 are turned-ON at the same duty cycle and the same instant. During this duty interval (D_sT), the inductor currents keep rising. When switches S_1 and S_2 are turned-OFF ($(1\text{-}D_s)T$), body diodes of the switches S_3 and S_4 are operated. In this boost mode of operation S_3 and S_4 switches are always turned-OFF which is clear from Figure 5.6(b).

The buck mode of operation is shown in Figure 5.7(a) and the operating waveform is given in Figure 5.7(b). In the buck mode, now V_H is an input supply and the load is connected at the low-voltage terminal (V_L). The quadratic buck operation will be achieved when the switches S_3 and S_4 operate at the same instant and same duty cycle. Body diodes of the switches S_1 and S_2 are turned-ON when switches S_3 and S_4

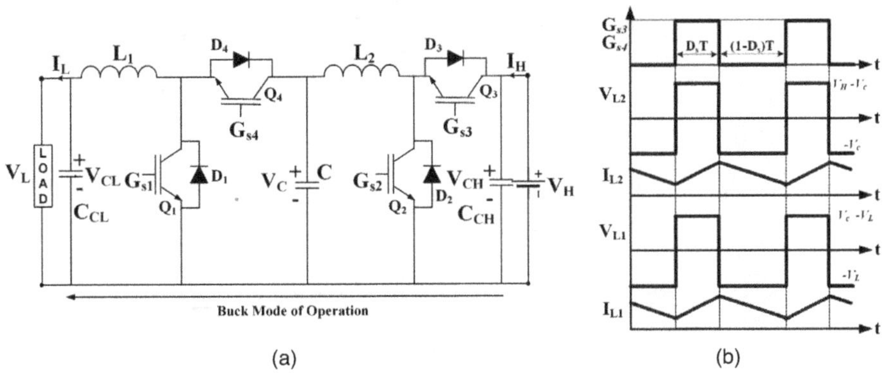

Figure 5.7 *Buck mode: (a) buck mode of operation, (b) operating waveform*

are turned-OFF. During this duty interval (D_sT), the inductor currents keep rising. In this mode of operation, switches, S_1 and S_2 remain turned-OFF as mentioned clearly in Figure 5.7(b).

5.4.2 *Steady-state analysis*

The boost and buck mode of operation are shown in Figures 5.6(a) and 5.7(a). The steady-state analysis of the boost mode and buck mode is analyzed with operating circuits. The inductor voltage and capacitor currents are used to analyze the steady-state performance of the converter.

5.4.2.1 Boost mode of operation

The boost mode operation is performed by connecting the load on the V_H terminal and battery supply is given from the terminal V_L. The operating waveforms are shown in Figure 5.6(b). The performance of the circuit during shoot-through and non-shoot-through is shown in Figure 5.8. From Figure 5.8(a) and (b), it is clear that switches S_1 and S_2 operate simultaneously.

The instantaneous voltage and current expressions of the energy storage elements can be derived from Figure 5.8(a) during shoot-through state. The inductor voltages and capacitor current expressions during shoot-through interval are (5.1)–(5.4).

$$L_1 \frac{di_{L1}}{dt} = v_L \tag{5.1}$$

$$L_2 \frac{di_{L2}}{dt} = v_C \tag{5.2}$$

$$C \frac{dv_C}{dt} = -i_{L2} \tag{5.3}$$

$$C_{CH} \frac{dv_{CH}}{dt} = -i_H = \frac{-v_{CH}}{R_H} \tag{5.4}$$

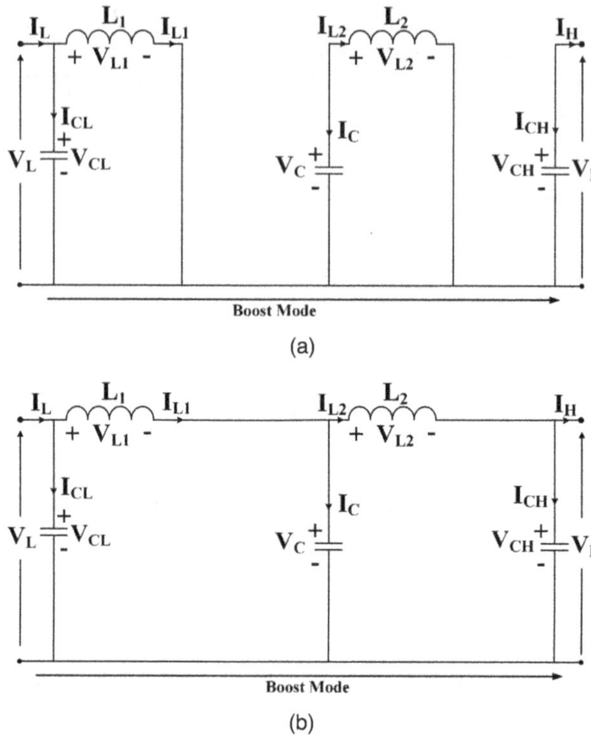

Figure 5.8 Boost mode of operation: (a) shoot-through, (b) non-shoot-through

where R_H is the load across the high-voltage terminal. During the non-shoot-through state following (5.5)–(5.8) are derived using Figure 5.8(b).

$$L_1 \frac{di_{L1}}{dt} = v_L - v_C \tag{5.5}$$

$$L_2 \frac{di_{L2}}{dt} = v_C - v_H \tag{5.6}$$

$$C \frac{dv_C}{dt} = i_{L1} - i_{L2} \tag{5.7}$$

$$C_{CH} \frac{dv_{CH}}{dt} = i_{L2} - i_H \tag{5.8}$$

By the principle of volt-second balance, the average steady-state DC voltage across the inductor is zero. The volt-second balance can be expressed for the inductor L_1 by (5.9).

$$V_{L1} * D_s + V_{L1} * (1 - D_s) = 0 \tag{5.9}$$

After putting the V_{L1} expression from shoot-through (5.1) and non-shoot-through (5.5) in (5.9). The capacitor voltage expression is given as (5.10).

$$V_c = \frac{V_L}{(1-D_s)} \tag{5.10}$$

Similarly, applying this principle to the inductor voltage waveform of L_2 we obtain (5.11).

$$V_H = \frac{V_c}{(1-D_s)} \tag{5.11}$$

By using (5.10) and (5.11), the voltage conversion ratio is given by (5.12)

$$V_H = \frac{V_L}{(1-D_s)^2} \tag{5.12}$$

This shows that the voltage conversion ratio of the converter quadratically depends on its duty ratio. The current expression is also calculated, by applying the charge-second balance principle on capacitors. The charge-second balance equation for capacitors C_{CH} is given by (5.13).

$$I_{CH} * D_s + I_{CH} * (1 - D_s) = 0 \tag{5.13}$$

After utilizing the I_{CH} expression from shoot-through (5.4) and non-shoot-through (5.8) in (5.13). The inductor current is simplified according to (5.14).

$$I_L = \frac{I_H}{(1 - D_s)^2} \tag{5.14}$$

Similarly, applying this principle to the capacitor current waveform of C, (5.15) is obtained.

$$V_H = \frac{V_L}{(1 - D_s)^2} \tag{5.15}$$

By using (5.14) and (5.15), the current conversion ratio is given by (5.16).

$$I_L = \frac{I_H}{(1 - D_s)^2} \tag{5.16}$$

where I_L is the input side current drawn from the supply and I_H is the load side current when operating in boost mode of operation. Moreover, it is evident from the transfer characteristic that the duty cycle can be varied theoretically from 0 to 1.

5.4.2.2 Buck mode of operation

The buck mode operation is performed by connecting the load on terminal V_L and supply is given through the high-voltage side. In EV application, the DC motor works in generating or breaking mode to charge the low-voltage battery. The shoot-through and non-shoot-through waveforms for the buck operation are shown in Figure 5.9(a) and (b). As depicted in Figure 5.9, in this mode of operation, the switches S_3 and S_4 operate simultaneously.

The instantaneous voltage and current expressions of the energy storage elements can be derived from Figure 5.9(a) during shoot-through state. The inductor voltages and capacitors current expression during the shoot-through interval is given as

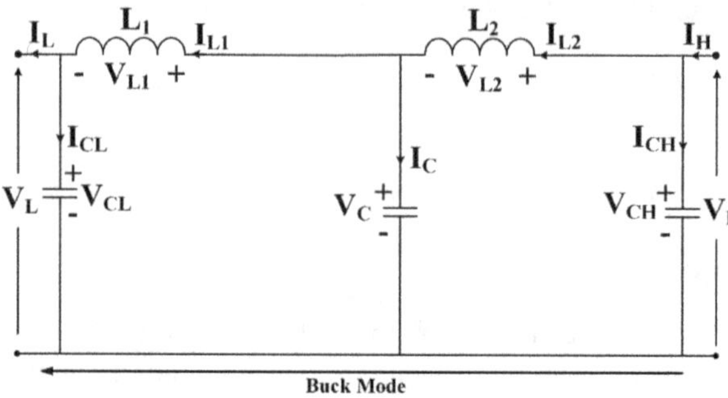

(a)

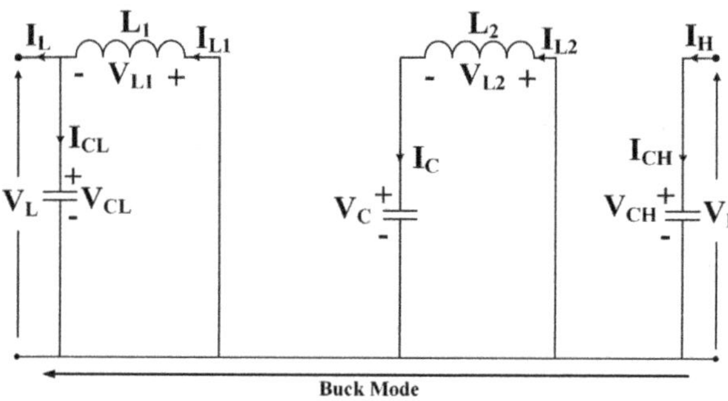

(b)

Figure 5.9 Buck mode of operation: (a) shoot-through and (b) non-shoot-through

$$L_2 \frac{di_{L2}}{dt} = v_H - v_C \tag{5.17}$$

$$L_1 \frac{di_{L1}}{dt} = v_C - v_L \tag{5.18}$$

$$C \frac{dv_C}{dt} = i_{L2} - i_{L1} \tag{5.19}$$

$$L_2 \frac{di_{L2}}{dt} = -v_C \tag{5.20}$$

where R_L is the load across the low-voltage terminal. During the non-shoot-through state, the following equations are derived using Figure 5.9(b)

$$L_2 \frac{di_{L2}}{dt} = -v_C \tag{5.21}$$

$$L_1 \frac{di_{L1}}{dt} = -v_L \tag{5.22}$$

$$C \frac{dv_C}{dt} = i_{L2} \tag{5.23}$$

$$V_L = D_s V_c \tag{5.24}$$

By the principle of volt-second balance, the average steady-state DC voltage across the inductor is zero. The volt-second balance can be expressed for the inductor L_2 given by (5.25)

$$V_{L2} * D_s + V_{L2} * (1 - D_s) = 0 \tag{5.25}$$

After putting the expression for V_{L2} from shoot-through (5.17) and non-shoot-through (5.21) in (5.25). The expression for capacitor V_c is given by (5.26).

$$V_c = D_s V_H \tag{5.26}$$

Similarly, applying this principle to the inductor voltage waveform of L_1, we get

$$V_L = D_s V_c \tag{5.27}$$

By using (5.26) and (5.27), the voltage conversion ratio is given by (5.28).

$$V_L = D_s^2 V_H \tag{5.28}$$

This shows that the voltage conversion ratio of the converter quadratically depends on its duty ratio. The expression for current is also calculated by applying the charge-second balance principle on capacitors. The charge-second balance equation for capacitors C is given by (5.29).

$$I_{L2} = D_s I_{L1} \tag{5.29}$$

After utilizing the expression for I_c from shoot-through (5.19) and non-shoot-through (5.23) in (5.2). The inductor current I_{L2} is given as (5.30).

$$I_{L2} = D_s I_{L1} \tag{5.30}$$

Similarly, applying this principle to the capacitor current waveform we obtain (5.31) and (5.32).

$$V_L = D_s^2 V_H \tag{5.31}$$

$$I_H = D_s I_{L2} \tag{5.32}$$

By using (5.30) and (5.32), the current conversion ratio is given by (5.33).

$$I_H = D_s^2 I_L \tag{5.33}$$

where I_H is the input current drawn from the supply and I_L is the load current when operating in buck mode of operations. Moreover, from the transfer characteristic, it can be observed that the duty cycle can be varied theoretically from 0 to 1.

5.4.3 Passive components design

The switches and passive components are the only components used in the converter. The rating of the switches is selected based on the peak voltage and peak current stress of the switch. In this case, the maximum voltage is at the V_H terminal and voltage stress on switches will be always less than V_H. However, the passive component values depend upon the switching frequency and the permissible ripples. The inductor and capacitor values are calculated according to (5.34) and (5.35).

$$L_i = \frac{D_s V_{Li}}{\Delta i_{Li}\% \, I_{Li} \, f_s} \tag{5.34}$$

$$C_i = \frac{D_s I_{ci}}{\Delta V_{ci}\% \, V_{ci} \, f_s} \tag{5.35}$$

where i =1 and 2. f_s is the switching frequency. The voltage across the inductors L_1 and L_2 are V_{L1} and V_{L2}, respectively. Moreover, i_{L1} and i_{L2} represent percentage ripple currents in L_1 and L_1 . V_{cH} and V_{cH} are the voltages across the capacitors C and C_{CH}, I_c and I_{CH} are the currents flowing through C and C_{CH}. V_c and V_{CH} represent percentage ripple voltages in C_1 and C_2 during D_s.

Using (5.34) and (5.35) the inductors and capacitors values are calculated in (5.36)–(5.39)

$$C = \frac{D_s I_{L2}}{\Delta V_c\% \, V_c f_s} = \frac{D_s I_H}{\Delta V_c\% \, (1 - D_s) \, V_c f_s} \tag{5.36}$$

$$L_2 = \frac{D_s V_c}{\Delta i_{Li}\% \, I_{L2} \, f_s} = \frac{D_s (1 - D_s) \, V_c}{\Delta i_{Li}\% \, I_H \, f_s} \tag{5.37}$$

$$C = \frac{D_s I_{L2}}{\Delta V_c\% \, V_c f_s} = \frac{D_s I_H}{\Delta V_c\% \, (1 - D_s) \, V_c f_s} \tag{5.38}$$

$$C_{CH} = \frac{D_s I_H}{\Delta V_{cH}\% \, V_{cH} \, f_s} = \frac{D_s I_H}{\Delta V_{cH}\% \, V_H \, f_s} \tag{5.39}$$

The inductor and capacitor design will be valid for boost mode as well as buck mode of operation as the values of currents and capacitor voltages will be same for specific duty cycle. Thus, the passive components may be designed based on the buck mode of operation and that will also be valid for the same voltage and current ripple in boost mode of operation.

The critical values for the continuous conduction mode (CCM) of inductors are given by (5.40) and (5.41).

$$L_{1CR} \geq \frac{D_s V_{L1}}{2 \, I_{L1} \, f_s} \tag{5.40}$$

$$L_{2CR} \geq \frac{D_s V_c}{2 \, I_{L2} \, f_s} \tag{5.41}$$

So, for the CCM the inductor values should be selected as per calculations of the critical inductance L_{1CR} and L_{2CR}.

5.4.4 Small-signal analysis

In this small-signal analysis, the small-signal AC variation of the signal is represented as (â) symbols. Moreover, the upper case represents the steady-state value and the lower case represents the instantaneous value. It is assumed that there are no-parasitic effects and switching frequency is much greater than the natural frequency of the converter. The boost mode and buck mode operating differential equations are utilized, using state-space averaging method [46] for the small-signal analysis.

5.4.4.1 Boost mode of operation

The equivalent circuits in shoot-through and non-shoot-through mode are given in Figure 5.8. For boost mode operation, the perturbation and linearization of the differential equation lead to the state-space model given by (5.42) the boost mode of operations.

$$\frac{d}{dt}\begin{bmatrix}\hat{i}_{L1}\\ \hat{i}_{L2}\\ \hat{v}_c\\ \hat{v}_H\end{bmatrix} = [A]\times\begin{bmatrix}\hat{i}_{L1}\\ \hat{i}_{L2}\\ \hat{v}_c\\ \hat{v}_H\end{bmatrix} + [B]^T\times[\hat{V}_L] + [E]^T\times[\hat{d}_s] \tag{5.42}$$

$$A=\begin{bmatrix} 0 & 0 & \frac{-(1-D_s)}{L_1} & 0\\ 0 & 0 & \frac{1}{L_2} & \frac{-(1-D_s)}{L_2}\\ \frac{1-D_s}{C} & \frac{-1}{C} & 0 & 0\\ 0 & \frac{1-D_s}{C_{CH}} & 0 & \frac{-1}{R_H C_{CH}} \end{bmatrix}$$

$$B=\begin{bmatrix}\frac{1}{L_1} & 0 & 0 & 0\end{bmatrix} \text{ and } E=\begin{bmatrix}\frac{V_c}{L_1} & \frac{V_H}{L_2} & \frac{-I_{L_1}}{C} & \frac{-I_{L_2}}{C_{CH}}\end{bmatrix}$$

The open-loop control-to-output transfer function of the boost mode of operation is given by (5.43):

$$\frac{\hat{v}_H}{\hat{d}_s} = \begin{bmatrix}0 & 0 & 0 & 1 & 0\end{bmatrix}[sI-A]^{-1}[U_1] \tag{5.43}$$

where $U_1 = \begin{bmatrix}\frac{V_c}{L_1} & \frac{V_H}{L_2} & \frac{-I_{L1}}{C} & \frac{-I_{L2}}{C_{CH}}\end{bmatrix}^T$.

By solving (5.43), the open-loop control-to-output transfer function is given by (5.44).

$$G_{\text{plant}} = \frac{\hat{v}_H}{\hat{d}_s} = \frac{N_1S^3+N_2S^2+N_3S+N_4}{D_1s^4+D_2s^3+D_3s^2+D_4s+D_5} \tag{5.44}$$

The numerator and denominator coefficients of (5.44) are as follows:

$$N_1 = CL_1L_2V_H$$

$$N_2 = -CL_1R_HV_L$$

$$N_3 = (1-D_s)^2 L_2V_H + L_1V_H$$

$$N_4 = -2R_HV_L(1-D_s)^2$$

$$N_1 = CL_1L_2V_H$$

$$D_2 = -\left(1 - D_s\right) CL_1L_2$$
$$D_3 = -\left(1 - D_s\right)^2 R_H \left(C_{CH}L_2 + CL_1\right) + C_{CH}L_1R_H \left(D_s - 1\right)$$
$$D_4 = -L_2 \left(1 - D_s\right)^3 + L_1 \left(D_s - 1\right)$$
$$D_2 = -\left(1 - D_s\right) CL_1L_2$$

5.4.4.2 Buck mode of operation

The shoot-through and non-shoot-through modes of buck operation are given in Figure 5.9. For buck mode of operation, the perturbation and linearization of the differential equation lead to the following state-space model given by (5.45).

$$\frac{d}{dt}\begin{bmatrix}\hat{i}_{L1}\\ \hat{i}_{L2}\\ \hat{v}_c\\ \hat{v}_H\end{bmatrix} = [A] \times \begin{bmatrix}\hat{i}_{L1}\\ \hat{i}_{L2}\\ \hat{v}_c\\ \hat{v}_H\end{bmatrix} + [B]^T \times [\hat{V}_L] + [E]^T \times [\hat{d}_s] \tag{5.45}$$

$$A = \begin{bmatrix} 0 & 0 & -\frac{1}{L_2} & 0 \\ 0 & 0 & \frac{D_s}{L_1} & -\frac{1}{L_1} \\ \frac{1}{C} & -\frac{D_s}{C} & 0 & 0 \\ 0 & \frac{1}{C_{CL}} & 0 & -\frac{1}{R_LC_{CL}} \end{bmatrix}$$

$$B = \begin{bmatrix} \frac{D_s}{L_2} & 0 & 0 & 0 \end{bmatrix} \text{ and } E = \begin{bmatrix} \frac{V_H}{L_2} & \frac{V_c}{L_1} & \frac{-I_{L1}}{C} & 0 \end{bmatrix}$$

$$\frac{\hat{v}_L}{\hat{d}_s} = \begin{bmatrix}0 & 0 & 0 & 1 & 0\end{bmatrix}\left[sI - A\right]^{-1}\left[U_1\right] \tag{5.46}$$

where $U_1 = \begin{bmatrix} \frac{V_H}{L_2} & \frac{V_c}{L_1} & \frac{-I_{L1}}{C} & 0 \end{bmatrix}^T$

$$G_{\text{plant}} = \frac{\hat{v}_L}{\hat{d}_s} = \frac{N_1'S^2 + N_2'S + N_3'}{D_1's^4 + D_2's^3 + D_3's^2 + D_4's + D_5'} \tag{5.47}$$

where

$$N_1' = CD_sL_2R_LV_H$$
$$N_2' = -D_sL_2R_LI_{L1}$$
$$N_3' = 2D_sR_LV_H$$
$$D_1' = CC_{CL}L_2L_1R_L$$
$$D_2' = C_{CL}L_2R_LD_s^2 + CL_2R_L + C_{CL}L_1R_L$$
$$D_3' = CL_2L_1$$
$$D_4' = L_1 + L_2D_s^2$$
$$D_5' = R_L$$

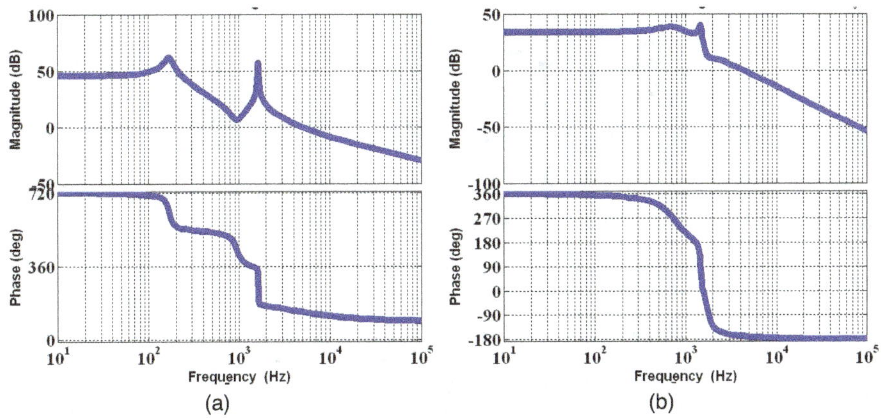

Figure 5.10 Bode plots of control-to-output transfer function of (a) quadratic boost and (b) quadratic buck

Bode plot of control-to-output transfer function is presented in Figure 5.10. Figure 5.10(a) shows the bode diagram for boost mode of operation and Figure 5.10(b) shows the bode diagram for buck mode of operation. The reference data used to obtain the bode plot are as follows: $V_L = 12$, $V_H = 48V$, $L_1 = 310$ μH, $L_2 = 360$ μH, $f_s = 100$ kHz, C = 47 μF, $C_{CH} = C_{CL} = 100$ μF.

5.5 Simulation and experimental verifications

The bidirectional quadratic converter is validated with the simulation and experimental results. The design parameters are mentioned previously. The bidirectional converter is designed for the 12 V to 48 V system. The active and passive components are fixed and only input/output terminals are changed for boost mode and buck mode of operations.

The results obtain using simulation are depicted in Figure 5.11. It can be clearly observed from Figure 5.11 that both switches S_1 and S_2 operate at same duty cycle. The voltage across the body diode of switches S_3 and S_4 is complimentary of the voltage across switches S_1 and S_2. Figure 5.12 shows the steady-state waveform for the boost mode of operation. It is observed that for the input voltage of $V_L = 12$ V and the output voltage is $V_H = 48$ V. Moreover, $V_c = 24$ V and the inductor currents are 8.35 A and 4.17 A. It is observed that the values obtained from the simulation match well with the calculated steady-state values.

Simulation results for buck mode of operation are depicted in Figure 5.13. It is evident from Figure 5.13 that switches S_3 and S_4 are switched-ON simultaneously. The peak voltage stress across the body diode of the switches S_1 and S_2 occurs during the non-shoot-through duty cycle. Figure 5.14 shows the steady-state waveform for the buck mode of operation. It is clearly observed from Figure 5.14 that for the input voltage of $V_H = 48$ V, the output voltage is $V_L = 12$ V. Moreover, $V_c = 24$ V

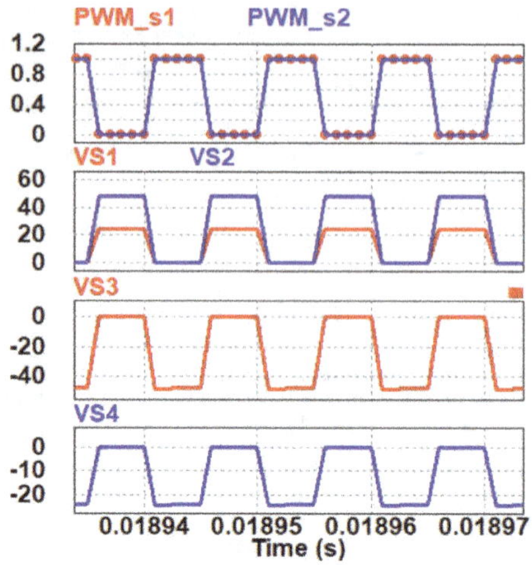

Figure 5.11 *Boost mode switching stress*

and the inductor currents are 8.35 A and 4.17 A. In this case too, the values obtained from the simulation match with the steady-state calculated values.

The bidirectional property of the quadratic converter is also verified experimentally using a digital signal processor TMS320F28335. The parameters used for experimental verification were the same as the ones used for simulation. The experimental results obtained for steady-state operation during boost mode are presented in Figure 5.15. Similar results as simulation were obtained experimentally too. This can be observed from Figure 5.15 where for V_L = 24 V (Ch1: 20V/div) and the obtained output voltage is V_H = 48 V (Ch4: 50V/div). The PWM signal given to switch S_1 with duty cycle D_s = 0.5 and the observed load current are also depicted

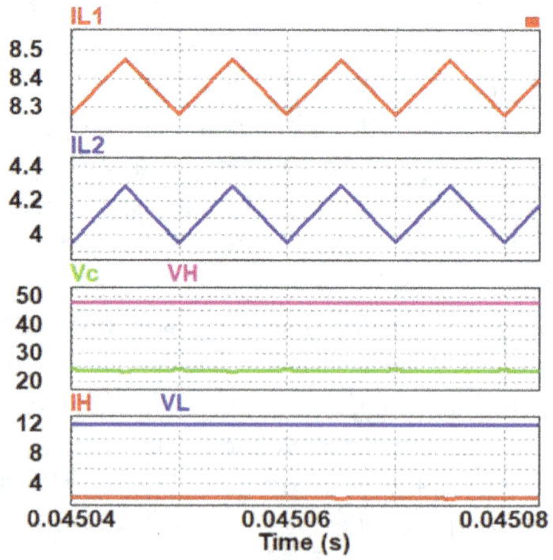

Figure 5.12 *Steady-state operation in boost mode*

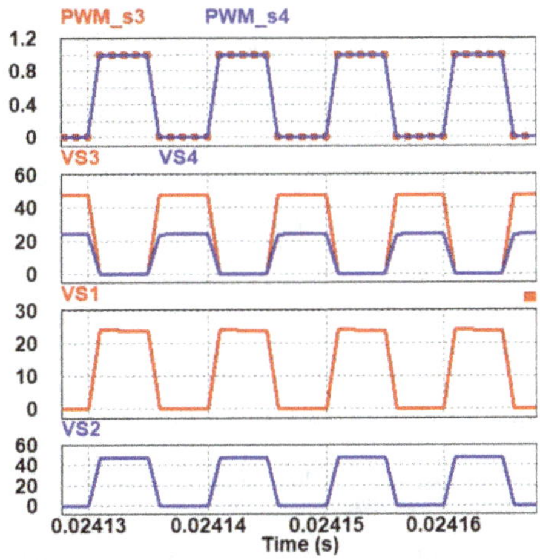

Figure 5.13 *Buck mode switching stress*

in Figure 5.15. Similarly, buck mode operation is depicted in Figure 5.16, where for the applied voltage V_H = 48 V (Ch1: 50V/div), the achieved output is V_L = 12 V (Ch3: 10V/div). Also, the voltage across switch S_3 and the supply current are depicted in Figure 5.16. It is observed that the experimental and simulation results are identical and match exactly with the theoretical steady-state analysis.

5.6 EV drives and control

The electric train drives use different drives system depending upon AC or DC drives. The selection of motor drive control depends upon the specific requirements

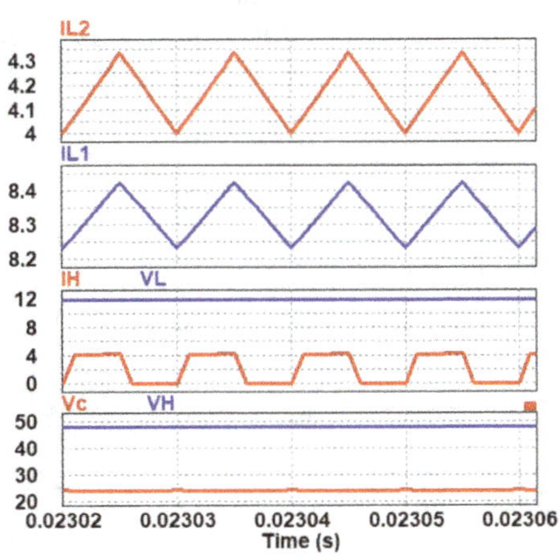

Figure 5.14 *Steady-state operation in buck mode*

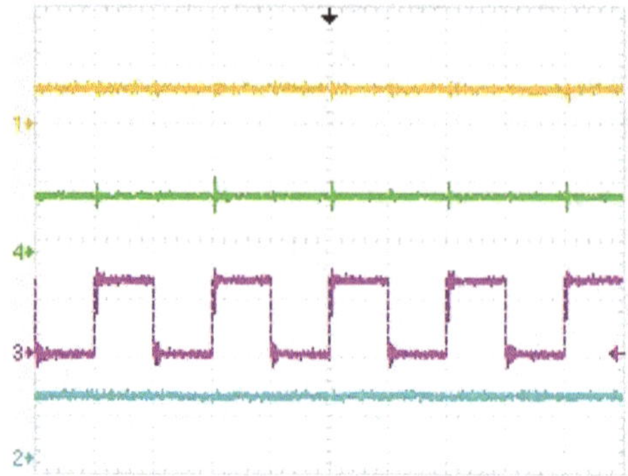

Figure 5.15 Steady-state experimental validation for boost mode of operations

of the converters and power ratings. The electric machines used for EV/HEV applications have some key features, advantages, and disadvantages [47–49].

The induction motor (IM), especially Cage IMs, is broadly recognized as the potential candidate for the electric propulsion of HEVs, due to low maintenance cost, robustness, ruggedness, and reliability. However, it also suffers from low inverter-usage factor and low efficiency [50, 51]. The permanent magnet synchronous motor has better performance as compared to the IM, but it suffers from eddy current loss at high speed [52]. The switched reluctance motor drives are also used prominently for HEV/EV applications. The key advantages of these drives are that they have rugged construction, good speed-torque characteristics, and simple control. Moreover, they can operate in long constant-power range. However, they suffer from toque ripples, EMI noise, and excessive current ripple [53–55]. The DC motors also have been prominent in electric propulsion due to their simple torque-speed controls. The voltage supply for this DC drives system is comparatively low than that of AC drives system. However, DC motor drives have low efficiency and higher maintenance cost, which is due to the occurrence of the mechanical commutator [56, 57]. The brushless DC (BLDC) motor utilizes a permanent magnet and it is

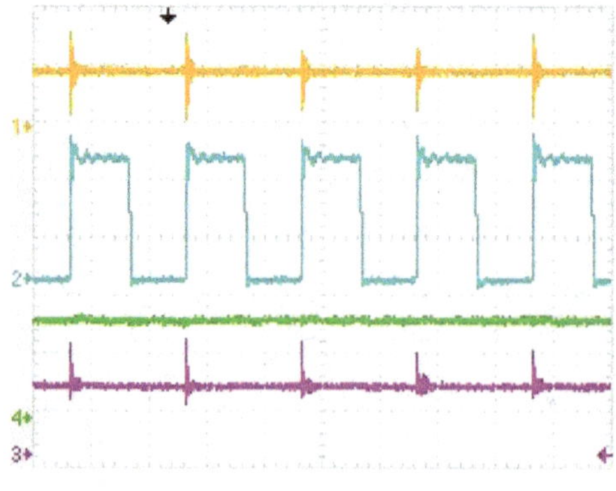

Figure 5.16 Steady-state experimental validation for buck mode of operations

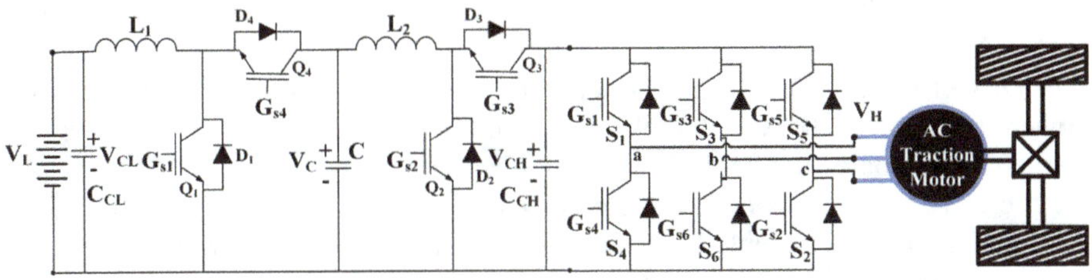

Figure 5.17 *Electric power train system with AC motor drive*

electrically commutated by power switches instead of brushes. Moreover, it offers the advantages such as high reliability, better efficiency, lightweight and smaller size, lower acoustic noise, better speed vs torque characteristics, and good dynamic response. Due to the above-stated advantages, the BLDC motors are more readily used compared to the brushed DC motor, synchronous motor, and IM [58–60].

Selection of machine drive system depends upon the EV design criteria, EV power requirement, battery management system, manufacturer utilized suitable AC drives or DC drives system. The discussed bidirectional converter can also be used for the AC motor drives, as shown in Figure 5.17. IM drive consists of different control schemes such as direct torque control, sensor-less control, voltage/frequency control, etc. Advance control techniques mentioned in many literature [61–63] are implemented on the AC drives represented in Figure 5.17.

The discussed bidirectional converter can also be used for the DC motor drives, as shown in Figure 5.18. For DC motor control different control techniques can be used to drive the motor [64–67]. These control techniques can be used for DC drives system shown in Figure 5.18. DC drives system is comparatively easy to control as compared to the AC motor drives.

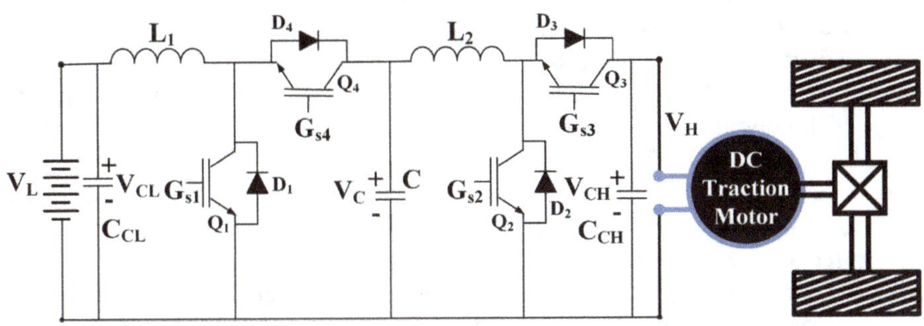

Figure 5.18 *Electric power train system with DC motor drive*

5.7 Conclusion

This chapter has presented a power converter classification, analysis, and design of power converter for HEV/EV applications. In this chapter, a brief review of power converter topologies used for HEV/EV applications is discussed. With the help of the literature review, desired characteristics of the power converter used for HEV/EV application are presented. The quadratic bidirectional converter is used as a design example to demonstrate the complete procedures for power converter design. The detailed operation of the quadratic bidirectional converter, steady-state analysis, small-signal modeling, and passive design analysis is carried out. Some experimental and simulation results are presented to exhibit the performance validation of a quadratic bidirectional converter. Finally, motor drive systems used in HEV/EV applications are discussed in brief.

References

[1] Chan C.C. 'The state of the art of electric, hybrid, and fuel cell vehicles'. *Proceedings of the IEEE*. 2007, vol. 95(4), pp. 704–18.

[2] Sabri M.F.M., Danapalasingam K.A., Rahmat M.F. 'A review on hybrid electric vehicles architecture and energy management strategies'. *Renewable and Sustainable Energy Reviews*. 2016, vol. 53, pp. 1433–42.

[3] Emadi A., Young Joo Lee., Rajashekara K. 'Power electronics and motor drives in electric, hybrid electric, and plug-in hybrid electric vehicles'. *IEEE Transactions on Industrial Electronics*. 2008, vol. 55(6), pp. 2237–45.

[4] Wang H., Liserre M., Blaabjerg F. 'Toward reliable power electronics: challenges, design tools, and opportunities'. *IEEE Industrial Electronics Magazine*. 2013, vol. 7(2), pp. 17–26.

[5] Dubey G.K. *Fundamentals of electrical drives*. India: Narosa Publishing House; 1995.

[6] Bose B.K. 'Modern power electronics and AC drives'. *Prentice-Hall of India*. 2002.

[7] Ahmad A., Singh R.K., Mahanty R. 'A novel non-isolated magnetically coupled based bidirectional quadratic converter'. 2016 IEEE International Conference on Power Electronics, Drives and Energy Systems (PEDES); Trivandrum; 2016. pp. 1–6.

[8] Liou W.R., Yeh M.L., Kuo Y.L. 'A high efficiency dual-mode buck converter IC for portable applications'. *IEEE Transactions on Power Electronics*. 2008, vol. 23(2), pp. 667–77.

[9] Gorji S.A., Sahebi H.G., Ektesabi M., Rad A.B. 'Topologies and control schemes of bidirectional DC–DC power converters: an overview'. *IEEE Access: Practical Innovations, Open Solutions*. 2019, vol. 7, pp. 117997–8019.

[10] Chakraborty S. 'DC-DC converter topologies for electric vehicles, plug-in hybrid electric vehicles and fast charging stations: state of the art and future trends'. *Energies*. 2019, vol. 12(1569), pp. 1–43.

[11] Forouzesh M., Siwakoti Y.P., Gorji S.A., Blaabjerg F., Lehman B. 'Step-up DC–DC converters: a comprehensive review of Voltage-Boosting techniques, topologies, and applications'. *IEEE Transactions on Power Electronics*. 2017, vol. 32(12), pp. 9143–78.

[12] Farooq. 'A review of non-isolated high step-down DC-DC converters'. *International Journal of Smart Home*. 2015, vol. 9, pp. 133–50.

[13] Du Y., Lukic S., Jacobson B., Huang A. 'Review of high power isolated bi-directional DC-DC converters for PHEV/EV DC charging infrastructure'. 2011 IEEE Energy Conversion Congress and Exposition; Phoenix, AZ; 2011. pp. 553–60.

[14] Li Y., Sathiakumar S. 'The improvement of cascaded converter design using quadratic boosting structure with high voltage gain and low duty cycle'. 2019 IEEE International Conference on Industrial Technology (ICIT); Melbourne, Australia; 2019. pp. 480–5.

[15] Ahmad A., Bussa V.K., Singh R.K., Mahanty R. 'Quadratic boost derived hybrid multi-output converter'. *IET Power Electronics*. 2017, vol. 10(15), pp. 2042–54.

[16] Singh R., Mishra S. 'A magnetically coupled feedback-clamped optimal bi-directional battery charger'. *IEEE Transactions on Industrial Electronics*. 2013, vol. 60(2), pp. 422–32.

[17] Rong-Jong W., Rou-Yong D. 'High step-up converter with coupled-inductor'. *IEEE Transactions on Power Electronics*. 2005, vol. 20(5), pp. 1025–35.

[18] Moradi S.H., Milimonfared J., Taheri M., Salehi S. 'High step-up soft-switched dual-boost coupled-inductor-based converter integrating multipurpose coupled inductors with capacitor-diode stages'. *IET Power Electronics*. 2015, vol. 8(9), pp. 1786–97.

[19] Zhao Y., Li W., He X. 'Single-phase improved active clamp coupled-inductor-based converter with extended voltage doubler cell'. *IEEE Transactions on Power Electronics*. 2012, vol. 27(6), pp. 2869–78.

[20] Yang Y., Ma J., Ho C.N-M., Zou Y. 'A new coupled-inductor structure for interleaving bidirectional DC-DC converters'. *IEEE Journal of Emerging and Selected Topics in Power Electronics*. 2015, vol. 3(3), pp. 841–9.

[21] Chub A., Vinnikov D., Blaabjerg F., Peng F.Z. 'A review of galvanically isolated impedance-source DC–DC converters'. *IEEE Transactions on Power Electronics*. 2016, vol. 31(4), pp. 2808–28.

[22] Muhammad M., Armstrong M., Elgendy M.A. 'A nonisolated interleaved boost converter for high-voltage gain applications'. *IEEE Journal of Emerging and Selected Topics in Power Electronics*. 2016, vol. 4(2), pp. 352–62.

[23] Hegazy O., Mierlo J. V., Lataire P. 'Analysis, modeling, and implementation of a multidevice interleaved DC/DC converter for fuel cell hybrid electric vehicles'. *IEEE Transactions on Power Electronics*. 2012, vol. 27(11), pp. 4445–58.

[24] Bussa V.K., Ahmad A., Singh R.K., Mahanty R. 'Interleaved hybrid converter with simultaneous DC and AC outputs for DC microgrid applications'. *IEEE Transactions on Industry Applications*. 2018, vol. 54(3), pp. 2763–72.

[25] Li S., Xie W., Smedley K.M. 'A family of an automatic Interleaved Dickson switched-capacitor converter and its ZVS resonant configuration'. *IEEE Transactions on Industrial Electronics*. 2019, vol. 66(1), pp. 255–64.

[26] Po-Wa L., Yim-Shu L., Cheng D.K.W., Xiu-Cheng L. 'Steady-state analysis of an interleaved boost converter with coupled inductors'. *IEEE Transactions on Industrial Electronics*. 2000, vol. 47(4), pp. 787–95.

[27] Bahrami H., Farhangi S., Iman-Eini H., Adib E. 'A new interleaved coupled-inductor nonisolated soft-switching bidirectional DC–DC converter with high voltage gain ratio'. *IEEE Transactions on Industrial Electronics*. 2018, vol. 65(7), pp. 5529–38.

[28] Ching T.W., Chan K.U. 'Review of soft-switching techniques for high-frequency switched-mode power converters'. 2008 IEEE Vehicle Power and Propulsion Conference; Harbin; 2008. pp. 1–6.

[29] Do H. 'A soft-switching DC/DC converter with high voltage gain'. *IEEE Transactions on Power Electronics*. 2010, vol. 25(5), pp. 1193–200.

[30] Faraji R., Farzanehfard H. 'Soft-switched nonisolated high step-up three-port DC–DC converter for hybrid energy systems'. *IEEE Transactions on Power Electronics*. 2018, vol. 33(12), pp. 10101–11.

[31] Sato Y., Uno M., Nagata H. 'Nonisolated multiport converters based on integration of PWM converter and phase-shift-switched capacitor converter'. *IEEE Transactions on Power Electronics*. 2020, vol. 35(1), pp. 455–70.

[32] Cai W., Yi F. 'An integrated multiport power converter with small capacitance requirement for switched reluctance motor drive'. *IEEE Transactions on Power Electronics*. 2016, vol. 31(4), pp. 3016–26.

[33] Shiluveru K., Singh A., Ahmad A., Singh R.K. 'Hybrid buck–boost multi-output quasi-Z-source converter with dual DC and single AC outputs'. *IEEE Transactions on Power Electronics*. 2020, vol. 35(7), pp. 7246–60.

[34] Prudente M., Pfitscher L.L., Emmendoerfer G., Romaneli E.F., Gules R. 'Voltage multiplier cells applied to non-isolated DC–DC converters'. *IEEE Transactions on Power Electronics*. 2008, vol. 23(2), pp. 871–87.

[35] Li S., Xie W., Smedley K.M. 'A family of an automatic Interleaved Dickson switched-capacitor converter and its ZVS resonant configuration'. *IEEE Transactions on Industrial Electronics*. 2019, vol. 66(1), pp. 255–64.

[36] Tang Y., Wang T., Fu D. 'Multicell switched-inductor/switched-capacitor combined active-network converters'. *IEEE Transactions on Power Electronics*. 2015, vol. 30(4), pp. 2063–72.

[37] Chen S., Lao M., Hsieh Y., Liang T., Chen K. 'A novel switched-coupled-inductor DC–DC step-up converter and its derivatives'. *IEEE Transactions on Industry Applications*. 2015, vol. 51(1), pp. 309–14.

[38] Mohammadzadeh Shahir F., Babaei E., Farsadi M. 'Voltage-lift technique based nonisolated boost DC–DC converter: analysis and design'. *IEEE Transactions on Power Electronics*. 2018, vol. 33(7), pp. 5917–26.

[39] Liu H., Li F., Ai J. 'A novel high step-up dual switches converter with coupled inductor and voltage multiplier cell for a renewable energy system'. *IEEE Transactions on Power Electronics*. 2016, vol. 31(7), pp. 4974–83.

[40] Ellabban O., Abu-Rub H. 'Z-source inverter: topology improvements review'. *IEEE Industrial Electronics Magazine*. 2016, vol. 10(1), pp. 6–24.

[41] Liu Y., Abu-Rub H., Ge B. 'Front-end isolated quasi-Z-source DC–DC converter modules in series for high-power photovoltaic systems—Part I: configuration, operation, and evaluation'. *IEEE Transactions on Industrial Electronics*. 2017, vol. 64(1), pp. 347–58.

[42] Abdelhakim A., Blaabjerg F., Mattavelli P. 'Modulation schemes of the three-phase impedance source inverters—Part I: classification and review'. *IEEE Transactions on Industrial Electronics*. 2018, vol. 65(8), pp. 6309–20.

[43] Ahmad A., Singh R.K., Lal N V. 'A family of enhanced voltage gain switched-Boost Impedance-source inverter topologies for renewable energy resources'. 2019 IEEE Energy Conversion Congress and Exposition (ECCE); Baltimore, MD, USA; 2019. pp. 4353–8.

[44] Hasan Babayi Nozadian M., Babaei E., Hosseini S.H., Shokati Asl E. 'Switched Z-source networks: a review'. *IET Power Electronics*. 2019, vol. 12(7), pp. 1616–33.

[45] Ahmad A., Singh R.K., Mahanty R. 'Bidirectional quadratic converter for wide voltage conversion ratio'. 2016 IEEE International Conference on Power Electronics, Drives and Energy Systems (PEDES); Trivandrum; 2016. pp. 1–5.

[46] Erickson R.W., Maksimovic D. *Fundamentals of power electronics*. 2nd ed. New Delhi, India: Springer International Edition; 2010.

[47] Xue X.D., Cheng K.W.E., Cheung N.C. 'Selection of electric motor drives for electric vehicles'. 2008 Australasian Universities Power Engineering Conference; Sydney, NSW; 2008. pp. 1–6.

[48] Xu W., Zhu J., Guo Y., Wang S., Wang Y., Shi Z. 'Survey on electrical machines in electrical vehicles'. 2009 International Conference on Applied Superconductivity and Electromagnetic Devices; Chengdu; 2009. pp. 167–70.

[49] Zhu Z.Q., Howe D. 'Electrical Machines and Drives for Electric, Hybrid, and Fuel Cell Vehicles'. *Proceedings of the IEEE*. 2007, vol. 95, pp. 746–65.

[50] Sung W., Shin J., Jeong Y. 'Energy-efficient and robust control for high-performance induction motor drive with an application in electric vehicles'. *IEEE transactions on vehicular Technology*. 2012, vol. 61(8), pp. 3394–405.

[51] Tripathi A., Narayanan G. 'Evaluation and minimization of low-order harmonic torque in low-switching-frequency inverter-fed induction motor drives'. *IEEE Transactions on Industry Applications*. 2016, vol. 52(2), pp. 1477–88.

[52] Pellegrino G., Vagati A., Boazzo B., Guglielmi P. 'Comparison of induction and PM synchronous motor drives for EV application including design examples'. *IEEE Transactions on Industry Applications*. 2012, vol. 48(6), pp. 2322–32.

[53] Yi F., Cai W. 'Modeling, control, and seamless transition of the bidirectional battery-driven switched reluctance motor/generator drive based on integrated multiport power converter for electric vehicle applications'. *IEEE Transactions on Power Electronics*. 2016, vol. 31(10), pp. 7099–111.

[54] Bostanci E., Moallem M., Parsapour A., Fahimi B. 'Opportunities and challenges of switched reluctance motor drives for electric propulsion: a comparative study'. *IEEE Transactions on Transportation Electrification*. 2017, vol. 3(1), pp. 58–75.

[55] Kachapornkul S., Jitkreeyarn P., Somsiri P., Tungpimolrut K., Chiba A., Fukao T. 'A design of 15 kW switched reluctance motor for electric vehicle applications'. 2007 International Conference on Electrical Machines and Systems (ICEMS); Seoul; 2007. pp. 1690–3.

[56] Chan T.F., Lie-Tong Y., Shao-Yuan F. 'In-wheel permanent-magnet brushless DC motor drive for an electric bicycle'. *IEEE Transactions on Energy Conversion*. 2002, vol. 17(2), pp. 229–33.

[57] Patterson D., Spee R. 'The design and development of an axial flux permanent magnet brushless DC motor for wheel drive in a solar powered vehicle'. *IEEE Transactions on Industry Applications*. 1995, vol. 31(5), pp. 1054–61.

[58] Shao J., Nolan D., Teissier M., Swanson D. 'A novel microcontroller-based sensorless brushless DC (BLDC) motor drive for automotive fuel pumps'. *IEEE Transactions on Industry Applications*. 2003, vol. 39(6), pp. 1734–40.

[59] Tae-Hyung K., Ehsani M. 'Sensorless control of the BLDC motors from near-zero to high speeds'. *IEEE Transactions on Power Electronics*. 2004, vol. 19(6), pp. 1635–45.

[60] Nian X., Peng F., Zhang H. 'Regenerative braking system of electric vehicle driven by brushless DC motor'. *IEEE Transactions on Industrial Electronics*. 2014, vol. 61(10), pp. 5798–808.

[61] Guidi G., Kubota H., Hori Y. Induction motor control for electric vehicle application using low resolution position sensor and sensorless vector control technique. Proceedings of Power Conversion Conference - PCC '97; Nagaoka, Japan; 1997. pp. 937–42.

[62] Wang F., Li S., Mei X., Xie W., Rodríguez J., Kennel R.M. 'Model-based predictive direct control strategies for electrical drives: an experimental evaluation of PTC and PCC methods'. *IEEE Transactions on Industrial Informatics*. 2015, vol. 11(3), pp. 671–81.

[63] Bose B.K., Patel N.R., Rajashekara K. 'A neuro-fuzzy-based on-line efficiency optimization control of a stator flux-oriented direct vector-controlled induction motor drive'. *IEEE Transactions on Industrial Electronics*. 1997, vol. 44(2), pp. 270–3.

[64] Zheng Wang., Chau K.T. 'Design, analysis, and experimentation of chaotic permanent magnet DC motor drives for electric compaction'. *IEEE Transactions on Circuits and Systems II: Express Briefs*. 2009, vol. 56(3), pp. 245–9.

[65] Lee W., Kim J.H., Choi W., Sarlioglu B. 'Torque ripple minimization control technique of high-speed single-phase brushless DC motor for electric Turbocharger'. *IEEE Transactions on Vehicular Technology*. 2018, vol. 67(11), pp. 10357–65.

[66] Pelczewski P.M., Kunz U.H. 'The optimal control of a constrained drive system with brushless DC motor'. *IEEE Transactions on Industrial Electronics*. 1990, vol. 37(5), pp. 342–8.

[67] Rodriguez F., Emadi A. 'A novel digital control technique for brushless DC motor drives'. *IEEE Transactions on Industrial Electronics*. 2007, vol. 54(5), pp. 2365–73.

Chapter 6

Reducing grid dependency of EV charging using renewable and storage systems

Viet Thang Tran[1], Md Rabiul Islam[1], Danny Sutanto[1], and Kashem M Muttaqi[1]

Electric vehicles (EVs) are increasingly deployed and may become the main private and public transportation mode in many countries in the future. This is the result of the clean energy programs promoted by many governments all over the world to deal with climate change, where incentives are offered for low- or zero-carbon emission vehicles to reduce the greenhouse gas emission from the road transport sector. Such incentives encourage automakers to increase EV production capacity. Following widespread deployment of EVs, it is envisaged that EV charging may become a dominant load in the power grid due to the high power required to charge the EV battery, particularly when EVs can be charged whenever they are parked at home or in parking lots. In such a scenario, the grid may experience an unpredictably large variation in the load demand that can lead to various technical issues, such as voltage regulation, harmonic contamination, and frequency variations. These will bring significant challenges to the operation and management of the power utilities. Therefore, to keep up with the increasing penetration of EVs, the distribution grid needs to be upgraded to increase its capacity to meet the new demand from EV charging. Unfortunately, such upgrades will take time and will slow down the deployment of EVs.

This chapter presents the solution of using solar photovoltaic (PV) and an energy storage unit to mitigate the effects of the high impact of EVs in all levels of charging from Level 1 to Level 3 of the EV charging system. The chapter is organized as follows: Section 6.1 presents an overview of the EV charging configurations. Section 6.2 presents the solution for mitigating the effects of high penetration of EVs with the Level 1 and Level 2 charging system. Section 6.3 proposes the solution for mitigating the effects of EVs with the Level 3 charging system. Conclusions are drawn in Section 6.4.

[1]School of Electrical, Computer and Telecommunications Engineering, University of Wollongong, Australia

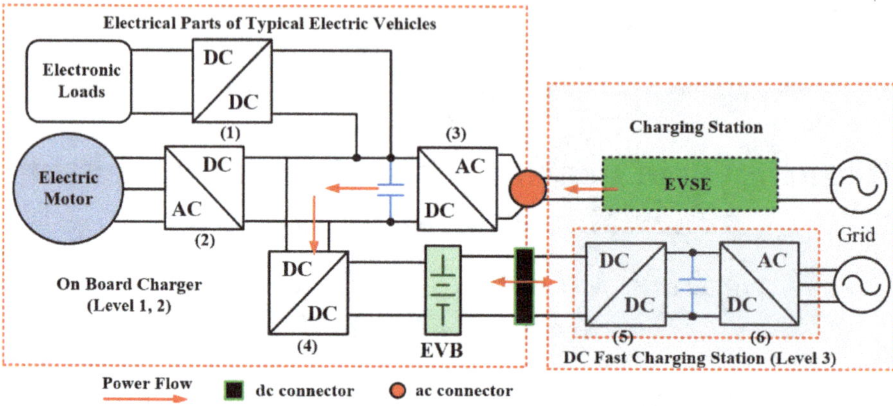

Figure 6.1 General configuration of an EV charging system

6.1 EV charging system

EV charging systems are increasingly being installed in public areas to facilitate the widespread use of EVs in the public domain [1, 2]. However, the high penetration of EVs can bring undesirable impacts on the distribution grid [3–7]. These may bring unexpected challenges for the power utilities for the operation and management of the grid system. Therefore, EV charging systems are continuously being improved to keep up with the increasing penetration of EVs. According to the charging standard [8], the maximum power at Level 1 and Level 2 charging is 19.2 kW. However, the power rating of a Level 1 portable charging cable that can be plugged directly into an electric power point is normally far lower than the standard maximum power allowable at Level 1, which is about 1–2.4 kW, which can comply with the maximum current rating of standard power points (10 A) in residential homes, offices, or shopping centers. Level 3 charging refers to a dc charging system that has a charging power rating ranging from 30 kW up to hundreds of kW.

6.1.1 EV charger topologies

In general, the EV chargers can be categorized as on-board and off-board chargers corresponding to Levels 1, 2, and 3 of the charging standards [9–19]. Figure 6.1 shows a general configuration of an EV charging system.

The on-board charger comprises the power converters shown as (3) and (4) in Figure 6.1, which should have high power density and high efficiency. The converter denoted as (3) in Figure 6.1 is normally a power factor correction (PFC) rectifier, which consists of an uncontrolled full-bridge rectifier and a power boost converter. The converter denoted as (4) in Figure 6.1 can be a boost converter or an isolated dc-dc converter such as the dual active bridge (DAB) converter or H-bridge inductor-capacitor-capacitor resonant converter [10]. Boost type converters are normally used on account of their simplicity in control, high efficiency, and capability of being easily modified into a bidirectional converter by replacing a switching device in place of a power diode [9, 15]. However, the lack of isolation in them results in difficulty to achieve high power transfer. To solve this issue, a DAB is a suitable converter as it has a high-frequency transformer acting as an isolator. This makes it able to achieve high power and bidirectional power flow. However, the control of DAB is more complicated because of the required high-computational algorithm [18, 19].

To control the charging procedure, electric vehicle service equipment (EVSE) is needed. The EVSE will have the function of turning on/off the ac power from the grid when the electric vehicle battery (EVB) gets fully charged and provide a charging fee or bill. The dc power in the dc link can also be used to supply the electronic loads in the car through the dc-dc converter denoted as (1) in Figure 6.1. The off-board charger is a Level 3 dc fast charger and is located in the charging station. The converter of the off-board charger, denoted as (5) in Figure 6.1, is a high-power dc-dc converter, which can be a multilevel converter or a cascaded converter. The converter denoted as (6) in Figure 6.1 is a three-phase PFC rectifier or a bidirectional converter that not only can charge the EV batteries but also can send power back to the grid.

6.1.2 EV charging/discharging strategies

The EV battery can be charged from or discharged to the distribution grid. Depending on the specific application, an EV charger can be designed to be either unidirectional or bidirectional. The current commercial EV charger is designed with a unidirectional converter. This means that EVs can be charged from the grid or renewable sources. The charging strategies for a unidirectional EV charger can be uncoordinated or coordinated. The uncoordinated charging strategy means that the EV can be charged at any time when it is plugged into the grid. This strategy has the potential to stress the distribution grid if a fleet of EVs is allowed to be charged at the same time. To mitigate the effects of EV charging on the grid, the coordinated charging strategy is proposed by researchers [20–29]. Several coordinated charging strategies have been proposed recently, such as the time-of-use-based charging [20], the real-time smart load management [22], and the distributed resources-based charging [23–29].

6.2 Integration of EV charging-home solar PV system

The integration of Level 1 and Level 2 EV charging into the home solar PV (EVC-HSP) system is shown in Figure 6.2. The power transfer between the HSP system and EV battery is implemented based on a power conditioning system (PCS), which consists of three power converters sharing a dc bus. The charging dc connector is based on the CHAdeMO standard as it allows EVB to be interfaced with the dc bus of the HSP system via a power dc-dc converter named EVB converter. The PV converter is deployed by a power boost converter to track the maximum power point (MPP) of the HSP system. The grid converter (GC) is implemented by a full-bridge converter to interface with the grid. It is noted that a conventional HSP system consists of a PV converter and GC. In this configuration, the GC is a unidirectional

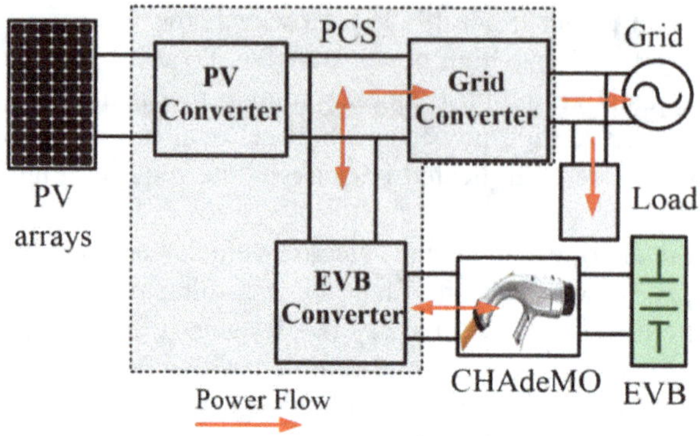

Figure 6.2 Configuration of the EVC-HSP system

converter that only injects the generated HSP power into the grid. However, in the ECV-HSP configuration, the GC is a bidirectional converter where the power can be transferred back and forth from the grid. By adding an EVB converter into the dc bus of the conventional HSP system, the ECV-HSP system is formed. The EVB converter is a unidirectional converter or bidirectional converter. If the EVB gets charged only from the HSP and grid, it can be designated as a unidirectional Boost converter. If the EVB has a function of supporting the grid, it can be designated as a bidirectional Buck/Boost converter. The output of the EVB converter is connected to the CHAdeMO connector. The mechanism of the handshaking signals of the CHAdeMO is presented in Figure 6.3. A control board is used to control the power flow of the EVC-HSP system. With this control mechanism, the effective power management of the EVC-HSP is obtained.

6.2.1 Operation modes of EVC-HSP system

The main idea for the EVC-HSP system is to utilize the HSP system to efficiently charge the EVB while still supporting the loads at home or grid in the case of EV traveling. This idea not only helps mitigate the effects of the high penetration of EVs but also helps increase the HSP system implementation. To achieve this, efficient operation modes of the EVC-HSP should be defined clearly. It has been well established that the HSP system will produce the highest energy around noon. The daily HSP power curve achieves a peak around 10 a.m. to 3 p.m. On the other hand, the home load demand curve often referred to as the "Duck curve", has the peak load around 5 p.m. to 9 p.m., with the lowest point of the curve occurring around noon. This reflects the basic idea that the EVB should be charged around noon and may discharge to support the home load and grid around the peak time from 5 p.m. to 9 p.m. The EVB may be charged from the grid at night when the electricity price is low. This also helps the grid become stable as of balancing the generation and power consumption from the grid. The effective operating modes of the EVC-HSP system can be defined as:

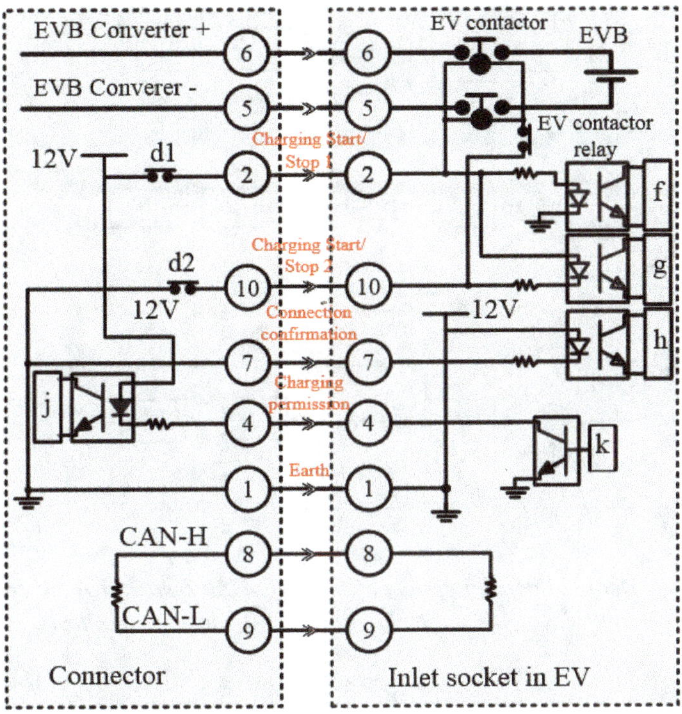

Figure 6.3 Mechanism handshaking signals of the CHAdeMo connector

Mode I: The EV is on the road and the home load is low. This mode may happen at any time during the daytime when people drive the EVs for doing their day-to-day works. The generated HSP power directly injected into the grid is given by (6.1), where η is the efficiency of the PCS, P_g is the grid power, and P_{pv} is the PV power:

$$P_g = \eta P_{pv} \tag{6.1}$$

Mode II: The EV is charged from the HSP power. Depending on the pre-defined maximum charging power, the EVB charging power reference P_{EVB}^{ref} is calculated according to (6.2) with the constraints of the maximum EVB state-of-charge (SoC_{max}).

$$P_{EVB}^{ref} = \begin{cases} -\frac{1}{\eta} P_{pv} \text{ if } P_{pv}(k) \leq P_{EVB}^{max},\ SoC \leq SoC_{max} \\ P_{EVB}^{max} \qquad \text{if } P_{pv} > P_{EVB}^{max},\ SoC \leq SoC_{max} \\ 0 \qquad\qquad \text{if } SoC > SoC_{max} \end{cases} \tag{6.2}$$

Mode III: The EVB is used to support the home load demand at peak time. This not only helps reduce the stress on the grid at the peak demand but also helps save the electricity bills for the EV owners. The EVB discharging power reference is calculated based on (6.3), where P_L is the home load power with the minimum EVB state-of-charge (SoC_{min}).

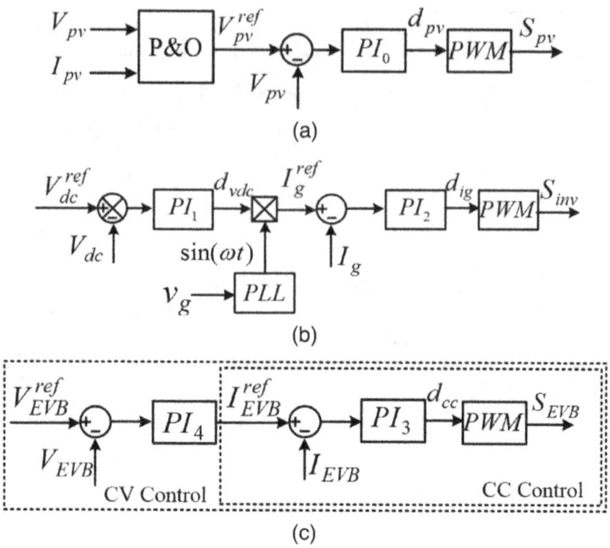

Figure 6.4 *Detailed control system of the EVC-HSP system: (a) PV converter control, (b) GC control, and (c) EVB converter control*

$$P_{EVB}^{ref} = P_L - P_{pv},\ SoC \geq SoC\,min \tag{6.3}$$

Mode IV: The EVB is used to support the grid with the required power P_g^{req}. The EVB discharging power reference is calculated based on (7.4).

$$P_{EVB}^{ref} = P_L + P_g^{req},\ SoC \geq SoC_{Res} \tag{6.4}$$

where SoC_{Res} is the reserved energy of the EVB. This mode can be achieved through the request from the grid operator, and it is a smart function of the smart grid where multi-sources are interconnected and cooperated.

6.2.2 Control strategy of EVC-HSP system

The power converter of the EVC-HSP system is controlled to achieve all operation modes. Each converter has its specific control structure as depicts in Figure 6.4. The output voltage of the HSP is tracked for the MPP of the HSP system; the HSP voltage reference is found by the perturbation and observation algorithm, and then a PI controller is designed to track the HSP voltage reference.

The output control signal d_{pv} of the EVB converter is expressed by (6.5), where k_{pv}^p and k_{pv}^i are proportional and integral parameters of the PI_0 controller, respectively. V_{pv} and V_{pv} are actual and reference voltages of the HSP system.

$$d_{pv} = \left(k_{pv}^p + \frac{k_{pv}^i}{s}\right)\left(V_{pv}^{ref} - V_{pv}\right) \tag{6.5}$$

The grid-connected converter is controlled through the dc bus V_{dc} to achieve a bidirectional power flow. The control for the GC is shown in Figure 6.4(b). The control signals of the GC, d_{vdc}, and d_{ig}, respectively, are calculated according to (6.6) and (6.7), where V_{dc}^{ref} is the voltage reference of the dc bus voltage and V_{dc}^{ref} is the grid current reference.

$$d_{vdc} = \left(k_{vdc}^{p} + \frac{k_{vdc}^{i}}{s}\right)\left(V_{dc}^{ref} - V_{dc}\right) \tag{6.6}$$

$$d_{ig} = \left(k_{ig}^{p} + \frac{k_{ig}^{i}}{s}\right)\left(I_{g}^{ref} - I_{g}\right) \tag{6.7}$$

The EVB is charged and discharged through the EVB converter. The control of the EVB converter is shown in Figure 6.4(c), where the constant current control is used for both charging and discharging processes. The current controller is following (6.8), where d_{cc} is the control signal, k_{cc}^{p} and k_{cc}^{i} are PI_3 controller parameters, and k_{cc}^{i} is the EVB charging/discharging current reference.

$$d_{cc} = \left(k_{cc}^{p} + \frac{k_{cc}^{i}}{s}\right)\left(I_{EVB}^{ref} - I_{EVB}\right) \tag{6.8}$$

The constant voltage (CV) control is used when the EVB reaches its gassing voltage and the EVB is kept to the EVB CV until the EVB is fully charged. The control variable d_{vc} of the voltage controller is the EVB current reference as given according to (6.9), where k_{vc}^{p} and k_{vc}^{i} are PI_4 parameters, respectively, and V_{EVB}^{ref} is the EVB voltage reference.

$$d_{vc} = \left(k_{vc}^{p} + \frac{k_{vc}^{i}}{s}\right)\left(V_{EVB}^{ref} - V_{EVB}\right) \tag{6.9}$$

6.2.3 Simulation results of EVC-HSP system

To validate the performance of the proposed EVC-HSP system and its control strategy, a detailed switching model of the entire EVC-HSP is simulated in MATLAB®/Simulink environment. The irradiance profile is selected to ensure that the EVC-HSP experiences all the operation modes depicted in Figures 6.5 and 6.6.

The green line is the PV power with a maximum power of 5 kW. The PV power has been varied through all the operation modes. All the energy generated by the PV system is utilized for charging to the EVB at the period from $t = 0$ s to $t = 1$ s with the maximum charging current. During the period from $t = 1$ s to $t = 2$ s, the PV irradiance is decreased; however, the EVB is nearing its fully charged condition, and hence it is required to limit the charging current. Therefore, the excess generated power from the HSP system will be injected into the grid. During the period from $t = 2$ s to $t = 3$ s, the EVB is fully charged, all the HSP power is injected into the grid. At the peak load, the EVB will support the load as it will be discharging its energy in the period from $t = 3$ s to $t = 4$ s, after discharging its energy and when the load demand is low, the EVB starts being charged as shown in the period from

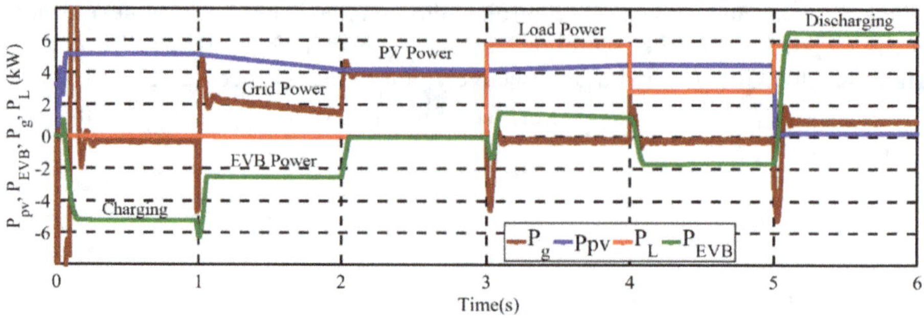

Figure 6.5 Simulation results of the EVC-HSP system in four operation modes

$t = 4$ s to $t = 5$ s. During the period from $t = 5$ s to $t = 6$ s, the EVB supports the grid during the peak load hours. The EVB power is shown in Figure 6.6.

The simulation results show that with the efficient energy management of the EVB and PV power, the EV owners will have the benefit of minimizing the electricity cost and the distribution grid can handle the increasingly high penetration of the PV power generation and high penetration of EVs as well. Figure 6.7 shows the state-of-charge (*SoC*) of the EVB through the different operation modes. The grid voltage and current waveforms of the operation modes are shown in Figure 6.8.

6.2.4 Experimental results of EVC-HSP system

A scaled laboratory prototype is built to test and debug the performance of the EVC-HSP system depicted in Figure 6.9. The test platform comprises an emulated 30 kW from the solar PV TerraSas emulator, a 10-kWh lead-acid battery, which is considered as the EV battery. Moreover, the power converters based on a 5-kW insulated-gate bipolar transistor (IGBT) module from SEMIKRON were also used. The control system is implemented digitally using a programmable platform from National Instruments called general-purpose inverter controller board. The maximum power generated by the HSP system is 275 W at the MPP voltage of 100 V. The dc-link

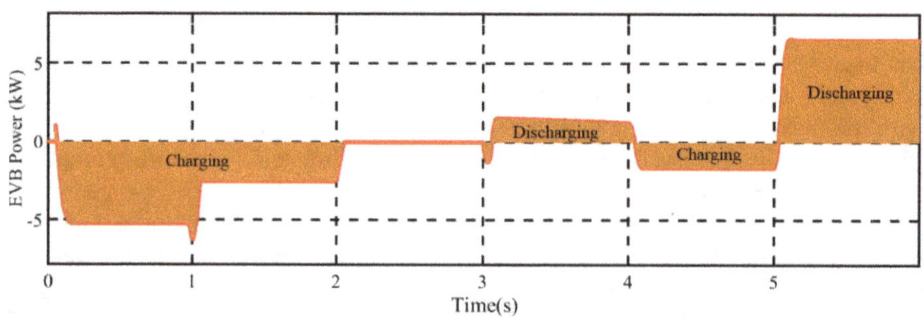

Figure 6.6 Simulation results of the EVB power in four operation modes

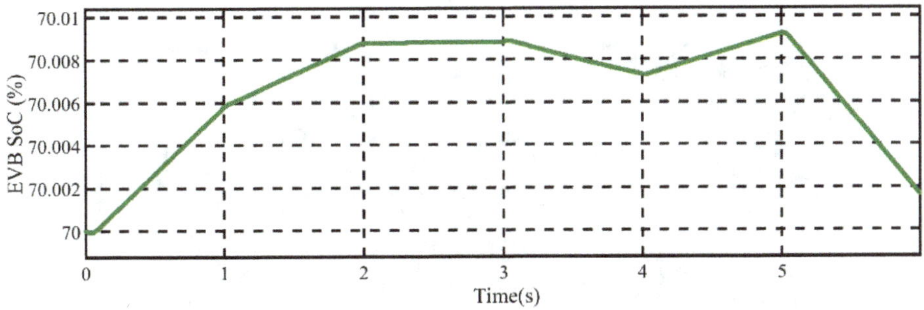

Figure 6.7 SoC, current, and voltage of the EVB in operation modes

voltage is maintained at 220 V with the grid voltage is set as 127 V. Two resistive loads of 60 Ω and 110 Ω, respectively, are considered as the home loads.

Figure 6.10 depicts the performance of the EVC-HSP system through the charging and the discharging procedure of the EV battery. Period (a) shows the charging of the EVB from the PV power and the period (b) shows the discharging of the EVB when it supports the local load. The EVB is charged in period (c) and it is discharged in period (d) when there is no PV power.

6.2.5 Summary designing of an EVC-HSP system

An EVC-HSP system for the integration of the HSP system and EV charging is presented. The energy management system successfully controls the power balance between the HSP source, EVB energy, the home load, and the grid demand to ensure the reduced dependency of the EVB charging power from the power grid. This will help to mitigate the uncertainty effects caused by the EVB charging. It is envisaged that the proposed system with its control algorithm will have a great potential to increase the deployment of EVs.

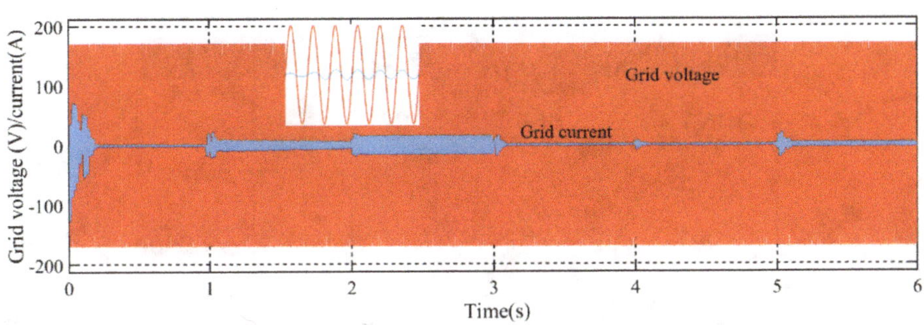

Figure 6.8 Grid voltage and current of the EVC-HSP system in four operation modes

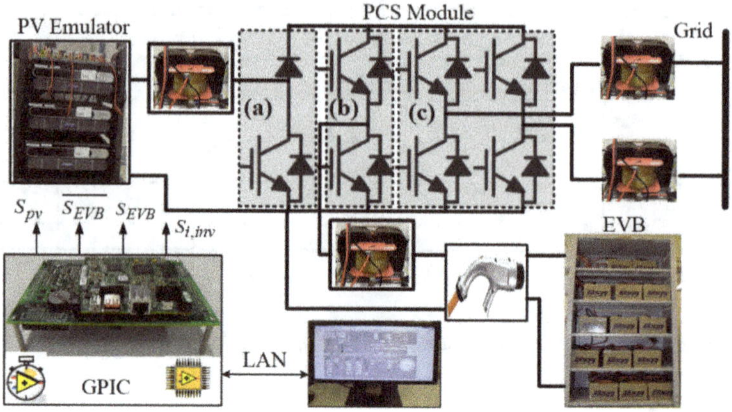

Figure 6.9 The detailed experimental setup of the laboratory scale of the EVC-HSP system

6.3 Level 3 – fast-charging infrastructure with solar PV and energy storage

To mitigate the effects of the EV fast-charging by reducing the dependence of the EV fast-charging on the grid, a fast-charging infrastructure (FCI) that can be integrated with an energy buffer unit (EBU) and a PV system is proposed as shown in Figure 6.11. The proposed FCI not only can help mitigate the effects of the high penetration of PV and EV into the distribution grid, but it can also play a key role in the development of the future smart grid. The four vital components of the FCI are grid converter, PV converter, EBU converter, and EVB converter. All these power

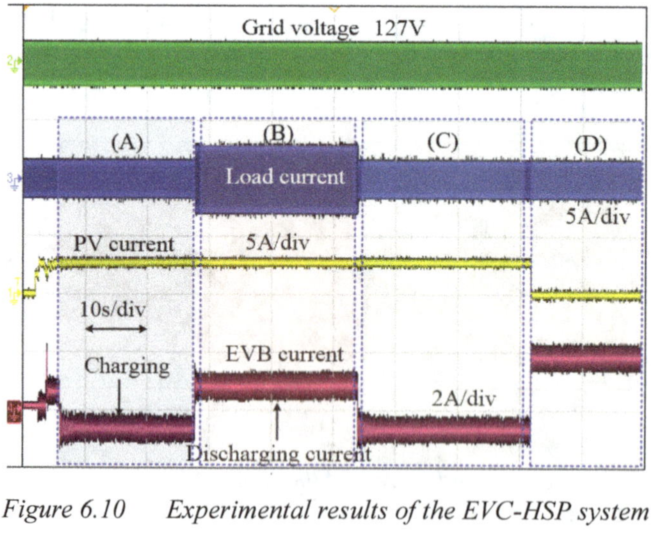

Figure 6.10 Experimental results of the EVC-HSP system

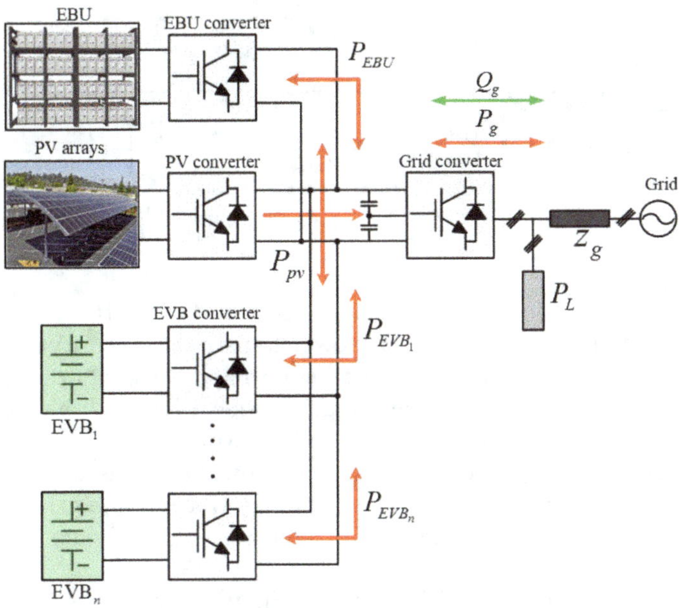

Figure 6.11 Infrastructure of a future EV fast-charging station

converters will share the same dc power bus. The number of EVs that can be charged concurrently depends on the number of EVB converters connected to the dc bus. The power flow in the proposed FCI is shown in Figure 6.11.

The PV power (P_{PV}) will flow into EBU, EVB, and the distribution grid. The EBU power (P_{EBU}), EVB power (P_{EVB}), and grid power (P_g) will flow back and forth between EBU, EVB, and the distribution grid because the EBU converter, EVB converter, and GC are bidirectional. As the power rating per charge ranges from several dozen kilowatts up to hundreds of kilowatts, if a few EVs are charged concurrently, the FCI power rating will be a few hundred kilowatts. The power rating is considered as the design criteria for FCI. To achieve reliable operation in any scenario, the power rating of the GC should be designed in such a way that it can handle the power flow in the FCI. The power rating of the GC depends on the power rating of the PV converter, EBU converter, and EVB converter, and it can be designed based on (6.10) or (6.11).

$$P_{GC}^{max} = P_{PV}^{max} + P_{EBU}^{max} + \sum_{i=1}^{i=n} P_{EVB}^{max} \tag{6.10}$$

$$P_{GC}^{max} = Max\left\{P_{PV}^{max},\ P_{EBU}^{max},\ \sum_{i=1}^{i=n} P_{EVB}^{max}\right\} \tag{6.11}$$

where P_{GC}^{max}, P_{PV}^{max}, P_{EBU}^{max}, and P_{EVB}^{max} are maximum power that GC, PV converter, EBU converter, and EVB converter can deliver, and n is the number of EVB converters.

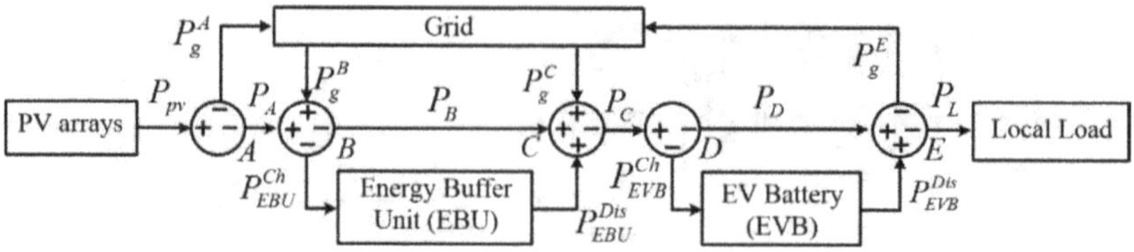

Figure 6.12 Power flow in the FCI

If the power rating of the GC is designed based on (6.10), the GC will be capable of operating for all types of power transfer between PV, EVB, EBU, and the distribution grid. In the worst case, during a blackout or peak demand, the power delivered to the distribution grid or local load will be the aggregation of the PV power, EVB power, and EBU power.

In the GC power rating designed based on (6.11), it is assumed that not all distributed sources can simultaneously inject power into the grid at the same time. The grid power rating can then be equal to the maximum power of the PV, EBU, and EVB.

The power flow in the FCI is shown in Figure 6.12. There are four power nodes: A, B, C, and D, which are defined as the aggregated power of each node. The positive sign (+) means the power flows into the node and the negative sign (−) means the power flows out from the node. The power at each node is expressed as (6.12).

$$\begin{cases} P_A = P_{PV} + \alpha_0 P_g^A \\ P_B = P_A + \alpha_1 P_g^B + \alpha_2 P_{EBU}^{Ch} \\ P_C = P_B + \alpha_3 P_g^C + \alpha_4 P_{EBU}^{Dis} \\ P_D = P_C + \alpha_5 P_{EVB}^{Ch} \\ P_L = P_D + \alpha_6 P_g^E + \alpha_7 P_{EBU}^{Dis} \end{cases} \tag{6.12}$$

The coefficients α_i (i=0:7) can be −1 or 0 based on the defined power modes. Nine operation modes are defined in such a way that EVB is preferentially charged from the PV and EBU power.

Mode 1: The FCI has no EV that needs to be charged, the generated PV power will be used to charge the EBU, the power flow at nodes A and B can be expressed as (6.13).

$$\begin{cases} P_A = P_{PV} & \alpha_0 = 0 \\ P_B = 0 & \\ P_A - P_{EBU}^{Ch} = 0 & \alpha_1 = 0;\ \alpha_2 = -1 \end{cases} \tag{6.13}$$

Mode 2: The EBU is fully charged, no EV needs to be charged. The generated PV power will be injected into the grid, and the power flow at the power nodes can be expressed as in (6.14).

$$\begin{cases} P_{PV} - P_g^A = 0 & \alpha_0 = -1 \\ P_A = 0,\ P_B = 0,\ P_C = 0,\ P_D = 0 \end{cases} \tag{6.14}$$

Mode 3: The EBU is fully charged, the EVB will be charged from the PV power only. The power at the power nodes can be expressed by (6.15).

$$\begin{cases} P_A = P_{PV} & \alpha_0 = 0 \\ P_B = P_A & \alpha_1 = 0;\ \alpha_2 = 0 \\ P_C = P_B & \alpha_3 = 0;\ \alpha_4 = 0 \\ P_C - P_{EVB}^{Ch} & \alpha_5 = -1 \end{cases} \tag{6.15}$$

Mode 4: PV power is insufficient, the EVB will be charged from the generated PV power and EBU power. In this case, EBU operates under the discharging mode. The power flow at the power nodes can be expressed by (6.16).

$$\begin{cases} P_A = P_{PV} & \alpha_0 = 0 \\ P_B = P_A & \alpha_1 = 0;\ \alpha_2 = 0 \\ P_C = P_B + P_{EBU}^{Dis} & \alpha_3 = 0;\ \alpha_4 = 1 \\ P_C - P_{EVB}^{Ch} = 0 & \alpha_5 = -1 \end{cases} \tag{6.16}$$

Mode 5: The EVB will get charged from the PV and grid. In this case, the output power from EBU is zero. In this mode, the powers at the nodes are given by (6.17).

$$\begin{cases} P_{PV} - P_g^A = 0 & \alpha_0 = -1 \\ P_A = 0,\ P_B = 0,\ P_C = 0,\ P_D = 0 \end{cases} \tag{6.17}$$

Mode 6: There is no PV power; the EVB will be charged from the EBU and the grid. The power at the power nodes can be expressed by (6.18).

$$\begin{cases} P_C = P_g^C + P_{EBU}^{Dis} & \alpha_3 = 1;\ \alpha_4 = 1 \\ P_C - P_{EVB}^{Ch} = 0 & \alpha_5 = -1 \end{cases} \tag{6.18}$$

Mode 7: The EBU stops discharging as it reaches the minimum *SoC*; the EVB gets charged from the grid power only. The power flow at the power nodes can be expressed by (6.19).

$$\begin{cases} P_C = P_g^C & \alpha_3 = 1;\ \alpha_4 = 0 \\ P_C - P_{EVB}^{Ch} = 0 & \alpha_5 = -1 \end{cases} \tag{6.19}$$

Mode 8: Local load support is required when the grid is disconnected. The power flow at the power nodes can be expressed by (6.20).

Table 6.1 Defined operation modes associated with α_i

Modes	α_0	α_1	α_3	α_4	α_5	α_6	α_7
1	0	0	−1	0	0	0	0
2	−1	0	0	0	0	0	0
3	0	0	0	0	−1	0	0
4	0	0	0	1	−1	0	0
5	0	0	1	0	−1	0	0
6	0	0	0	1	−1	0	0
7	0	0	1	0	−1	0	0
8	0	0	0	1	0	0	1
9	0	0	0	1	0	−1	1

$$\begin{cases} P_A = P_{PV} & \alpha_0 = 0 \\ P_B = P_A & \alpha_1 = 0;\ \alpha_2 = 0 \\ P_C = P_B + P_{EBU}^{Dis} & \alpha_3 = 0;\ \alpha_4 = 1 \\ P_D = P_C & \alpha_5 = 0 \\ P_L = P_D + P_{EVB}^{Dis} & \alpha_6 = 0,\ \alpha_7 = 1 \end{cases} \tag{6.20}$$

Mode 9: Grid support is required during the peak demand. The power flow at the power nodes can be expressed by (6.21).

$$\begin{cases} P_A = P_{PV} & \alpha_0 = 0 \\ P_B = P_A & \alpha_1 = 0;\ \alpha_2 = 0 \\ P_C = P_B + P_{EBU}^{Dis} & \alpha_3 = 0;\ \alpha_4 = 1 \\ P_D = P_C & \alpha_5 = 0 \\ P_D - P_g^E + P_{EVB}^{Dis} = 0 & \alpha_6 = -1,\ \alpha_7 = 1 \end{cases} \tag{6.21}$$

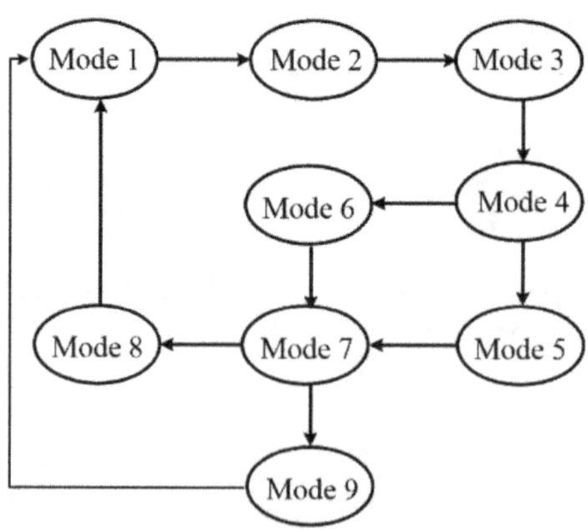

Figure 6.13 Mode transition of the FCI

The summary of the operation modes of the FCI presented above is shown in Table 6.1.

The transition between the modes is shown in Figure 6.13 shows that $\alpha_i = 0$ at all modes, this is because the FCI will have the functionality of changing the EBU from the grid when the electricity price is low. However, in this chapter, this mode is not considered, as the idea is to utilize the generated PV power as much as possible.

6.3.1 Power converter for FCI

The current advancement in the EV battery and semiconductor device technologies allows the EV to travel a long distance of up to 400 km, which requires the EVB capacity to be a few hundred kilowatts. Therefore, the power rating of the FCI needs to be high enough to deliver the high power to an EV with less charging time. By considering the required high power, the converter types need to be chosen carefully. The proposed power converter of the FCI is shown in Figure 6.14.

6.3.1.1 Grid converter

The GC will play a key role in the FCI to achieve reliable operation. As discussed in the previous analysis, the power rating of the GC should be chosen based on (6.12) or (6.13). The three-level neutral point clamped (3L-NPC) topology is chosen as it shows the benefit of having a lower current total harmonic distortion, consequently facilitating the reduced volume of grid filter. The 3L-NPC inverter has been chosen for the GC as it has been proven for its high frequency, high voltage, and high power in the power system applications with the power ranging up to a few hundred kilowatts per single topology. In the past, only power IGBTs can handle this level of power; however, with the fast development of the silicon carbide metal-oxide-semiconductor field-effect transistors (MOSFETs), having the voltage and current rating about 1 kV and 100 A, respectively, the required level of power can be achieved. Also, using this topology, the power level can be upgraded easily by connecting more MOSFETs in parallel at each leg of the 3L-NPC topology. The GC based on 3L-NPC topology is shown in Figure 6.14.

6.3.1.2 PV converter

The PV boost converter is to track the maximum power from the PV arrays. However, for a high-power application, the boost converter will require a big inductor, which will result in having a high-power loss thereby reducing the overall efficiency. To overcome this disadvantage, an interleaved boost converter is utilized for the proposed PV converter. This helps increase system efficiency by reducing the core size of the inductor. Also, the inductor current ripple is reduced significantly. The PV converter is shown in Figure 6.14.

6.3.1.3 EBU and EVB converter

The DAB converter is the best candidate for the EV charger. Taking advantage of its many capabilities, the DAB converter is used as the power converter to control the

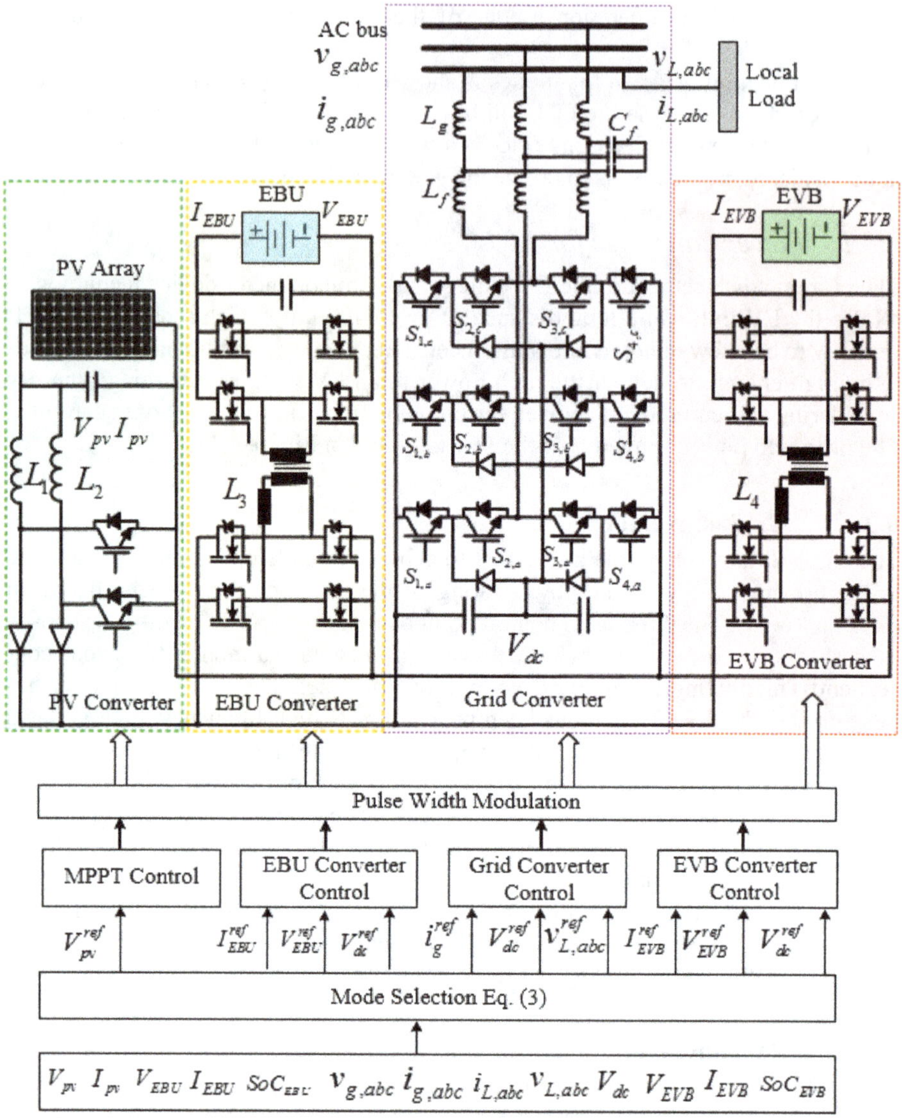

Figure 6.14 Power converter and control diagram of the FCI

charging and discharging of the EBU and EVB. The EBU and EVB converters are shown in Figure 6.14.

6.3.2 Control diagram for FCI

Based on the defined operation modes, the control flowchart of the FCI is shown in Figure 6.15. The control operation of the FCI will follow the control flowchart, as shown in Figure 6.15. The transition between the defined modes is based on the

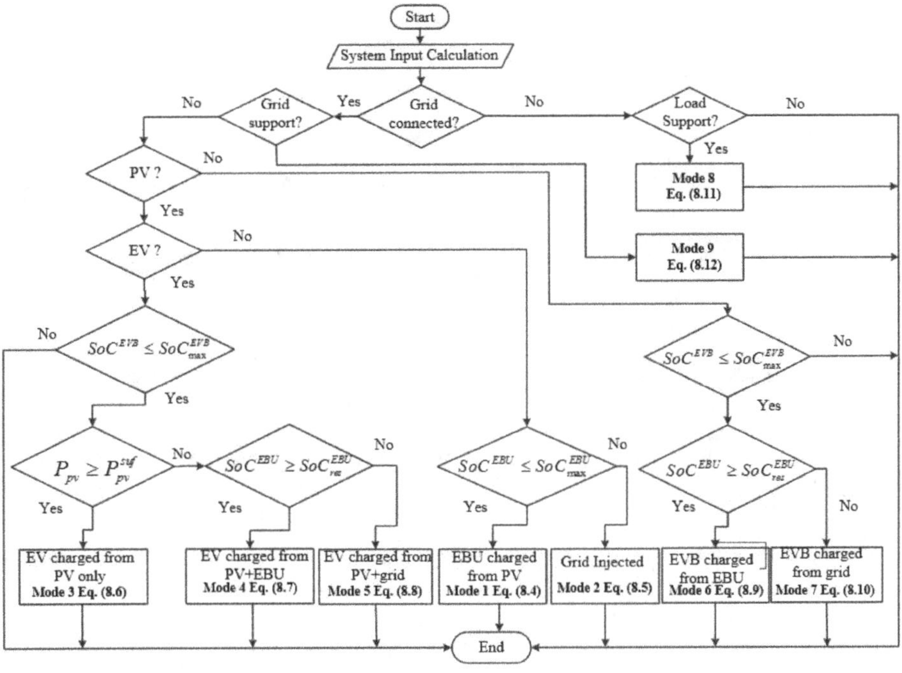

Figure 6.15 Control flowchart of the FCI

status of the grid, PV power, EBU power, and EVB power. The detailed control of each power converter is discussed in this section.

6.3.2.1 Maximum power point tracking control

The control scheme of the PV converter is shown in Figure 6.16. There is a two-loop control: the outer loop for the PV voltage control V_{pv} and the inner loop for the boost inductor current control $I_{L,}$ which is calculated according to (6.22).

$$I_L = I_{L_1} + I_{L_2} \tag{6.22}$$

where I_{L1} and I_{L2} are inductor currents of the interleaved boost converter.

The two power switches will be controlled compensatively. The inductor reference current is calculated based on (6.23), where K_p^1 and K_i^1 are parameters of the voltage controller $C_{Vpv}(s)$.

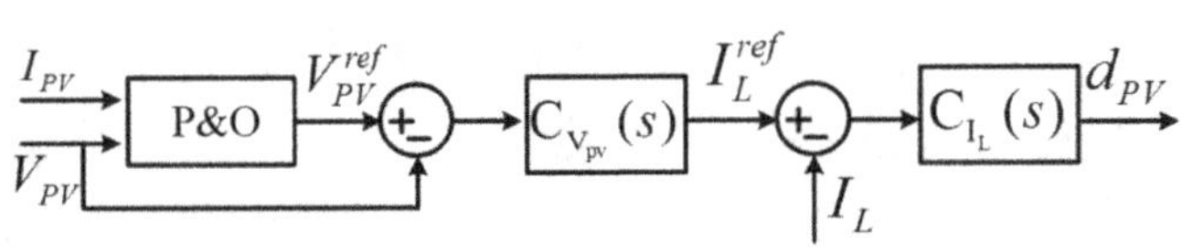

Figure 6.16 MPPT control of the PV converter

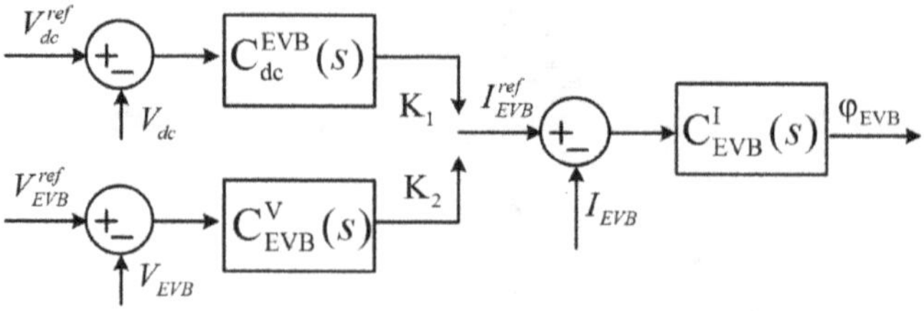

Figure 6.17 The control diagram of the EVB converter

$$I_L^{ref} = K_p^1\left(V_{pv}^{ref} - V_{pv}\right) + \frac{K_i^1}{s}\left(V_{pv}^{ref} - V_{pv}\right) \tag{6.23}$$

The duty cycle d_{pv} is calculated based on (6.24), where K_p^2 and K_p^2, are parameters of the current controller $C_{IL}(s)$. The duty cycle of the compensated switch K_p^2 is calculated based on (6.25).

$$d_{pv} = K_p^2\left(I_L^{ref} - I_L\right) + \frac{K_i^2}{s}\left(I_L^{ref} - I_L\right) \tag{6.24}$$

$$\bar{d}_{pv} = 1 - d_{pv} \tag{6.25}$$

6.3.2.2 Electric vehicle battery control

The EVB converter control structure is depicted in Figure 6.17. Based on the selected modes as shown in the flowchart in Figure 6.15, the EVB converter has three control schemes. The controlled variables are calculated by (6.26)–(6.28).

There are three combinations of the operation of switches K_1 and K_2 corresponding to the three control schemes of the EVB converter. The first scheme is the EVB control scheme given by (6.26). This control is used during both the charging and discharging modes of EVB with the switches K_1 and K_2 opened. The second scheme given by (6.26) and (6.27) is for EVB charging at a CV, where K_1 is opened and K_2 is closed. The third control scheme represented by (6.26) and (6.28) involves the control of dc-link voltage. This control scheme is chosen in the local support mode (Mode 8), where the EVB power is used to supply the local load. In this scheme, K_1 is closed and K_2 is opened.

$$I_{EVB}^{ref} = K_p^3\left(V_{EVB}^{ref} - V_{EVB}\right) + \frac{K_i^3}{s}\left(V_{EVB}^{ref} - V_{EVB}\right) \tag{6.26}$$

$$\varphi_{EVB} = K_p^4\left(I_{EVB}^{ref} - I_{EVB}\right) + \frac{K_i^4}{s}\left(I_{EVB}^{ref} - I_{EVB}\right) \tag{6.27}$$

$$I_{EVB}^{ref} = K_p^5\left(V_{dc}^{ref} - V_{dc}\right) + \frac{K_i^5}{s}\left(V_{dc}^{ref} - V_{dc}\right) \tag{6.28}$$

where $K_p^3, K_i^3, K_p^4, K_i^4, K_i^5$, and K_p^5 are parameters of C_{EVB}^V, C_{EVB}^I, and C_{dc}^{EVB} controllers, respectively.

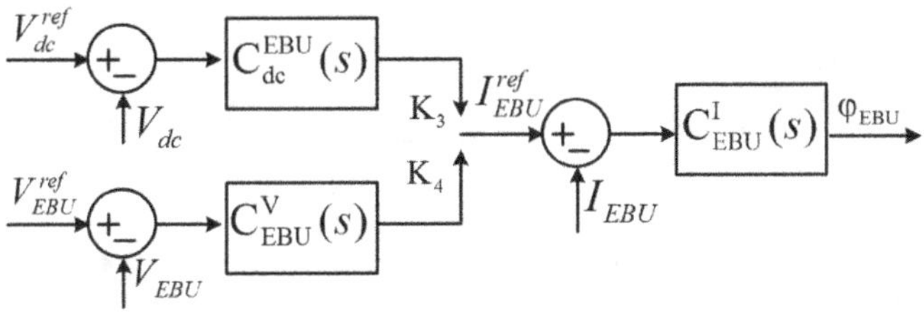

Figure 6.18 The control diagram of the EBU converter

6.3.2.3 Energy buffer unit control

The control of the EBU converter is similar to the control of the EVB converter, as shown in Figure 6.18. Based on the selected modes shown previously in the flowchart of Figure 6.14, the EBU converter has three control schemes. The controlled variables are calculated by (6.29)–(6.31).

$$I_{EBU}^{ref} = K_p^6 \left(V_{EBU}^{ref} - V_{EBU}\right) + \frac{K_i^6}{s}\left(V_{EBU}^{ref} - V_{EBU}\right) \tag{6.29}$$

$$\varphi_{EBU} = K_p^7 \left(I_{EBU}^{ref} - I_{EBU}\right) + \frac{K_i^7}{s}\left(I_{EBU}^{ref} - I_{EBU}\right) \tag{6.30}$$

There are three combinations of the operation of switches K_3 and K_4 corresponding to three control schemes of the EBU converter. The first control represented by (6.29) is the EBU current control. This control scheme is used during both the charging and discharging modes of EBU and the switches K_3 and K_4 remain open during the scheme. The second control scheme given by (6.30) and (6.31) is chosen when the EBU is being charged in the CV mode, in this scheme, where K_3 is opened and K_4 is closed. The third control scheme given by (6.29) and (6.31) is associated with the control of dc-link voltage. This control scheme is chosen for the local support mode (Mode 8) where the EBU supplies power to the local load. In this scheme, K_3 is closed and K_4 is opened.

$$I_{EBU}^{ref} = K_p^8 \left(V_{dc}^{ref} - V_{dc}\right) + \frac{K_i^8}{s}\left(V_{dc}^{ref} - V_{dc}\right) \tag{6.31}$$

where $K_P^6, K_i^6, K_P^7, K_i^7, K_i^8$, and K_p^8 are parameters of C_{EBU}^V, C_{EBU}^I, and C_{dc}^{EBU} controllers, respectively.

6.3.2.4 Grid converter control

The GC has two control schemes, the grid-connected and the standalone mode as shown in Figure 6.19. The overall control system of the GC is shown in Figure 6.19 where the space vector modulation is chosen as the modulation method due to its lowest harmonic performance. In the grid-connected mode, the GC will be an active rectifier when it delivers power from the distribution grid to EVB and will be a grid inverter when it delivers power from PV, EBU, and EVB to the distribution grid. In this mode, a two-loop control is deployed, the inner control loop controls the active

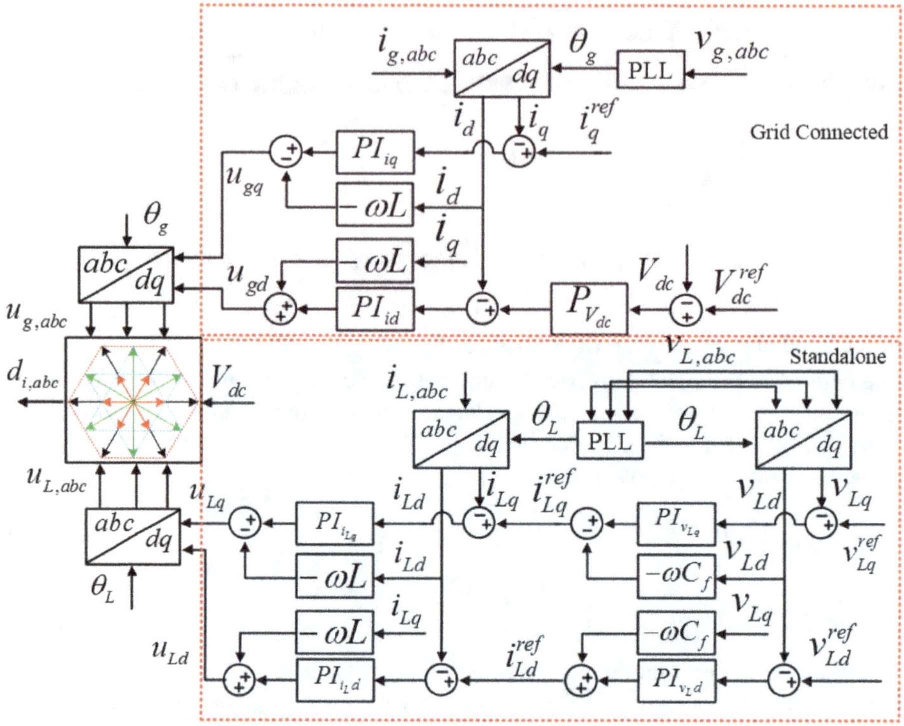

Figure 6.19 The control diagram of the GC converter

and reactive power, and the outer loop controls the dc-link voltage V_{dc}. All the controllers are designed in the *d-q* frame as given by (6.32)–(6.34) with i_d, where i_d is the grid current in the *d-q* frame, d_{vdc} is the control variable of the dc-link voltage control, and u_{gd} and u_{gq} are the control variables of the grid control in the *d-q* frame.

$$d_{vdc} = \left(k_{vdc}^{p} + \frac{k_{vdc}^{i}}{s}\right)\left(V_{dc}^{ref} - V_{dc}\right) \tag{6.32}$$

$$u_{gd} = \left(k_{d}^{p} + \frac{k_{d}^{i}}{s}\right)\left(i_{d}^{ref} - i_{d}\right) \tag{6.33}$$

$$u_{gq} = \left(k_{q}^{p} + \frac{k_{q}^{i}}{s}\right)\left(i_{q}^{ref} - i_{q}\right) \tag{6.34}$$

In the standalone mode, the GC acts as an inverter that converts the dc power to ac power to supply the local load. This is represented as mode 8, shown in the flowchart of Figure 6.15. The detailed control of the GC in the standalone mode is shown in Figure 6.19. The outer loop is the output voltage of the GC, $v_{L,abc}$. All the controllers are designed in the *d-q* frame with i_{Ld} and i_{Lq} are the load current in the *d-q* frame, and v_{Ld} and v_{Lq} are the load voltage in the *d-q* frame.

$$i_{Ld}^{ref} = \left(k_{Ld}^{p} + \frac{k_{Ld}^{i}}{s}\right)\left(v_{Ld}^{ref} - v_{Ld}\right) \tag{6.35}$$

$$i_{Lq}^{ref} = \left(k_{Lq}^{p} + \frac{k_{Lq}^{i}}{s}\right)\left(v_{Lq}^{ref} - v_{Lq}\right) \tag{6.36}$$

$$i_{Lq}^{ref} = \left(k_{Lq}^{p} + \frac{k_{Lq}^{i}}{s}\right)\left(v_{Lq}^{ref} - v_{Lq}\right) \tag{6.37}$$

$$SoC^{EVB}(t) = SoC_{0}^{EVB} + \frac{1}{C^{EVB}}\int_{t_0}^{t} i_{EVB}(t)\,dt \tag{6.38}$$

6.3.3 Simulation results for FCI

To validate the performance of the proposed FCI and its control system, a detailed model of all power converter switches is programmed in the MATLAB/Simulink environment. Two lead-acid battery banks rated 10 kWh are modeled in Simulink. These battery banks are considered as the EVB and EBU, respectively. It is noted that there is no limit in the power rating of the FCI in the simulation environment. The main reason to choose the power rating of 10 kWh is to match with the experimental system that will be implemented at the laboratory scale. The *SoC* of the EVB and EBU can be calculated as given in (6.39) and (6.40):

$$SoC^{EVB}(t) = SoC_{0}^{EVB} + \frac{1}{C^{EVB}}\int_{t_0}^{t} i_{EVB}(t)\,dt \tag{6.39}$$

$$SoC^{EBU}(t) = SoC_{0}^{EBU} + \frac{1}{C^{EBU}}\int_{t_0}^{t} i_{EBU}(t)\,dt \tag{6.40}$$

where C_0^{EVB} and C_0^{EBU} are the initial *SoC* at $t = t_0$, C is the nominal EVB and EBU capacities, and $i_{EVB}(t)$ and $i_{EBU}(t)$ are the EVB and EBU currents at time t. The simulation results of modes 4–7 are shown in Figures 6.20 and 6.21.

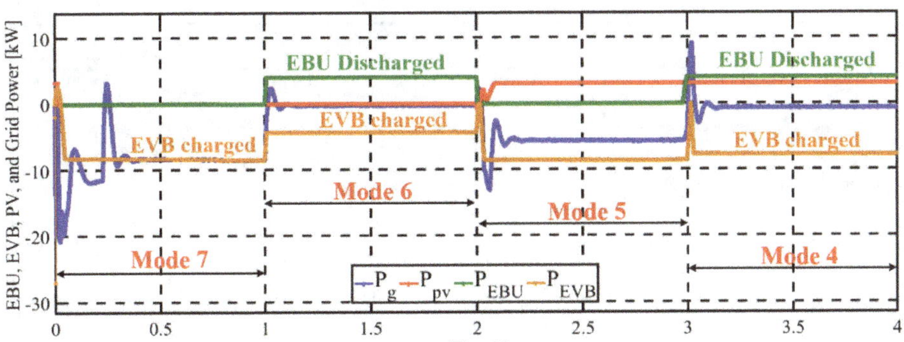

Figure 6.20 Simulation results of the transition between modes 4, 5, 6, and 7

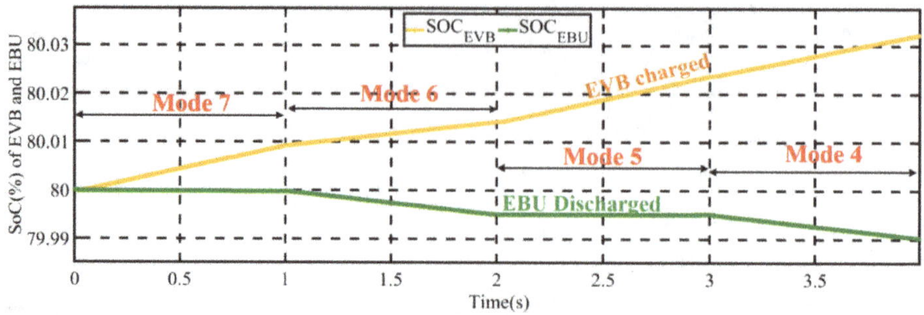

Figure 6.21 Simulation results of the SoC changing in modes 4, 5, 6, and 7

In mode 7, the EVB is charged from the grid only, which starts from t = 0 s to t = 1 s. Mode 6 starts from t = 1 s to t = 2 s when the EVB is charged directly from the EBU without the PV power while there is no requirement for grid support. Mode 5 starts from t = 2 s to t = 3 s, which shows the charging power of the EVB is taken from the grid and the PV while the EBU is assumed to have entered the low *SoC* mode. The period from t = 3 s to t = 4 s represents mode 4, where the EVB is charged from the EBU and PV power. The simulation results of the modes 1, 2, 3, 8, and 9 are shown in Figures 6.22 and 6.23.

In mode 1, the EBU is charged directly from the PV power only, which starts from t = 0 s to t = 1 s. Mode 2 starts from t = 1 s to t = 2 s when the EVB is absent or not in the operating modes, where the PV power is injected into the grid. Mode 3 starts from t = 2 s to t = 3 s, where the charging power of the EVB is drawn from PV only. The period from t = 3 s to t = 4 s represents mode 8, which is the load support mode where the grid is disconnected. All the distributed powers will support the local load. The period from t = 4 s to t = 5 s represents mode 9, which is the grid support mode, where the grid is at the peak demand and needs support. All the distributed powers will inject the power into the grid as per request.

The simulation results show that with the proposed efficient energy management of the distributed powers and grid power, the grid will not be tremendously

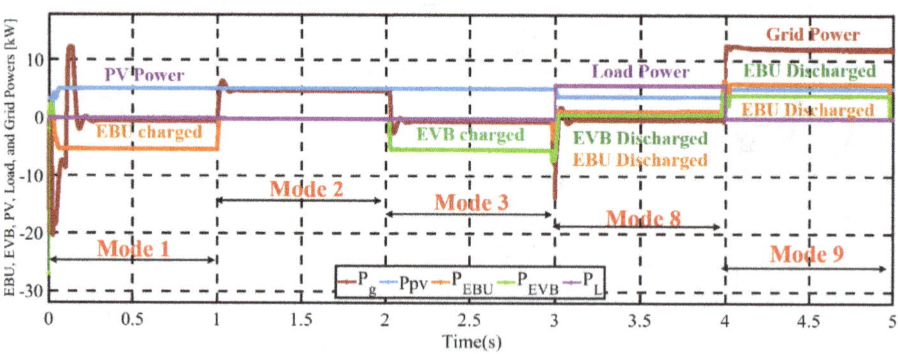

Figure 6.22 Simulation results of the transition between modes 1, 2, 3, 8, and 9

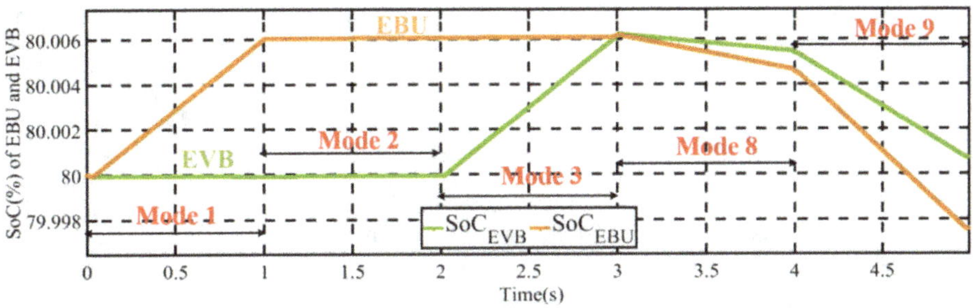

Figure 6.23 Simulation results of the SoC changing in modes 1, 2, 3, 8, and 9

affected by the EV fast-charging. The distribution grid can also handle the increasingly high penetration of EVs. The simulation results show that, in a future smart grid, where the grid is to be controlled for being smarter, the proposed energy management will be more effective as the PV power generation and load demand curves can be forecasted dynamically. This ensures that the EVB power reference can be produced more accurately.

6.3.4 Summary designing of an FCI

An FCI using solar PV and an energy storage unit to mitigate the effects of the high impact of EVs in all levels of charging from Level 1 to Level 3 of the EV charging system has been proposed. The power converters of the proposed FCI have been selected carefully, and the performance of the proposed FCI has been investigated in detail. The control system has been designed based on the operation modes, which are defined based on all possibilities of the exchanged power among the EVB, EBU, PV, and the distribution grid. The status of all sources will define each operation mode accordingly. The control of each power converter utilizes programmable switches to direct the control scheme to the corresponding mode. The simulation results of all operation modes of the overall system, showing the transition operations of the various operation modes, demonstrate the robustness of the proposed FCI.

6.4 Conclusions

In this chapter, an overview of Levels 1, 2, and 3 EV charging systems has been presented. The investigation of the power converters and charging/discharging strategies for the EV is shown. To reduce the effects of EV charging on the distribution grid, an EVC-HSP system and FCI using solar PV and an energy buffer unit have been presented. This not only helps to mitigate the effects of the potential impact of the integration of EVs in the distribution grid but also helps to mitigate the impact of high penetration of PV systems. The proposed designs of the power converters and controllers in this chapter can serve as a reference to help engineers who are working in the field of renewable energy to choose suitable configurations and power converters for their systems.

References

[1] Tran V.T., Islam M.R., Muttaqi K.M., Sutanto D. 'An on-board V2X electric vehicle charger based on amorphous alloy high-frequency magnetic-link and SiC power devices'. *IEEE Industry Applications Society Annual Meeting*. USA: Baltimore, MD; 2019. pp. 1–6.

[2] Lutsey N. *Power play: how goverments are spurring the electric vehicle industry, the inter. coun. clean transport [online]*. Available from https://theicct.org/sites/default/files/publications/EV_Government_WhitePaper_20180514.pdf [Accessed 20 Feb 2020].

[3] Clement-Nyns K., Haesen E., Driesen J. 'The impact of charging plug-in hybrid electric vehicles on a residential distribution grid'. *IEEE Transactions on Power Systems*. 2010, vol. 25(1), pp. 371–80.

[4] Qian K., Zhou C., Allan M., Yuan Y. 'Modeling of load demand due to EV battery charging in distribution systems'. *IEEE Transactions on Power Systems*. 2011, vol. 26(2), pp. 802–10.

[5] Yilmaz M., Krein P.T. 'Review of the impact of vehicle-to-grid technologies on distribution systems and utility interfaces'. *IEEE Transactions on Power Electronics*. 2013, vol. 28(12), pp. 5673–89.

[6] Dubey A., Santoso S. 'Electric vehicle charging on residential distribution systems: impacts and mitigations'. *IEEE Access*. 2015, vol. 3, pp. 1871–93.

[7] Farzin H., Moeini-Aghtaie M., Fotuhi-Firuzabad M. 'Reliability studies of distribution systems integrated with electric vehicles under battery-exchange mode'. *IEEE Transactions on Power Delivery*. 2016, vol. 31(6), pp. 2473–82.

[8] International S. *SAE's J1772 'combo connector' for ac and dc charging advances with IEEE's help [online]*. Available from http://articles.sae.org/10128/ [Accessed 20 Feb 2020].

[9] Kisacikoglu M.C., Kesler M., Tolbert L.M. 'Single-phase on-board bidirectional PEV charger for V2G reactive power operation'. *IEEE Transactions on Smart Grid*. 2015, vol. 6(2), pp. 767–75.

[10] Haoyu W., Dusmez S., Khaligh A. 'Design and analysis of a full-bridge LLC-based PEV charger optimized for wide battery voltage range'. *IEEE Transactions on Vehicular Technology*. 2014, vol. 63(4), pp. 1603–13.

[11] Musavi F., Edington M., Eberle W., Dunford W.G. 'Evaluation and efficiency comparison of front end AC-DC plug-in hybrid charger topologies'. *IEEE Transactions on Smart Grid*. 2012, vol. 3(1), pp. 413–21.

[12] Shi C., Wang H., Dusmez S., Khaligh A. 'A SiC-based high-efficiency isolated onboard PEV charger with ultrawide DC-link voltage range'. *IEEE Transactions on Industry Applications*. 2017, vol. 53(1), pp. 501–11.

[13] Restrepo M., Morris J., Kazerani M., Canizares C. 'Modeling and testing of a bidirectional smart charger for distribution system EV integration'. *IEEE transactions on smart grid.* 2016,1.

[14] Gautam D.S., Musavi F., Edington M., Eberle W., Dunford W.G. 'An automotive onboard 3.3-kW battery charger for PHEV application'. *IEEE Transactions on Vehicular Technology*. 2012, vol. 61(8), pp. 3466–74.

[15] Kesler M., Kisacikoglu M.C., Tolbert L.M. 'Vehicle-to-grid reactive power operation using plug-in electric vehicle bidirectional offboard charger'. *IEEE Transactions on Industrial Electronics*. 2014, vol. 61(12), pp. 6778–84.

[16] Zhang J., Lai J.-S., Kim R.-Y., Yu W. 'High-power density design of a soft-switching high-power bidirectional dc–dc converter'. *IEEE Transactions on Power Electronics*. 2007, vol. 22(4), pp. 1145–53.

[17] Buja G., Bertoluzzo M., Fontana C. 'Reactive power compensation capabilities of V2G-enabled electric vehicles'. *IEEE Transactions on Power Electronics*. 2017, vol. 32(12), pp. 9447–59.

[18] Tan L., Wu B., Yaramasu V., Rivera S., Guo X. 'Effective voltage balance control for bipolar-DC-bus-fed EV charging station with three-level DC–DC fast charger'. *IEEE Transactions on Industrial Electronics*. 2016, vol. 63(7), pp. 4031–41.

[19] Rivera S., Wu B. 'Electric vehicle charging station with an energy storage stage for split-DC bus voltage balancing'. *IEEE Transactions on Power Electronics*. 2017, vol. 32(3), pp. 2376–86.

[20] Cao Y., Tang S., Li C. 'An optimized EV charging model considering TOU price and SOC curve'. *IEEE Transactions on Smart Grid.* 2012, vol. 3(1), pp. 388–93.

[21] Richardson P., Flynn D., Keane A. 'Optimal charging of electric vehicles in low-voltage distribution systems'. *IEEE Transactions on Power Systems*. 2012, vol. 27(1), pp. 268–79.

[22] Deilami S., Masoum A.S., Moses P.S., Masoum M.A.S. 'Real-time coordination of plug-in electric vehicle charging in smart grids to minimize power losses and improve voltage profile'. *IEEE Transactions on Smart Grid.* 2011, vol. 2(3), pp. 456–67.

[23] Chukwu U.C., Mahajan S.M. 'V2G parking lot with PV rooftop for capacity enhancement of a distribution system'. *IEEE Transactions on Sustainable Energy*. 2014, vol. 5(1), pp. 119–27.

[24] Alam M.J.E., Muttaqi K.M., Sutanto D. 'Effective utilization of available PEV battery capacity for mitigation of solar PV impact and grid support with integrated V2G functionality'. *IEEE Transactions on Smart Grid.* 2016, vol. 7(3), pp. 1562–71.

[25] Foster J.M., Trevino G., Kuss M., Caramanis M.C. 'Plug-in electric vehicle and voltage support for distributed solar: theory and application'. *IEEE Systems Journal*. 2013, vol. 7(4), pp. 881–8.

[26] Badawy M.O., Sozer Y. 'Power flow management of a grid tied PV-battery system for electric vehicles charging'. *IEEE Transactions on Industry Applications*. 2017, vol. 53(2), pp. 1347–57.

[27] Islam M.R., Rahman M.F., Xu W. *Advances in solar photovoltaic power plants: Green Energy and Technology Series*. Germany: Springer-Verlag GmbH, Heidelberg; 2016.

[28] Abdelrazek S.A., Kamalasadan S. 'Integrated pv capacity firming and energy time shift battery energy storage management using energy-oriented optimization'. *IEEE Transactions on Industry Applications*. 2016, vol. 52(3), pp. 2607–17.

[29] Tran V.T., Islam M.R., Muttaqi K.M., Sutanto D. 'An efficient energy management approach for a solar-powered EV battery charging facility to support distribution grids'. *IEEE Transactions on Industry Applications*. 2019, vol. 55(6), pp. 6517–26.

Chapter 7

Optimal charge control strategies of EVs for enhancement of battery life and lowering the charging cost

A Zahedmanesh[1], KM Muttaqi[1], and D Sutanto[1]

Abstract

Given the significant advantages of electric vehicles (EVs), such as their high efficiency and their environmental benefits, EVs are considered as a promising alternative to conventional passenger vehicles. The incentive policies for purchasing EVs in many countries, the increasing cost reduction of the EV batteries, and the increased EV driving distance suggest a more widespread utilization of EVs in the near future.

However, there are still some concerns against the increased connections of EVs to the distribution power systems, such as the potential increase of the electric load demand and the additional power quality issues that can be produced by the converters of the charging devices. Without any coordination, there is a potential of having an unexpected peak demand if the user can charge the EV without considering the power system conditions that can cause grid congestion or reduce the grid efficiency. An effective charge/discharge control strategy can be devised to mitigate the unexpected peak demand due to EV charging, while at the same time the EV batteries can also be used to support the power grid based on vehicle-to-grid concept.

Further, it is necessary to consider the lifespan of the EV batteries, because the EV batteries are an expensive and a key part of both battery electric vehicles (BEVs) and plug-in hybrid electric vehicles (PHEVs). The additional and repetitive charging/discharging cycles required for the V2G services can affect the expected lifetime of the EV batteries. In order to account for the lifetime of the EV batteries, the degradation cost of the batteries needs to be considered for designing their energy management (EM) systems. An optimal control strategy for the enhancement of battery life and the reduction of the operational cost, for the case when the EVs are connected in power networks is therefore needed.

[1]School of Electrical, Computer and Telecommunications Engineering, University of Wollongong, Australia

In this chapter, discussions on the lifespan of the lithium-ion (Li-ion) batteries are initially provided and different methods that can be used for the battery thermal management system (BTMS) of EVs for the enhancement of battery life are introduced. In addition, various reported strategies for the optimal integration of EVs in the power systems are reviewed. Subsequently, an optimal EV charging control (EVCC) strategy is proposed for the EM of the EVs in a parking lot to lower the operational cost for the EVs, while considering the lifespan of the EV batteries. The operation of EVs in a parking lot to provide the V2G services is then investigated through different scenarios. The simulation results show that the proposed EVCC can minimize the charging cost of the EVs and provide effective V2G services, and when the degradation cost of the EVs is considered, a further cost reduction can be achieved.

7.1 Introduction

The significant challenges of conventional transportation, such as the large dependency of passenger vehicles on fossil fuels, necessitate more sustainable and safer alternative approaches for transportation. In comparison with conventional transportation systems, electric vehicles (EVs) are viable solutions, as they can enable daily commuting with adequate speed and with relatively very high efficiency. Moreover, because of the environmental superiority of EVs, they can provide a more sustainable future alongside renewable energy sources. Recent technological improvements in the manufacturing of EVs, most importantly, the technological advancement and the cost reduction in the development of EV batteries, further support the use of EVs. The application of EVs instead of fossil fuel passenger vehicles is encouraged in many countries through either legislation or incentive policies [1]. The number of electric cars on the road exceeded 5.1 million in 2018, and it is expected that the integration of EVs into the power grid will grow extensively in the near future [1].

Nevertheless, different challenges remain to be investigated for the transition to widespread application of EVs. Initially, suitable policies and incentive instruments are essential to encourage patrons for the utilization of EVs instead of the less-expensive conventional types of vehicles. In this context, the increase of the capacity of EV batteries and the deployment of recharging infrastructures for high-speed and convenient EV charging can address the anxiety of the EV owners for daily commuting. It is also necessary to maximize the monetary benefits for the stakeholders including the owners of EVs and the electricity providers by reducing the cost for EV charging and by minimizing the expenditure due to the battery degradation and EV battery replacement.

Another challenge for the widespread application of EVs is the need for additional substantial energy to supply the EV batteries from the distribution power systems. An uncoordinated integration of EVs can significantly increase the load demand, raise the power systems losses and result in the overloading of power systems components at the distribution levels. Therefore, it is essential to investigate the impacts of the EVs' connection on the distribution power systems so that the constraints of the electric grids are maintained within the required limits.

Consequently, innovative and more comprehensive strategies are required for the optimal integration of EVs in distribution grids at medium voltage (MV) and low voltage (LV) levels that consider the concerns of EV owners and distribution system operators. To this aim, the characteristics of the connected EVs as electrical loads acting as mobile energy storage systems, the uncertainties of EVs' travel times, and the constraints of the power system should be appropriately considered. Furthermore, EV owners should be persuaded to actively cooperate in adopting such coordinating measures.

Many reported research studies are concerned with control strategies for the optimal charging and discharging of EVs in power systems [2–5]. One obvious approach to lower the electricity cost and to avoid the negative impacts of the EV's integration is to charge EVs during off-peak hours based on electricity tariffs or other pricing schemes [6]. Besides the strategies for the scheduled EV charging, the concept of the vehicle-to-grid (V2G) system has been also proposed [7]. Generally, the EVs are parked most of the time; for example, in Australia, the total average of travel period for passenger vehicle fleet is 69 minutes and it can be assumed that most cars are parked and not in operation for about 95% of the time [8]. Conforming to the concept of V2G, in addition to the managed charging, some of the stored energy in the EV batteries can be delivered to the electricity grids during peak demand when the vehicles are connected to power systems, while ensuring that there is sufficient charge for transport. Consequently, EVs with V2G capabilities can provide different services including:

- shaving the peak load [9],
- improving the power quality of the distribution electricity networks [10],
- reducing the intermittency of renewable energy sources (RESs) and providing better integration of EVs and RESs [11],
- minimizing power losses [12],
- maximizing the profits of entities and reducing the charging costs to provide financial benefits [13],
- increasing the reliability and balancing the power supply and demand in the distribution power systems by providing ancillary services, namely voltage regulation, frequency regulation, spinning reserve, and operating in backup uninterruptible power supply systems during power blackouts [14].

There are still some barriers to the wide implementation of V2G systems. First, V2G systems require a bidirectional power flow between the vehicle and the grid, and consequently bidirectional battery chargers are essential. In addition to bidirectional power electronic converters, suitable control systems and monitoring infrastructure should be installed for the V2G applications. Moreover, advanced bidirectional communication systems for delivering the control commands between different entities are necessary.

Another issue is that the frequent charging and discharging cycles in the V2G system, that is required for the ancillary services, can affect the lifetime of EV batteries. While the battery packs of the EVs are initially designed for transportation, the utilization of battery packs for V2G applications, which entails constant charging/discharging

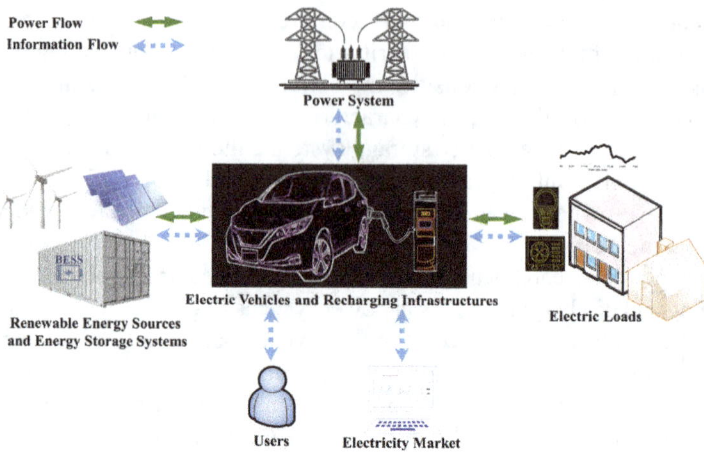

Figure 7.1 Configuration for the integration of EVs with different entities in power system

cycles, can significantly lessen the battery lifespan. Moreover, the numerous charging/discharging of EVs can degrade the lifespan of other charging facilities and vehicle components. Further, the stakeholders, such as the EV owners, need to be persuaded to participate in the V2G applications, which can be attained through economic incentives such as lowering the electricity cost to the entities. Suitable control strategies are needed to be devised for the optimal integration of EVs at distribution power systems that can allow the EV to provide V2G services and lower the charging cost, while at the same time enhance the battery life of the EVs. In Figure 7.1, the configuration for the integration of EVs with different entities in power systems is illustrated.

7.2 Integration of EVs in power systems

EVs can be classified as follows:

- Battery electric vehicles (BEVs), e.g., Nissan LEAF, Fiat 500e, Tesla Model S.
- Plug-in hybrid electric vehicle (PHEV), e.g., Ford Fusion Energi, Chevrolet Volt.
- Hybrid electric vehicle (HEV), e.g., Toyota Prius, Ford Escape Hybrid, Ford Fusion Hybrid.

In this chapter, the focus is on the vehicles, which can be plugged into power systems, and the term, EV, used in this chapter refers to the first two groups.

Many organizations have prepared standards and guidelines for the connection of EVs to the electricity grids through the charging infrastructure; among them are the SAE J1772 standard developed by the Society of Automotive Engineers (SAE) and the IEC 61851 standard developed by the International Electro-technical

Table 7.1 Electrical ratings for charging of EV based on SAE J1772

Charging method	Nominal supply voltage (V)	Maximum current (A)	Charger type
AC Level 1	120 V-AC Single-phase	12	On-board
	120 V-AC Single-phase	16	
AC Level 2	208–240 V AC Single-phase	≤80	On-board (with electric vehicle supply equipment (EVSE))
DC Level 1	50–1 000 V-DC	80	Off-board
DC Level 2	50–1 000 V-DC	400	Off-board

Commission (IEC). As this chapter deals with conductive charging, some of the main definitions in the SAE J1772 standard are presented here.

According to SAE J1772, the conductive system architecture is suitable for use with electrical ratings as specified in Table 7.1.

7.2.1 EV chargers

Based on the definition of the SAE J1772 standard, the EV charger is an electric device that converts the AC to a regulated DC energy to charge the battery pack of the EVs. The EV chargers can be categorized as follows:

- Based on the location of the charging unit, the EV charger can be (i) on-board or (ii) off-board.
- Based on the power flow, the EV chargers can be (i) unidirectional or (ii) bidirectional.

The EV on-board chargers are embedded inside the EVs for Levels 1 and 2 charging, and the off-board chargers are outside the EV, which provides the energy conversion. Due to the limitations of weight, space, and power rating, high-power DC chargers cannot be used in the vehicle as on-board chargers.

In addition to the main charger, other charging components may be utilized for AC charging, including the EVSE with its connector. For Level 2 charging using an EVSE, the AC power is transferred from the grid side to EVs through the AC lines of the connector, and the control pilot of the connector exchanges information between the EVSE and EV. Unlike the on-board chargers, the off-board chargers can provide high-charging/discharging power that can exceed 100 kW and it is possible to directly charge or discharge the EVs through the off-board chargers without the restriction of size and weight. The off-board chargers, such as DC fast charging, are mostly used for charging with a higher rating.

As indicated before, the V2G technologies require bidirectional power flow and consequently bidirectional battery chargers are required. Despite the requirement of bidirectional chargers, at present most of the charging equipment, including the on-board chargers in the EVs, can only provide unidirectional power flow, which means that it is only possible to charge EVs from the grid and not possible to send the power back to the grid. The constant current/constant voltage charging method is widely utilized for EVs [15].

Other classifications have been also considered based on the topology of the converters in the battery chargers, consequently EV chargers can be categorized as below:

- two-stage or single-stage chargers,
- isolated or non-isolated chargers.

7.2.2 EV batteries

The EV battery is the key and a high-price component of the EVs that is indispensable for running both the BEVs and PHEVs [16–19]. In automotive applications, the batteries are primarily needed to provide continuous power to the electric traction motor of the EVs. However, the EV battery capacity can be utilized in the V2G applications when the EV is parked and connected to the power grid.

The EV batteries can be categorized based on their technologies and chemical characteristics. Instead of the battery types that were originally used in automotive applications, e.g., the lead–acid and nickel–metal hydride batteries, currently, rechargeable storage systems based on Li-ion batteries are widely employed in the EVs. This is due to the outstanding features of Li-ion batteries, namely high energy density, less self-discharge rate, long cycle life, low maintenance, improved safety, and suitable cycle life [16, 20].

The main elements of the Li-ion batteries include the electrolyte, the positive and negative electrodes, and the separator, where the elements are embedded as a package with different types of battery cells, such as the cylindrical cells, the prismatic cells, and the pouch cells [21]. The Li-ion batteries in the automotive application generally utilize the cylindrical cells. For the cathodes, the LiFePO4 material is often adopted, due to its safe application and long cycle life, as well as the ternary lithium due to its capacity advantages. For the anodes, silicon metal oxides and graphite are generally adopted [22].

The cost of EV Li-ion batteries per kWh can be dependent on different factors, such as the energy density or W-H/kg rating, the lifecycle, and their performance at different temperatures [17]. The expected development of the battery pack technologies and the increasing production capacity of battery manufacturing plants show that the EV battery cost can be decreased to 105–120 $/kWh in the near term [1].

The EV batteries are constantly being discharged or charged during accelerating or regenerative energy braking modes. Additional and repetitive charging/discharging cycles of the EV battery occur when the EV batteries are used for V2G services. This influences the chemical structure and the internal parameters of the Li-ion batteries and may cause the degradation of the EV battery [16, 17]. The degradation of the battery can be indicated based on two indices. The first index is the aging of the cycle life or the decrease of the number of cycles that can be initially provided by the battery [23]. The other index is the usable energy of the battery, where a portion of the capacity of the battery will gradually fade [17, 23]. Obviously, for V2G services, in addition to the expenditures of the purchased and sold energy in the electricity markets, the economical and reliable scheduling of the EV batteries entails taking the battery degradation cost into account. For example, it may not be feasible nor economical for an EV battery with high degradation cost to be discharged to provide its stored energy to the grid, despite the high electricity prices in the market [24].

In the literature, different indices have been defined to represent the critical states of a battery; among them, the definitions of the state of charge (SOC), the state of health (SOH), and the state of energy (SOE) are mentioned in this chapter due to their importance for the optimal charge control strategies of EVs [20].

SOC: The SOC can be defined as the portion of the available capacity (Q_a) in Ah, to the nominal capacity of the battery (Q_r) in Ah. Various approaches have been introduced for the estimation of the SOC, e.g., the coulomb counting method and the open-circuit voltage (OCV)-based method [20, 25].

Based on the coulomb counting method, the SOC can be represented as in (7.1):

$$soc_t = soc_{t0} + \frac{1}{3\,600Q_r} \cdot \int_{t0}^{t} (\eta \cdot I_b) \cdot dt \tag{7.1}$$

where soc_t is the SOC value at time t (s), η is the coulombic efficiency, and η (A) is the battery current, respectively [20, 25].

SOE: The SOC can express the capacity state of the battery; however, due to different battery voltages during the discharge process, the same amount of SOC change at different voltages results in unequal energy variations [25]. Hence, the SOE can be utilized rather than the SOC to better represent the available energy stored in the EV batteries; the SOE can be mathematically presented as (7.2) [25]:

$$soe_t = soe_{t0} + \frac{1}{3\,600E_r} \cdot \int_{t0}^{t} (P_b) \cdot dt \tag{7.2}$$

where soe_t is the SOE value at time t (s), E_r is the nominal level of battery energy (Wh), and P_b (W) denotes the charging/discharging power, respectively.

SOH: Another index is the SOH, which expresses the condition of the EV battery compared to its primary and ideal conditions. The two general methods to measure the SOH of the Li-ion batteries are based on (i) the capacity fade or (ii) the internal resistance [26–28].

7.2.2.1 Battery modeling

A stringent and compressive analysis of an EV battery in V2G services necessitates the adoption of an accurate battery modeling. Authentic battery models (BMs) are initially essential to tune the parameters of the electric circuits of EV battery chargers. The BMs are also needed to estimate the battery voltage and current and determine the control variables, e.g., the charging power, for the optimal charge control strategies of the EVs. Moreover, the BMs can be adopted to investigate the battery capacity fade and lifetime and assign the degradation cost of the EV batteries, accordingly [16].

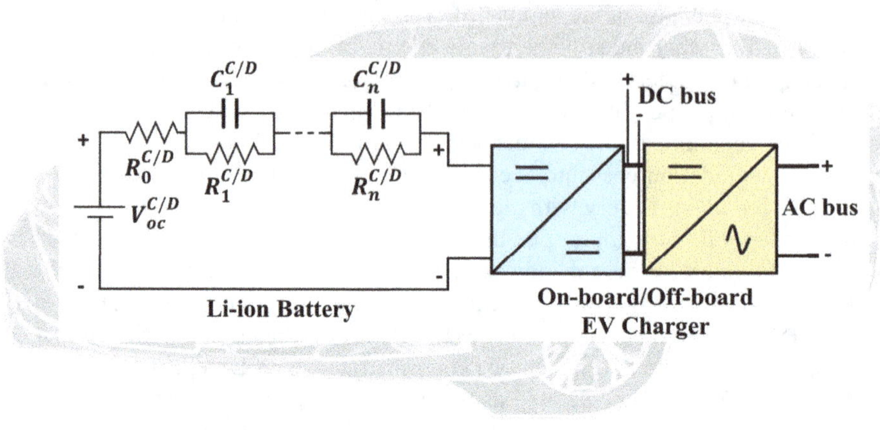

Figure 7.2 ECM for Li-ion batteries in EVs

The electric circuit model (ECM) shown in Figure 7.2 can be intuitively utilized for the dynamic and transient studies of EV batteries during the charging and discharging process. The parameters of the model in Figure 7.2 can be defined as multivariable functions of the SOC, the depth-of-discharge (DOD), the charge rate (C_r), the discharge rate (D_r) of the battery, the temperature, and the aging levels of the batteries [16, 20].

In Figure 7.2, $V_{OC}^{c/d}$ is the OCV and $R_0^{c/d}$, $R_1^{c/d}$, $C_1^{c/d}$, $R_n^{c/d}$ and, $C_n^{c/d}$ are the parameters of ECMs.

In order to extract the parameters of ECMs, different approaches such as the pulse charge/discharge tests or the iteration-based optimization techniques have been proposed [16, 20].

7.2.2.2 Battery degradation cost

The degradation of Li-ion batteries needs to be considered for the effective and reliable participation of EVs in the V2G services [29]. The degradation of batteries can notably attenuate the main operation of the BEVs for transport applications. Also, the battery degradation can increase the emissions from the PHEVs because of the additional pressure on the combustion engines of that types of the EVs. The cycling conditions and the state parameters have major effects on the degradation of the Li-ion batteries. More particularly, the energy capacity, the cycle life, and subsequently the degradation of Li-ion batteries mainly depend on the temperature and the application of battery, and also the DOD, the SOC, the C_r, and the D_r [30].

Temperature can have drastic impacts on the capacity and the cycle life of the battery [30], influencing the chemical nature of Li-ion battery. More power can be obtained from a Li-ion battery at high temperatures, which is interpreted as the increased battery capacity at high temperatures. However, when a Li-ion battery is used at high temperatures, the expected cycle life of the battery decreases significantly. The datasheet of Li-ion batteries provided by the battery manufacturers can be used to obtain the regression models that show the dependence of the battery capacity and the cycle life on the ambient temperature.

It has been shown that the cycle life of the Li-ion battery is vigorously sensitive to the value of DOD, as such the critically high values of the DOD can eventuate a lower cycle life for a battery [30]. Empirical analysis can be adopted in order to identify the non-linear relationship between the DOD and the cycle life of the Li-ion battery. Taking the cycling data of different types of Li-ion batteries into account, logarithmic or exponential models have been used for curve fitting to demonstrate the effects of the DOD on the battery cycle life [17, 30]. The extreme high or low values of the SOC and rise in the internal impedance of the Li-ion batteries can eventually result in the battery capacity fade [23]. Meanwhile, the SOC can be expressed based on the amount of the DOD, for instance, according to a definition for the DOD, one complete cycle of the battery is expressed as the discharging of the battery from (SOC = 1) to (SOC = 1−DOD) followed by charging the battery again to SOC = 1 [30].

The value of C_r that is more than 1 can diminish the energy capacity of Li-ion batteries; nonetheless, the overall effects of the C_r/D_r can be negligible [23].

The impacts of every charging and discharging cycle on the battery degradation for a Li-ion battery can be analyzed irrespective of the previous battery cycles [23]. In that case, it is plausible to account for the cost of the battery degradation in terms of the price per transferred battery energy, e.g., \$/kWh, for the optimal charge control strategies of the EVs. The dominant components of the degradation cost for Li-ion batteries can be the DOD and the temperature [23].

The cost function in (7.3) can be adapted to account for the degradation cost (C_{deg}) of the Li-ion batteries in the optimal charge/discharge control strategies [17, 19] and [30].

$$C_{deg} = 0.5 \frac{C_{bat} \cdot cap_r + C_{lab}}{cyc_r \cdot cap_r \cdot s_{tem}^{cap} \cdot s_{dod}^{cyc} \cdot s_{tem}^{cyc}} \tag{7.3}$$

where C_{bat} is the cost for purchasing the Li-ion battery of EVs (\$/kWh), cap_r is the rated battery capacity (kWh), C_{lab} is the labor cost for replacing the battery (\$), and cyc_r is the reference cycle life of the Li-ion battery, respectively.

In (7.3), the effects of the DOD on the cycle life and the effects of the ambient temperature on the cycle life and the battery capacity are regarded as s_{dod}^{cyc}, s_{tem}^{cyc}, and s_{tem}^{cap}, respectively, and can be written as [30]

$$s_{dod}^{cyc} = \frac{cyc_{dod}}{cyc_r} \tag{7.4}$$

$$s_{tem}^{cyc} = \frac{cyc_{tem}}{cyc_r} \tag{7.5}$$

$$s_{tem}^{cap} = \frac{cap_{tem}}{cap_r} \tag{7.6}$$

where cyc_{dod} demonstrates the battery cycle life as a non-linear function of the DOD, and cyc_{tem} and cap_{tem} represent the cycle life and the capacity of the battery as a function of the temperature, respectively.

As stated in (7.3), a precise estimation of the Li-ion battery degradation cost necessitates the use of a non-linear function; however, this increases the burden for the mathematical computation of cost functions. Linear regression analysis that represents the sensitivity of the battery cost to the amount of energy used during each cycle is also introduced in studies [31, 32].

7.2.2.3 BTMS for enhancement of battery life

The working environment of the EV battery can seriously affect the service life and the safety of the EVs and for the enhancement of battery life, it is necessary to reduce the negative effects of battery temperature variations and inconsistencies. The arrangement of the Li-ion battery cells in the high-power EV automotive battery cases [33] can affect the heat dissipation of battery cells. The appropriate temperature range for the EV batteries to enhance the battery life cycle is normally between 25°C and 40°C [33], where higher- or lower-temperature operation can reduce the battery life cycle. Operating the batteries in intense high temperatures, if left unmanaged, can cause overheating and can lead to undesired extreme consequences. Conversely, extreme low-temperature operation can reduce the battery capacity. Hence, the BTMS of the EV battery is essential. The BTMS is a technology to control the temperature of the battery within an optimal range considering the electro-chemical feature of the battery cells and the mechanism of heat transfer.

The BTMS of the EVs are designed to fulfill two main tasks. The first task is to reduce the likelihood of drastic changes in battery temperature and improve safety. The thermal management system should not only cause the battery to dissipate heat in high-temperature conditions but also heat the battery in low-temperature conditions, to ensure that the battery is operating at the right temperature. The second task is to reduce the temperature difference between the battery cells and maintain a uniform battery temperature. The unequal temperatures for the operation of battery cells for long periods of time can increase the capacity difference between cells. For the EV battery cells, which are connected in series, typically the capacity of some battery cells reduces gradually. The capacity of the EV battery depends on the capacity of the worst battery cell, and the safety of the battery depends on the most unstable cell. Therefore, a BTMS is essential for the sustainable operation of EV battery packs. The main methods employed in the BTMS include air, liquid, and phase change material (PCM) cooling.

BTMS using air cooling: The BTMS that is based on air cooling employs natural convection and forced convection. The natural-cooled BTMS is a passive system without any external equipment, and it depends on the battery surface radiation and heat transfer between the air and the battery. However, in the forced convection BTMS, additional equipment, such as fan or radiator, is used for the ventilation of the EV batteries, while air cooling can also be provided using the air conditioning system of the EVs or another refrigeration system.

BTMS using liquid cooling: Because of the specific heat capacity of liquids, which is greater than that of air, the heat transfer efficiency of the liquid cooling is higher and the temperature of the EV battery pack can decrease faster, where the coolant liquid can be ethylene glycol aqueous solution or hydraulic oil. However, the liquid cooling system has a more complex structure and requires a very high level of strength, which makes it more difficult to build and maintain. For example, there is a risk of fluid leakage, which can cause a short circuit in the system as well as accidents. Two types of liquid cooling are used: direct and indirect contact cooling systems. The direct contact cooling system is achieved by directly submerging the battery pack in liquid and it entails the insulation of the system. In the indirect contact type, the battery and the liquid are divided by the metal, and also the metal can be covered by thermal interface materials (TIMs) for increasing the heat transfer between the metal and the battery. In the liquid cooling systems, the sealing of the battery pack and the insulation of the system need to be considered.

BTMS using PCM cooling: Recently, the PCM technology is increasingly used for BTMS. PCM is an effective way to manage the heat transfer of the EV battery packs, because the implementation and the structure of the PCM are reasonably uncomplicated and the maintenance cost is economical. PCM is a substance that changes its phase and can dissipate or absorb energy at a specific temperature. The inherent temperature changes of the PCM make it suitable to be used for the temperature control of the battery pack in EVs. No additional energy consumption is needed in the EVs to help the BTMS using the PCM cooling system. Despite the substantial advantages, there are still some limitations such as volume changes for the use of the PCM cooling in the BTMS.

In order to improve the functionality of the BTMS for the thermal management of the EV battery pack, rather than merely adopting just one cooling technology, different cooling technologies can be combined. As an example, the combined technology of air cooling or liquid cooling with heat pipes and PCM is proposed in [33]. The advantages and drawbacks of different cooling technologies for BTMS are summarized in Table 7.2.

Table 7.2 Comparison of different cooling systems for the BTMS of battery pack

	Advantages	Drawbacks
Air cooling	Simple structure, economical cost, low weight, ventilation of hazardous gases, safety	Affected by ambient temperature, slow cooling performance
Liquid cooling	Consistent temperature control for different cells, high heat transfer capacity	High weight and cost, risk of leakage, requirement for additional equipment, e.g., pump
PCM cooling	Uniform temperature control for battery cells, relatively simple structure and implementation, low maintenance cost	Low thermal conductivity, poor fluidity, volume changes

7.3 Charge/discharge control strategies of EVs

Unlike some other non-responsive loads, nearly all EVs can operate as controllable loads [34, 35]. In addition, for the case of EVs with bidirectional charging equipment, the stored energy in EVs can be used for providing ancillary services. EVs can support the following services in V2G applications:

1. coordination of the charging power rate (e.g., for valley filling),
2. reactive power compensation (e.g., for voltage regulation), and
3. provision of active power (e.g., for peak shaving).

It is worthy to note that the first two tasks can be also achieved using unidirectional charging equipment with limited modifications. The first service can be interpreted as smart charging, for example, the charging of EVs can be implemented during off-peak times, while conversely during peak times smart charging may be achieved by peak shaving. EVs can also support reactive power compensation considering that the converters in the EV recharging equipment are capable of reactive power control (RPC) at the connection points to the grids. The RPC is a feasible service for the EV charging equipment since it does not intervene with the primary operation of the recharging infrastructures for replenishing the EVs. The RPC can be also less complex than the active power management because it can be provided and managed, independently for each time-step of the horizon for optimal charging strategies. In addition, both the first two tasks, including the charging management and the reactive power compensation, can be attained without significantly influencing the EV batteries' charging and also their degradation. Nevertheless, the monetary benefits from these tasks are limited.

Generally, in the simplest form, a pulse width-modulated (PWM) rectifier followed by a bidirectional DC–DC converter can be used for providing active power in V2G applications. With such a topology, the charger can regulate the EVs' charging/discharging power rate during the plug-in times of the EVs.

The adoption of all these three functionalities for EVs necessitates the application of suitable control strategies for charging/discharging of EVs in the distribution electric systems. These control strategies pursue different objectives, where the control objectives can be (i) technical to achieve voltage and frequency regulation, peak load shaving and valley filling, mitigating the intermittency of renewables, reducing energy losses, etc., (ii) economic to achieve charging cost reduction, reduction of fossil fuel consumption, aggregator profit maximization, maximizing the satisfaction of EV owners, etc., or (iii) a combination of both of them.

7.3.1 Configuration for the optimal charging/discharging strategies of EVs

The optimal charge or discharge of EVs can be achieved using different structures or hierarchies for control systems, namely (1) centralized, (2) decentralized, and (3) distributed.

Centralized control: In the centralized control method, a central entity (e.g. the EV fleet operator) programs the optimal charging/discharging schedule and all individual EVs adopt the coordination commands, considering the technical and economic constraints of the systems and the preferences of EV owners. Therefore, in this approach, the structure of decision-making is straightforward and if full knowledge of the system constraints is known, then the optimal charging/discharging of EVs and the economic and technical objectives can be achieved with minimum concerns. Consequently, many research works have used the centralized control method [36]. Nevertheless, the effective implementation of the centralized method entails the complete cognition of EVs characteristics, including their technical specifications such as the charging/discharging rates, the capacity of EV batteries, the ratings of the EV chargers, and the arrival/departure time of EVs. Moreover, the constraints of power systems, such as the topology of distribution networks and the load demand of other loads should be determined by the central control unit. Also, when the cost optimization objectives are taken into consideration, probabilistic approaches should be adopted to cope with the dynamic electricity prices or tariff zones. Consequently, centralized control methods require extra computational intensity and additional communication systems and this issue restricts the merits of these control methods.

Decentralized control: In addition to the centralized method, decentralized control methods are widely adopted in previous literature. In the decentralized control method, the computational intelligence is distributed between each individual EV or EV aggregator, and the decision-making is executed by EVs locally without any global awareness of other units' actions. Based on previous studies, two decentralized control approaches can be adopted for EVs, namely (1) the optimized control and (2) the droop control [37].

In the optimized control method, individual EV coordinates its charge/discharge rate according to the commands received from other entities, such as EV aggregator or the distribution system operator (DSO). Hence, this technique still needs the exchange of information regarding electricity prices, EVs trips, or grid constraints between the DSO or the EV aggregator and the EVs. In the droop control method, EVs will modify the charging or discharging schedule according to the local measurements, for example, EVs can be utilized for voltage or frequency regulation only based on local measurement without communication systems.

Distributed control: The distributed method can be interpreted as a hierarchical decentralized control or a connective control between the decentralized and centralized methods. In this architecture no central control unit is utilized but intermediate layers exist in the control structure and different entities participate in the control process. As a result, the decision-making is implemented between different intelligent entities according to the hierarchy of the control system, and accordingly the charge/discharge of EVs is coordinated. This structure reduces the data exchange between aggregation entities and EVs and improves the simplicity and scalability of the system. A widely adopted approach for distributed coordination is the multi-agent system (MAS). In MAS, to reduce the complexity of control problems, the autonomous entities (agents) undertake subtasks while they cooperate with each other in an environment.

In addition to the structure of the control systems, the strategies for the integration of EVs in distribution power networks can be classified based on several aspects, for example, the chronology for scheduling the EVs' integration in power systems. In this context, the reported approaches on optimal charging/discharging control of EVs can be categorized as (i) single-level method, e.g., day ahead or real-time schedule of EVs or (ii) multi-level method, e.g., two-level schedule of EVs. In multi-level methods, the integration of the EVs is scheduled or modified at different chronological levels, while the coordination measure can pursue similar or independent control objectives at every temporal level. Subsequently, the hierarchy, the constraints, or the length of temporal intervals would be specified based on the control objectives of multi-level methods.

In order to solve the optimization problems of charging/discharging control strategies of EVs, considering the objective functions and the constraints of the optimization, different mathematical techniques have been adopted in previous works, which can include linear programming, e.g., mixed-integer linear programming (MILP), non-linear programming, e.g., quadratic programming or other mathematical methods, e.g., game theory, and also population-based metaheuristic optimization methods, e.g., genetic algorithm or particle swarm optimization (PSO).

7.3.2 Development of the analytical models of EVs

Currently, the economic and technical impacts of the integration of EVs in power systems can be negligible. Nevertheless, the integration of EVs with high penetration levels has been foreseen for the near future; thus, comprehensive models including acquisitive models of EVs, analytical models for the economic participation of the EVs in restructured electricity markets, and test frameworks of MV and LV systems need to be developed. To this aim, the following requirements should be fulfilled.

The electric models of EVs should pervasively illustrate their electric characteristics (which is considered unconventional), namely

- the power consumption behavior of EVs while connected through charging equipment.
- the power flow dynamics of charging equipment (which can be unidirectional/bidirectional).
- the capacity of EVs' batteries and rate of change of the SOC while charging/discharging.
- the upward and downward acceptable charging/discharging power at different SOC levels (these limits will be derived based on the former characteristics or historical data).

Furthermore, in order to illustrate the availability and variability of EVs, the following features should be considered:

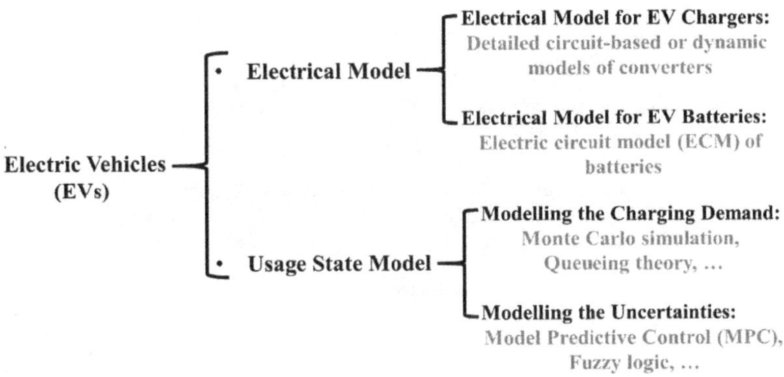

Figure 7.3 *Framework for the modeling and the analysis of the optimal integration of EVs in a power system*

- the connection/disconnection time;
- the frequency of use; and
- the initial and the desired SOC levels.

Comprehensive and detailed models should be also developed for power grids where these models should provide the capability of systems under different conditions, for instance, during different states (transients, short and long duration), and for different topologies (e.g., unsymmetrical or unbalanced three-phase grids). Besides, the developed models are required to be capable of providing the study of the reciprocal impacts of different electrical elements in distribution power systems. The developed models should also be capable of illustrating the economic operation of the entities in power systems. The developed simulation frameworks are subsequently used for (1) studying the impacts of integration of EVs in power systems and also (2) verifying the effectiveness of proposed solutions for the optimal integration of EVs in MV and LV distribution electric systems.

In Figure 7.3, the framework for the modeling and the analysis of the optimal integration of EVs in a power system is depicted.

7.4 Optimal control strategy for integration of EVs to enhance battery life and lower the charging cost

The constraints of distribution systems and the concerns of the EV owners for the integration of EVs can be realized by utilizing the concepts of the optimal economic operation of the power systems. In this context, lowering the operational cost for running EVs, which includes both the charging cost and the expenditure of the EV battery, can further support the prospect for the extensive application of EVs. The schematic diagram for the optimal charging/discharging control system of

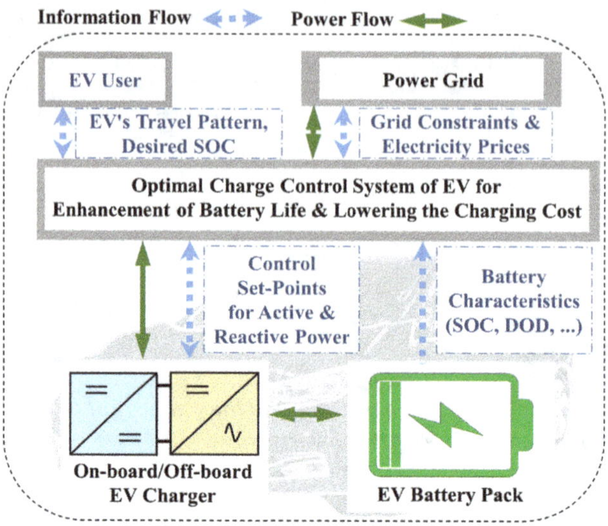

Figure 7.4 Schematic diagram for optimal charge control system of EV for enhancement of battery life and lowering the charging cost

EVs for the enhancement of battery life and lowering the charging cost is shown in Figure 7.4.

7.4.1 Optimal EV charging control strategy

In this chapter, an optimal EV charging control (EVCC) strategy is proposed in order to lower the charging cost for the EVs in a parking lot. The structure for the EVs' integration in the parking lot is shown in Figure 7.5. The parking lot in Figure 7.5 is in a workplace area, where the EV owners generally arrive in the morning and connect the EVs for charging upon their arrival to the workplace. In order to provide V2G services, bidirectional converters are considered for the EV recharging facilities. As shown in Figure 7.5, solar photovoltaic (PV) panels are also integrated to supply electrical energy to the workplace area. In this case, EVs can provide flexibile services to facilitate the integration of PV solar systems, so that an EV battery can operate as a mobile storage system to mitigate the variability of the energy produced by the PV. Meanwhile, in that case the energy to run the EVs can be mainly provided by environmentally friendly energy resources.

It can be regarded that the whole system shown in Figure 7.5 constitutes a distributed power plant (DPP) that is scheduled by a DPP controller. The main task of the DPP controller is the energy management (EM) of the DPP or to organize the participation of the DPP in the electricity market and to supply the power demand of the EVs and the other electric consumers, namely the electric appliances in the buildings. All the indices, sets, parameters, and variables of the EVCC are presented in Table 7.3. The configuration for the power transfer and the information exchange for the DPP is shown in Figure 7.6.

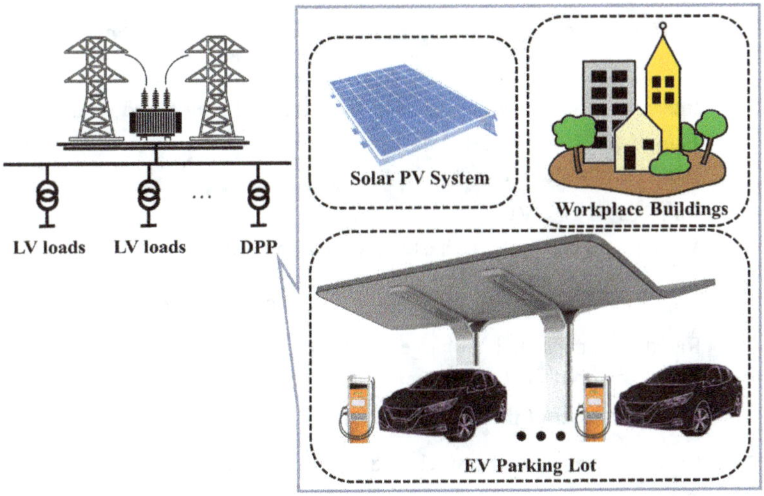

Figure 7.5 Structure for the integration of EVs in the DPP

The cost of the electricity can be generally defined as the quadratic function in (7.7).

$$O_1 = \sum_{\kappa \in k} \xi_{\kappa}^{\alpha} \cdot \left(E_{\kappa}^{tot}\right)^2 + \xi_{\kappa}^{\beta} \cdot \left(E_{\kappa}^{tot}\right) + \xi_{\kappa}^{\gamma} \tag{7.7}$$

$$k = \begin{bmatrix} \kappa & \kappa+1 & \ldots & |k|-1 & |k| \end{bmatrix} \tag{7.8}$$

It is also essential to account the degradation cost of the EV batteries as given in (7.9).

$$O_2 = \sum_{\kappa \in k} \sum_{v \in V} C_{v,\kappa}^{deg} \cdot P \cdot v, \kappa|\kappa| \tag{7.9}$$

$$P_{v,\kappa} = \eta_c \cdot l_{v,\kappa} \cdot P_{v,\kappa}^{c} - \frac{1}{\eta_d} \cdot \left(1 - l_{v,\kappa}\right) \cdot P_{v,\kappa}^{d} \tag{7.10}$$

where in (7.9), $C_{v,\kappa}^{deg}$ is the degradation cost of the Li-ion battery in EV v at time-step κ, as previously introduced in (7.3).

The utility function in (7.11) is also defined to account the convenience of the EV owners by hastening the charging process of the EVs.

$$O_3 = \sum_{\kappa \in k} \sum_{v \in V} \left(\frac{k_v}{2}\right) \cdot \left(S_v^{nom} - P_{v,\kappa}^{c}\right)^2 \tag{7.11}$$

The multi-objective function for the EVCC is defined as (7.12).

$$O = min\left(\varpi_1 \cdot O_1 + \varpi_2 \cdot O_2 + \varpi_3 \cdot O_3\right) \tag{7.12}$$

Subject to:

$$E_{\kappa}^{tot} = P_{\kappa}^{tot} \cdot |\kappa|, \forall \kappa \in k \tag{7.13}$$

Table 7.3 Indices, sets, parameters, and variables defined for the EVCC

$\kappa, \lvert\kappa\rvert, \kappa$	Index, length, and set for the time-steps for a 12-hour scheduling of the DPP
$\lvert k\rvert$	Length of the optimization horizon
v, V	Index, set for the EVs
soc_v^{ini}	Initial SOC for EVv
k_v	Willingness parameter to hasten the charging process for EVv
N_v	The number of chargers in the parking lot
$\eta_{c/d}$	Efficiency of the EV chargers for charging/discharging
E_κ^{tot}	Rated power for the EV chargers in the parking lot (kVA)
N_v	Rated power of the transformer of the DPP (kVA)
E_κ^{tot}	Energy transfer of the DPP at time-step κ (kWh)
$\xi_\kappa^{\alpha,\beta,\gamma}$	Cost parameters for the purchased energy in the electricity market
P_κ^{tot}	Total power transfer of the DPP at time-step κ (kW)
$soc_{v,\kappa}$	SOC for EV $P_{v,\kappa}^{c/d}$ at time-stepκ
$P_{v,\kappa}^{c/d}$	Charging/discharging power of EV v at time-step κ (kW)
cap_v^{nom}	Nominal battery capacity of EV v (kWh)
P_κ^{dem}	Purchased, sold power of the DPP in the electricity market (kW)
P_κ^{dem}	Power demand of the DPP at time-step κ (kW)
P_t^{pv}	Generated power of the solar PV panels at time-step $l_{v,\kappa}$ (kW)
$l_{v,\kappa}$	Binary variable to avoid the charging/discharging of EV v at the same time
l_κ^g	Binary variable to avoid the purchasing/selling energy at the same time
$C_{v,\kappa}^{deg}$	Degradation cost of the Li-ion batteries of EV v at time κ (\$)

$$P_\kappa^{tot} = \left(\sum_{v\in V} P_{v,\kappa}\right) + P_\kappa^{dem} - P_\kappa^{pv}, \forall\kappa \in k \tag{7.14}$$

$$P_\kappa^{tot} + P_\kappa^{buy} - P_\kappa^{sell} = 0, \forall\kappa \in k \tag{7.15}$$

$$0 \le |P_\kappa^{tot}| \le S_{tra}^{nom}, \forall v \in V, \forall\kappa \in k \tag{7.16}$$

$$soc_{v,\kappa+1} = soc_{v,\kappa} + \frac{1}{cap_v^{nom}} \cdot |\kappa| \cdot P_{v,\kappa}, \forall v \in V, \forall\kappa \in k \tag{7.17}$$

$$dod_{v,\kappa} = 1 - soc_{v,\kappa}, \forall v \in V, \forall\kappa \in k \tag{7.18}$$

$$soc_{v,\kappa=\kappa_v^{dep}} = 1, \forall v \in V \tag{7.19}$$

$$0 \le |P_{v,\kappa}| \le S_v^{nom}, \forall v \in V, \forall\kappa \in k \tag{7.20}$$

The energy transfer of the DPP as given in (7.13) and (7.14) includes the charging demand of the EVs, the load demand of the workplace buildings, and the produced energy by the PV system. The constraint in (7.15) is to balance the energy transfer of the DPP with the purchased and sold energy in the electricity market. The total

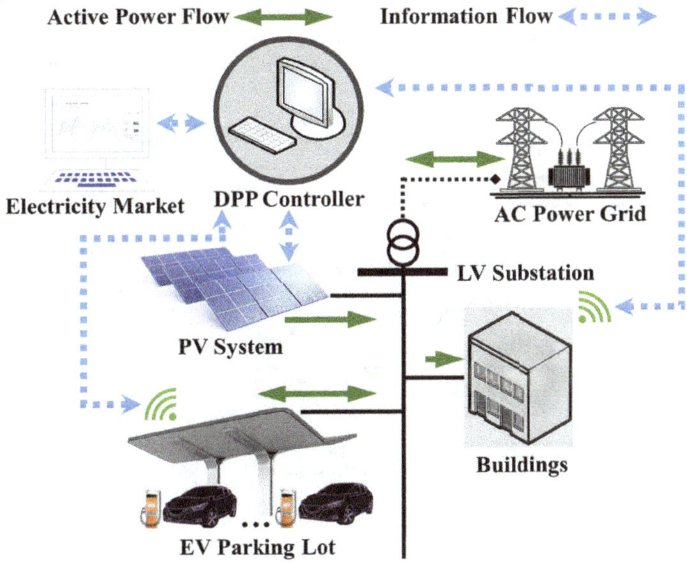

Figure 7.6 Configuration for the power flow and the information exchange for the DPP

power transfer of the DPP and the power transfer of the EV chargers should not exceed their permissible bounds as presented in (7.16) and (7.20). The variation of the SOC and the DOD of the EVs can be expressed as (7.17) and (7.18), while it is essential to fulfill the required SOC of the EVs before the departure time of the EVs, κ_v^{dep}, as given in (7.19).

7.4.2 Simulation results and discussions

Three cases are considered in order to assess the effectiveness of the proposed EVCC to schedule the EM of the DPP, namely (i) the *base* case, where the proposed EVCC is not applied for the EM of the DPP, (ii) case 1, where the proposed EVCC is applied for the EM of the DPP, without considering the EV batteries' degradation cost in (7.9), and (iii) case 2, where the proposed EVCC is applied for the EM of the DPP and the EV batteries' degradation cost is also taken into account.

The charging demand and the connection times or equivalently the plug-in/plug-off times of the EVs are modeled using the approach developed in [34], where the nominal battery capacity for all the EVs in the parking lot is determined as 33 kWh. In total, 18 EV chargers are regarded to be supplied in the parking lot. The rated power of the transformer for supplying the DPP, the maximum generated power of the solar PV panels, the rated power, and the efficiency for charging/discharging using the EV chargers are 800 kVA, 250 kW, 7.2 kW, and 0.9, respectively. The cost for purchasing the Li-ion battery of the EVs is assigned as 120 \$/kWh, the cost parameters in (7.7) are extracted from the data available by the electricity market operators in Australia [38].

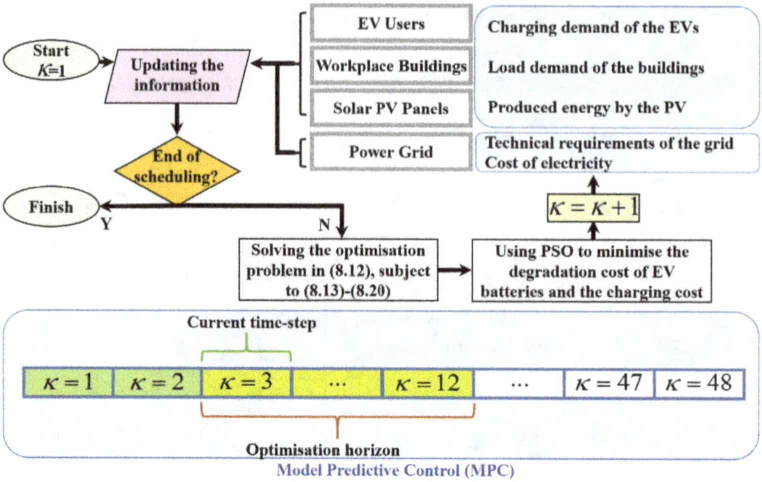

Figure 7.7 Flowchart for solving the optimization problem

In the case that the battery management system of the battery has the temperature controller or the BTMS, the effects of temperature can be omitted [20]. Therefore, in (7.9) only the impact of the DOD is taken into consideration, where a convex function is extracted for the relationship between the DOD of the EV batteries and the cycle life [19].

In order to mitigate the forecast errors of the PV generation and the cost parameters, model predictive control (MPC) is used [30]. The length of each time-step and length of the optimization horizon for the MPC are assigned as 30 minutes and 5 hours, respectively. Based on the concept of the MPC, at the beginning of each time-step, the DPP controller assigns the purchased/sold energy for the optimization horizon. While, at each time-step, an updated forecast of the produced power of the PV system and the cost parameters are available based on the historical data for the next optimization horizon. By using this approach, the charging/discharging of the EVs can be controlled according to the real-time data for time-step κ and the forecasted data for the optimization horizon. The multi-objective optimization problem in (7.12) was solved using PSO. The flow chart for describing the steps involved in solving the optimization problem in (7.12) is shown in Figure 7.7.

In Figure 7.8, the generated power of the PV system and the load demand of the workplace buildings for a sunny day in February are shown. The load modeling technique described in [10] and the actual load demand from commercial areas in the Sydney region are considered to replicate the load demand of DPP [38]. The produced energy of the PV system at different times is modeled based on the method discussed in [10], while the maximum solar irradiance level is derived based on the statistical data and the expected capacity of PV generation in Australia [39].

Figure 7.9 shows the power transfer of the parking lot in the DPP during a 12-hour period that begins at 5:00, for the *base* case, case 1, and case 2.

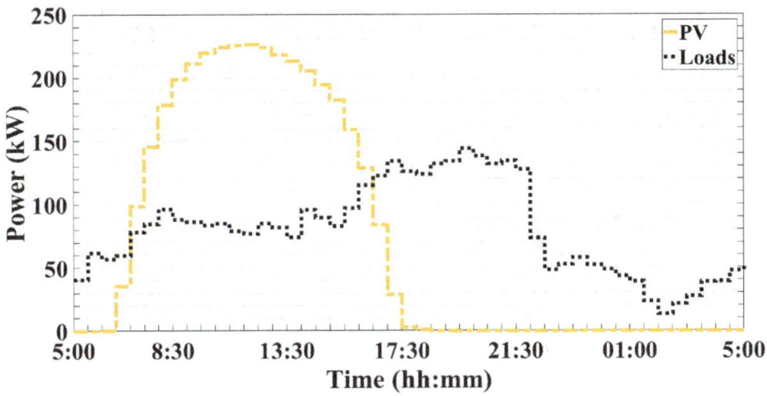

Figure 7.8 *Generated power of the PV system and the load demand of the workplace buildings*

In Figure 7.9, the EVs are only charged upon the connection to the EV chargers irrespective of the cost of energy; on the contrary, in case 2, besides the managed charging, the stored energy in the EV batteries is transferred to the power grid in order to maximize the revenue for the DPP controller, while the charging load to provide the required SOC value at the departure times of the EVs is mainly shifted to the late working hours. Moreover, in Figure 7.9, an entirely different profile has been obtained, when the battery degradation cost is considered for case 2.

Particularly, when the degradation of the EV batteries is regarded, it can cause less deep charging and limited numbers of charging/discharging cycles for the EV batteries, as presented for case 2 in Figure 7.9.

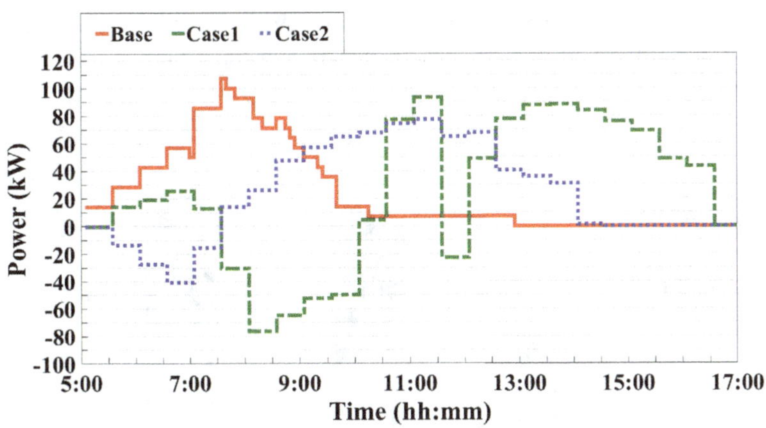

Figure 7.9 *Total power transfer of the parking lot in the DPP during a 12-hour period for the base case, case 1, and case 2*

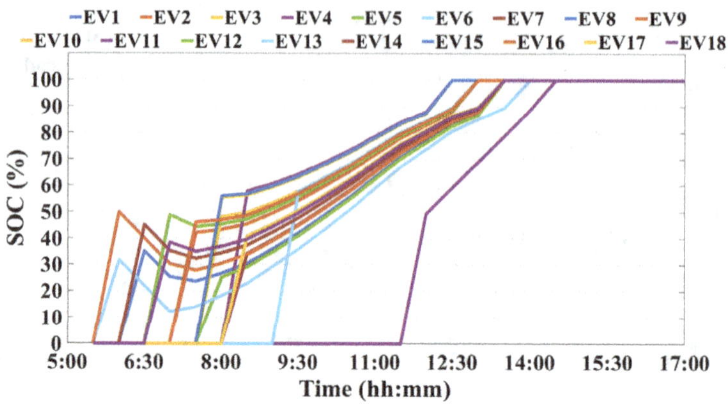

Figure 7.10 The change of the SOC values of the connected EVs for case 2

In Figure 7.10, the change of the SOC values for all the connected EVs is illustrated for case 2.

For case 2, EVs are discharged during the early times in the morning, when the PV produced power is lower, as depicted in Figure 7.10. Subsequently, the EVs are gradually charged in order to fulfill their required charging demand.

Figure 7.11 shows the total profit of the DPP controller subsequent to the participation of the DPP in the wholesale electricity market. Furthermore, the total cost for the integration of the EVs in the parking lot that includes the cost of transferred energy for fully charging the EVs and the degradation cost of the EV batteries is given in Figure 7.12.

In Figure 7.11, the profit of the VPP controller is maximum when the proposed EVCC is applied to the DPP, while the profit for the participation of the DPP in the wholesale electricity market is even higher for case 2, when the degradation cost of the EVs is considered.

Similarly, as expected the charging cost of the EVs is much lower for both, case 1 and case 2, when the EVCC is applied for the EM of the DPP, while both the cost

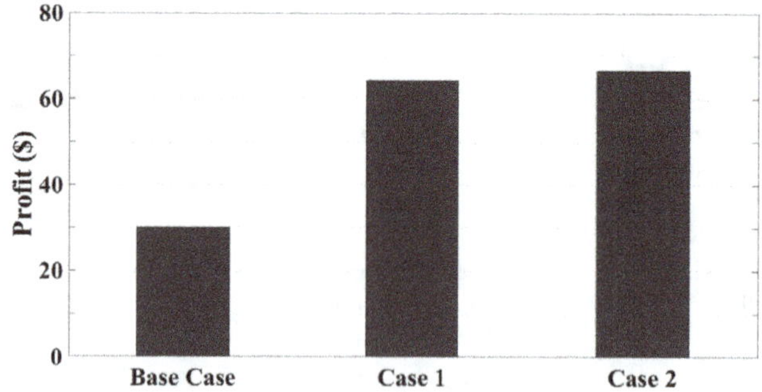

Figure 7.11 The total profit of the DPP controller

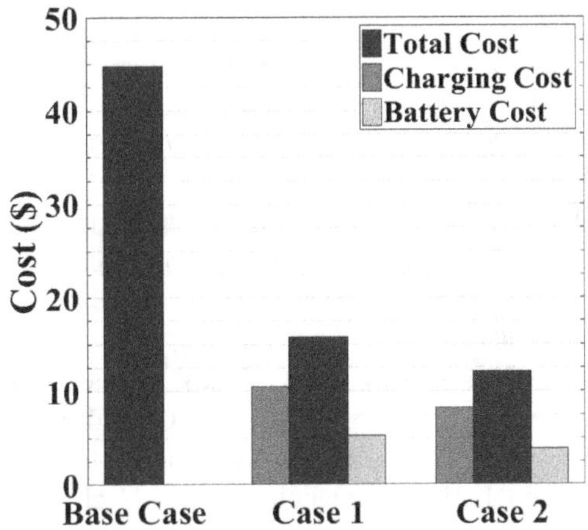

Figure 7.12 The total charging cost of the EVs in the parking lot

of purchased energy to charge the EVs and the degradation cost of the EVs are less, when the degradation cost is considered for the EM of the DPP in case 2, as depicted in Figure 7.12.

Clearly, the minimization of the cost function in (7.9) can result in less degradation of the EV batteries. As a result, the proposed EVCC strategy can lower the charging cost and enhance the lifespan of the EV batteries.

7.5 Conclusion

Suitable strategies for the optimal operation of the EVs when connected to the grids are required, not only to mitigate the negative impacts of the mass application of EVs on power networks but also to utilize them to provide ancillary services using the concept of V2G. Such strategies need to consider the concerns of the EV owners and also the electric and economic requirements of the power systems. Since the EV battery system is an expensive and substantial part of both BEVs and PHEVs, it is also necessary to account for the lifespan of the EV batteries.

In this chapter, the requirements for the optimal control strategies for the enhancement of EV battery life and the lowering of the operational cost when integrating EVs in power systems are first reviewed. Strategies for the optimal integration of EVs in power systems and discussions regarding the lifespan of the EV batteries are then presented. Moreover, an optimal EVCC strategy is proposed for the EM of a DPP comprising a parking lot, a workplace building, and a solar PV system. The operation of EVs in the DPP has been investigated for different case studies. The simulation results have clearly shown that the proposed EVCC can minimize the charging cost of the EVs and the degradation cost of the EV batteries.

References

[1] Outlook G.E.V 2019. *Scaling Up the Transition to Electric Mobility [online].* 2019. Available from https://www.iea.org/reports/global-ev-outlook-2019 [Accessed January 2020].

[2] Rahmani-Andebili M., Fotuhi-Firuzabad M. 'An adaptive approach for PEVs charging management and reconfiguration of electrical distribution system penetrated by renewables'. *IEEE Transactions on Industrial Informatics.* 2018, vol. 14(5) 2001–10.

[3] Chandra Mouli G.R., Kefayati M., Baldick R., Bauer P. 'Integrated pv charging of eV Fleet based on energy prices, V2G, and offer of reserves'. *IEEE Transactions on Smart Grid.* 2017, vol. 10(2), pp. 1313–25.

[4] Weckx S., Driesen J. 'Load balancing with EV chargers and PV inverters in unbalanced distribution grids'. *IEEE Transactions on Sustainable Energy.* 2015, vol. 6(2) 635–43.

[5] Wang S., Bi S., Zhang Y.-J.A., Huang J. 'Electrical vehicle charging station profit maximization: admission, pricing, and online scheduling'. *IEEE Transactions on Sustainable Energy.* 2018, vol. 9(4), pp. 1722–31.

[6] Deng R., Yang Z., Chow M.-Y., Chen J. 'A survey on demand response in smart grids: mathematical models and approaches'. *IEEE Transactions on Industrial Informatics.* 2015, vol. 11(3), pp. 570–82.

[7] Ortega-Vazquez M.A. 'Optimal scheduling of electric vehicle charging and vehicle-to-grid services at household level including battery degradation and price uncertainty'. *IET Generation, Transmission & Distribution.* 2014, vol. 8(6), pp. 1007–16.

[8] Ustan T.S., Zayegh A., Ozansoy C. 'Electric vehicle potential in Australia: its impact on smart grids'. *IEEE Industrial Electronics Magazine.* 2013, vol. 60(4), pp. 15–25.

[9] Wang Z., Wang S. 'Grid power peak shaving and Valley filling using vehicle-to-grid systems'. *IEEE Transactions on Power Delivery.* 2013, vol. 28(3), pp. 1822–9.

[10] Zahedmanesh A., Sutanto D., Muttaqi K.M. 'Analyzing the impacts of charging plug-in electric vehicles in low voltage distribution networks: A case study of utilization of droop charging control system based on the SAE J1772 Standard'. 2017 Australasian Universities Power Engineering Conference (AUPEC); Melbourne, VIC; 2017. pp. 1–6.

[11] Zahedmanesh A., Muttaqi K.M., Sutanto D. 'Alleviation of voltage variations introduced by unbalanced allocation of single-phase loads in a distribution network integrated with PVs and PEVs'. 2020 IEEE International Conference on Power Electronics, Smart Grid and Renewable Energy (PESGRE2020); Cochin, India; 2020. pp. 1–6.

[12] Masoum A.S., Deilami S., Moses P.S., Masoum M.A.S., Abu-Siada A. 'Smart load management of plug-in electric vehicles in distribution and residential networks with charging stations for peak shaving and loss minimisation

considering voltage regulation'. *IET Generation Transmission & Distribution.* 2015, vol. 5(8), pp. 877–88.

[13] Yan Q., Zhang B., Kezunovic M. 'Optimized operational cost reduction for an EV charging station integrated with battery energy storage and PV generation'. *IEEE Transactions on Smart Grid.* 2019, vol. 10(2), pp. 2096–106.

[14] Chandra Mouli G.R., Kefayati M., Baldick R., Bauer P. 'Integrated PV charging of EV fleet based on energy prices, V2G, and offer of reserves'. *IEEE Transactions on Smart Grid.* 2019, vol. 10(2), pp. 1313–25.

[15] Haidar A.M.A., Muttaqi K.M. 'Behavioral characterization of electric vehicle charging loads in a distribution power grid through modeling of battery chargers'. *IEEE Transactions on Industry Applications.* 2016, vol. 52(1), pp. 483–92.

[16] Thirugnanam K., E. R. J. T. P., Singh M., Kumar P. 'Mathematical modeling of Li-ion battery using genetic algorithm approach for V2G applications'. *IEEE Transactions on Energy Conversion.* 2014, vol. 29(2), pp. 332–43.

[17] David A.O., Al-Anbagi I. 'EVs for frequency regulation: cost benefit analysis in a smart grid environment'. *IET Electrical Systems in Transportation.* 2017, vol. 7(4), pp. 310–17.

[18] Hoke A., Brissette A., Smith K., Pratt A., Maksimovic D. 'Accounting for Lithium-ion battery degradation in electric vehicle charging optimization'. *IEEE Journal of Emerging and Selected Topics in Power Electronics.* 2014, vol. 2(3), pp. 691–700.

[19] Ginigeme K., Wang Z. 'Distributed optimal vehicle-to-grid approaches with consideration of battery degradation cost under real-time pricing'. *IEEE Access.* 2020, vol. 8, pp. 5225–35.

[20] Li K., Wei F., Tseng K.J., Soong B.-H. 'A practical lithium-ion battery model for state of energy and voltage responses prediction incorporating temperature and ageing effects'. *IEEE Transactions on Industrial Electronics.* 2018, vol. 65(8), pp. 6696–708.

[21] Zwicker M.F.R., Moghadam M., Zhang W., Nielsen C.V. 'Automotive battery pack manufacturing – a review of battery to TAB joining'. *Journal of Advanced Joining Processes.* 2020, vol. 1(1–3), p. 100017.

[22] Madani S., Swierczynski M., Kear S. 'A review of thermal management and safety for lithium ion batteries'. 2017 Twelfth International Conference on Ecological Vehicles and Renewable Energies (EVER); Monte Carlo; 2017. pp. 1–20.

[23] Ju C., Wang P., Goel L., Xu Y. 'A two-layer energy management system for microgrids with hybrid energy storage considering degradation costs'. *IEEE Transactions on Smart Grid.* 2018, vol. 9(6), pp. 6047–57.

[24] Ortega-Vazquez M.A. 'Optimal scheduling of electric vehicle charging and vehicle-to-grid services at household level including battery degradation and price uncertainty'. *IET Generation, Transmission & Distribution.* 2014, vol. 8(6), pp. 1007–16.

[25] Hu X., Feng F., Liu K., Zhang L., Xie J., Liu B. 'State estimation for advanced battery management: key challenges and future trends'. *Renewable and Sustainable Energy Reviews.* 2019, vol. 114, p. 109334.

[26] Sarrafan K., Sutanto D., Muttaqi K.M., Town G. 'Accurate range estimation for an electric vehicle including changing environmental conditions and traction system efficiency'. *IET Electrical Systems in Transportation*. 2017, vol. 7(2), pp. 117–24.

[27] Sarrafan K., Sutanto D., Muttaqi K.M. 'An electric circuit based EV battery model for runtime prediction and state of charge tracking'. 2017 IEEE Transportation Electrification Conference (ITEC-India); Pune; 2017. pp. 1–6.

[28] Sarrafan K., Muttaqi K.M., Sutanto D. 'Real-time state-of-charge tracking embedded in the advanced driver assistance system of electric vehicles'. *IEEE Transactions on Intelligent Vehicles*. 2020, vol. 5(3), pp. 497–507.

[29] De Gennaro M., Paffumi E., Martini G., Giallonardo A., Pedroso S., Loiselle-Lapointe A. 'A case study to predict the capacity fade of the battery of electrified vehicles in real-world use conditions'. *Case Studies on Transport Policy*. 2020, vol. 8(2), pp. 517–34.

[30] Zia M.F., Elbouchikhi E., Benbouzid M. 'Optimal operational planning of scalable DC microgrid with demand response, islanding, and battery degradation cost considerations'. *Applied Energy*. 2019, vol. 237(2), pp. 695–707.

[31] Zahedmanesh A., Muttaqi K.M., Sutanto D. 'A sequential decision-making process for optimal technoeconomic operation of a grid-connected electrical traction substation integrated with solar PV and BESS'. *IEEE Transactions on Industrial Electronics*. 2020, vol. 68(2), pp. 1353–64.

[32] Sarker M.R., Pandžić H., Sun K., Ortega-Vazquez M.A. 'Optimal operation of aggregated electric vehicle charging stations coupled with energy storage'. *IET Generation, Transmission & Distribution*. 2018, vol. 12(3), pp. 1127–36.

[33] Liu H. 'Thermal management system design for densely-packed EV battery'. *M.Sc. thesis, School of Electrical, Computer and Telecommunications Engineering*. University of Wollongong; 2020.

[34] Zahedmanesh A., Muttaqi K.M., Sutanto D. 'Direct control of plug-in electric vehicle charging load using an in-house developed intermediate control unit'. 2018 IEEE Industry Applications Society Annual Meeting (IAS); Portland, OR; 2018. pp. 1–9.

[35] Zahedmanesh A., Muttaqi K.M., Sutanto D. 'Direct control of plug-in electric vehicle charging load using an in-house developed intermediate control unit'. *IEEE Transactions on Industry Applications*. 2019, vol. 55(3), pp. 2208–18.

[36] Zahedmanesh A., Muttaqi K.M., Sutanto D. 'A consecutive energy management approach for a VPP comprising commercial loads and electric vehicle parking lots integrated with solar PV units and energy storage systems'. 2019 1st Global Power, Energy and Communication Conference (GPECOM); Nevsehir, Turkey; 2019. pp. 242–7.

[37] García-Villalobos J., Zamora I., Junquera I., San Martín J.I., Eguía P. 'Decentralized control techniques for plug-in electric vehicles in MV/LV distribution networks'. International Conference on Renewable Energies and Power Quality (ICREPQ'16); Madrid (Spain); 2016. pp. 365–70.

[38] Australian Energy Market Operator (AEMO). Available from https://www.aemo.com.au/ [Accessed January 2020].

[39] Solar Report Jan. 2018 Australian Energy Council. Available from https://www.energycouncil.com.au/media/11188/australianenergy-council-solar-report-january-2018.pdf [Accessed January 2020].

Chapter 8

Energy management strategies in microgrids with EV and wind generators

Payam Teimourzadeh Baboli[1], *Alireza Sheikhi Fini*[2], *and Mostafa Sedighizadeh*[3]

In this chapter, different energy management strategies of microgrid (MG) operation in grid-connected or stand-alone mode are investigated. The problem of optimal energy management could be solved in two timescales.

The first timescale includes the optimal operation of an MG and the outputs of the problem could be the optimal scheduling of controllable resources in the MG, such as distributed generations (DGs), energy storage systems (ESSs), and amount of transferred power with the upstream network. The second timescale in MG energy management is solving its control problem, which deals with the control strategies of inverter-based devices to improve power exchange and enhance MG power quality. The state of the art of MG energy management would be thoroughly investigated in both timescales. Not only the current state of energy management strategies, methods, and algorithms would be critically compared to each other but the future prospects will also be highlighted.

Finally, an energy management strategy is proposed for a real MG using the day-ahead concept. The elements considered in the MG are photovoltaic (PV) panel, wind turbine (WT), electric vehicle (EV), ESSs, controllable DG, and transformer that connect the MG with the upstream network and provide the transfer of electrical power with the main grid. The objective of this case study is to optimize the economic dispatch of energy within the MG, considering the uncertainties in the predictions of electricity demand, solar radiation, wind speed, behavior of EVs, and the corresponding state of charge (SOC) for a given period of time.

8.1 Introduction

The traditional view of power systems has been reformed by advances in technology together with economic and environmental incentives. The most of US electrical

[1]Energy Division, OFFIS, Institute for Information Technology, Oldenburg, Germany
[2]Power System Operation and Planning Research Department, Niroo Research Institute, Iran
[3]Faculty of Electrical Engineering, Shahid Beheshti University, Iran

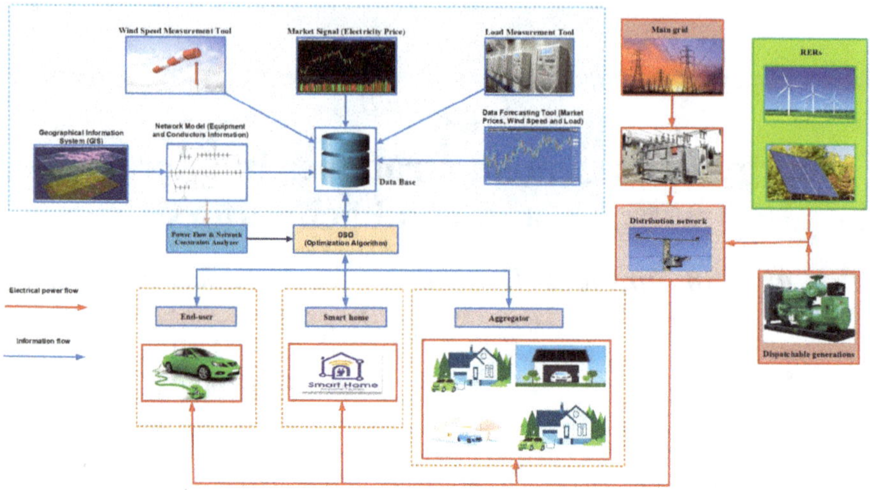

Figure 8.1 Topology of MG with EV integration

power grid has been built in the 1930s. Therefore, five huge blackouts occurred in the past 40 years due to exhaustion and overstressed US infrastructure electrical power grid [1]. To overcome these challenges, MG as a relatively emerging technology and capable solution can solve the problems pertaining to current electrical power infrastructure and guaranteeing the reliability of the electrical energy supply.

Technically defining, the main infrastructure of MG is a low-voltage distribution network that is supplied by point of common coupling via a distribution substation. MGs include a diversity of electrical devices that are DGs, ESSs, and controllable loads such as EVs [2]. Regarding the features and penetration level of distributed energy resources (DERs) and ESSs within MG, the energy management system (EMS) can be meaningfully dissimilar from a traditional power system. A typical MG can operate in two operational modes that are interconnected mode and in islanded (autonomous) mode [2].

In fact, MG-EMS is a control software that optimally distributes the electrical power output among the DER and ESS units, economically supplies the controllable/uncontrollable loads, and automatically enables MG to switch among interconnected and islanded modes based on the real-time (RT) operating conditions of MG [3].

Figure 8.1 indicates a typical hierarchal control of an MG equipped with EVs. In general, an advanced MG-EMS has to operate and organize a diversity of DGs, ESSs, and loads to ensure quality, reliability, sustainability, and enviro-friendly of the supplied energy as cost-effective.

8.2 Day-ahead MG EMS considering EVs

Designing and analyzing EV optimal charging strategies is one of the main prerequisites for implementing day-ahead EMS in the MG equipped by EVs. By properly modeling the charge/discharge of EVs, the problem of optimizing the

EMS of the MG can be simplified and at the same time, useful insight can be provided for the planning and operation of the EV. Figure 8.1 also shows how EVs interact with the MG and its operator in terms of both electrical power flow and information flow. Because MG operators, EV aggregators, smart homes, and EV users make their decisions about participating in the day-ahead market based on the information exchanged, they must have access to this information without any restrictions.

Various studies have examined the effects of merging EVs into EMS from different perspectives. For example, the Electric Power Research Institute in the United States categorizes how to intelligently charge EVs in an MG based on driver behavior [4]. In another study [5], three charging modes, i.e., charging at home, charging on the streets, and devoted charging stations were considered, and MG scheduling costs in vehicle to grid (V2G) and grid to vehicle modes were extracted and compared. In [6], a field study in North America determined that the present distribution network infrastructure and facilities respond to the use of lightweight EVs with high penetration in MG operation considering RESs. In other studies, scheduling of MGs equipped with EVs was studied as numerous scenarios and it can be seen that in these scenarios, MG scheduling by considering EVs can reduce costs and increase network intelligence. However, it can be noted that robust scheduling that can lead to the best results in all operating modes requires the creation of the necessary MG infrastructure to support EV charging/discharging program, which requires long-term MG scheduling using a standard approach taking into account the cost of its implementation.

8.2.1 Effects of EV's charging/discharging strategies on the EMS

Optimally implementation of EMS is extensively affected by EV's charging/discharging strategies which are chosen by players of the electricity market that are MG operators, EV aggregators, smart homes, and EV users. Therefore, the following subsections deal with EV's exchanging electrical power strategies from the aforementioned player's point of view.

8.2.1.1 MG operator strategy

From the MG operator's point of view, one of the most important goals of using EVs is to perform the peak shaving mechanism and smooth the load curve. In this regard, the work [7] has been studied such a strategy with two scenarios of charge at low load and discharge at peak load in order to determine the effects of charge/discharge EVs on reducing operating costs and greenhouse gas (GHG) emissions throughout the Northern Ireland and Republic of Ireland electricity market. In [8], MG optimal day-ahead management program is formulated as an optimization problem. The objectives are operational cost and GHGs emission minimization, subject to several technical and economic constraints. The MG includes EVs which are charged/discharged in some controlled and uncontrolled patterns.

8.2.1.2 Aggregator strategy

The aggregator can implement more flexible and practical charging scheduling by choosing the right strategy. In fact, the aggregator manages the participation of EVs in MG day-ahead energy planning [9, 10] by directly controlling network-connected EV chargers. The aggregator uses the MG satisfactorily by dividing the load and, if possible, prolonging the charging time. Numerous studies have been conducted in this regard. For example, the work [11] has proposed two methods of planning based on active strategy and proactive strategy for EV aggregators. In the active strategy, the maximum power that the aggregator can transfer to the EVs simultaneously is set to the upper limit, with each additional load higher than this delayed. The total load in the proactive strategy is formerly calculated and, after that, distributed into a number of charging times. Instead of scheduling EV loads, some authorities have focused on load distribution strategies and charge efficiencies. For example, in [12] the charge rate is considered as a criterion for determining the power and sequence of charge and has shown that the efficiency-based strategy works better than the non-efficiency-based strategy. Of course, choosing this strategy in planning may deprive the user of the freedom to charge/discharge.

8.2.1.3 Smart home strategy

A residential smart MG includes several smart homes that optimally manage their energy. The EMS of a smart home manages the time slot of the appliances and determines the economic dispatch of home dispatchable/non-dispatchable DGs as well as EVs as active loads.

The work [13] studies the stochastic energy management of smart homes with WT as renewable energy source (RES) and EVs as prosumer units considering demand response programs (DRPs). The work [14] proposes optimal operation of grid-interactive homes including PVs and EVs with considering DRPs. The objectives of the optimization problem are minimizing the operational cost of the smart-home owners, subject to several techno-economic-enviro constraints.

8.2.1.4 User strategy

Another option for pursuit aggregator commands is to use a user strategy in which EV owners charge their EVs via a smart device. This intelligent device determines the charging intervals based on the user's behavior and choice so that the load curve becomes smoother when the load of the house is changing. These types of charging strategies are only suitable for local charging scheduling issues. These strategies are designed from the user, smart home, aggregator, and distribution system operator (DSO) point of view. Normally, in MG-EMS, tasks are performed as optimization problems to minimize costs or maximize user and DSO welfare. Consequently, end user strategies similar to aggregator and DSO strategies do not simultaneously consider the network requirements and needs of the EV owner. In this regard, smarter strategies should be considered to solve the problem of EV charging planning by considering the objective functions and appropriate constraints [15].

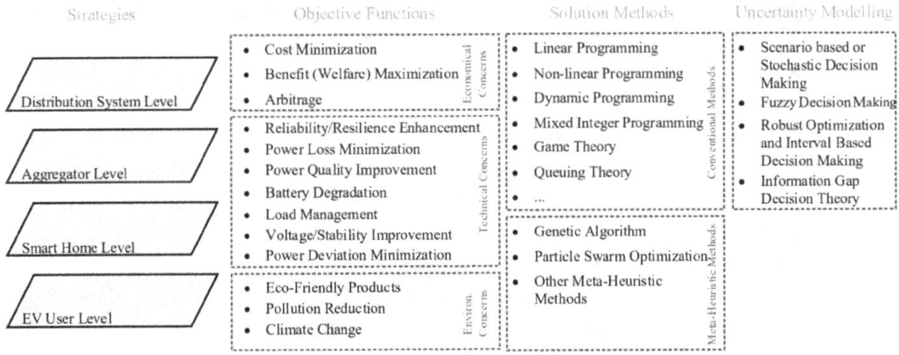

Figure 8.2 EV scheduling, strategies, objectives, and methods

8.2.2 Objective functions and constraints for MG-EMS equipped EVs

As mentioned, the charging program should be formulated as an optimization problem to achieve the desired goals, taking into account the relevant constraints. To solve this optimization problem, there are several methods that can be referred to as traditional analysis methods such as linear programming (LP), nonlinear programming (NP), dynamic programming (DP), and the like. In recent years, the use of meta-heuristic algorithms (MAs) such as genetic algorithms (GAs), particle swarm optimization (PSO), etc., which are inspired by nature, has become common to solve the problems caused by linearization of goals and optimization constraints. The following are the common objectives and constraints used to plan the charging of EVs in the day-ahead market.

8.2.2.1 Objectives

As mentioned, four levels of MG operator, aggregator, smart home, and end user can be considered to determine the charge/discharge of day-ahead planning strategy of EVs. These four levels can have different optimization goals and constraints, which are classified into three general categories: economic, technical, and environmental, as shown in Figure 8.2. In the following sections, the objectives related to these three categories are introduced.

A. Economic objectives

A.1 Cost minimization

One of the most significant objectives in a day-ahead EMS of an MG is to minimize costs that consist of costs of power generation, operation and investment costs, and costs pertaining to demand side. However, the aforementioned objectives are limited to numerous constraints, i.e., fuel constraints, emissions constraints, etc. It can be noted that generation cost involves the cost of fossil fuel, cost of non-dispatchable generations, and cost of EVs. There are several researches in this area, for example, the work [16] performs day-ahead EMS for an MG operator aimed at minimization of cost considering internal and external agents. Internal agents are internal resources and external agents are other MGs in the same network, aggregators, or EVs that can interact with the MG through a local energy market. The work [17] proposes a stochastic-based optimal MG-EMS with high penetration level of RESs and EVs. This model involves the high portable charging demands of EVs together with the stochastic behavior of RESs. An optimal V2G scheme is organized to alleviate the negative impacts of EVs on the operation of the individual MGs and minimization of the investment and operation costs.

A.2 Benefit (welfare) maximization

EV owner participants try to maximize their revenue (welfare) via V2G service. Recently, some researches pay attention to this topic. In [18], an optimal EV charging strategy has been presented by dynamic programming approach. Its objective was maximizing benefits for EV owners. The work [19] presents an aggregator strategy whose objective is benefit maximization by selling energy, regulation, and reserve. The proposed strategy helps the MG operator by filling valley and peak shaving and consequently reducing the operational cost while maximizing EV owner's welfare by charging freedom and further financial benefit.

A.3 Arbitrage

The work [20] presents an optimal charging schedule of EVs taking energy arbitrage and MG operation cost into account. The objectives of the optimization problem are to maximize EV owners' energy arbitrage benefit and to decrease peak demand, power loss, and transformer aging. The work [21] proposes an analytical model including energy arbitrage and other objectives that are operation cost and power loss for day-ahead energy management in an MG. The results show a suitable pricing scheme for energy arbitrage, EV charging, and power losses.

B. Technical objectives

B.1 Reliability/resiliency enhancement

EVs as active loads can collaborate with DRPs to enhance MG reliability. The work [22] develops a stochastic analytical reliability and adequacy model for MGs considering EVs. It optimally determines the number of EVs' states that simultaneously assures the speed and accuracy of the calculation of reliability indices. The work [23] introduces a stochastic general model for MG reliability evaluation considering EVs by proposing a new state matrix that are determined by segmentation concept and graph theory.

EVs as a type of ESSs can make the MGs more resilient against the grid blackout caused by natural disasters. The work [24] studies the resiliency of smart homes against natural disasters that result in an MG outage. Dissimilarity in the pattern of availability of EVs in each smart home leads the owners to be benefitted from the EVs charging/discharging hours. The contributions of EVs in the energy management of smart homes are as a peer-to-peer operation of the smart homes, EV charging/discharging, partial charge ability, load curtailment, and load adjustment. The minimization of energy cost and maximization of energy resilience under natural disasters are objectives of the proposed model. In this model, resiliency is introduced as critical load restoration and minimum energy loss under various power outages. In [25], EVs are fast EMSs that operate as a backup capacity to enhance the flexibility and resiliency of an MG. The proposed EMS model is formulated as an optimization problem. The objectives of this optimization problem are maximization of flexibility and resiliency of the MG as well as system cost reduction.

B.2 Power loss minimization

Regarding the International Energy Agency report, between 2% and 56% of electricity is lost annually through transmission lines and distribution feeders [26]. However, excess energy loss occurs during EV integration. The work [27] studies the V2G schemes to evaluate their impacts on the several technical factors of MG, such as electrical power loss. The efficacy of electric vehicles in minimizing power loss and improving ancillary services such as frequency and voltage control was taken into account.. Regarding rising development of MG and EVs, the work [28] tries to enhance the energy management performance of multiple MG systems considering the stochastic behavior of EVs charging. It proposed an optimization model to minimizing the electrical power losses, operating costs, etc.

B.3 Power quality improvement

MGs faced severe intermittency and uncertainty due to the increased penetration level of renewable generations and EVs. Therefore, numerous operational indices such as power quality are threatened. In order to manage RESs and EVs, and alleviate the unpredicted behavior of such sources and loads, the work [29] proposes a capable and adequate ancillary services to maintain the MG reliability and the harmonic supports to ensure power quality. In [30], integrated EVs charging is modeled considering uncertainties of RESs. The aim is to enable the design of an EMS that improves voltage quality and harmonic distortion in spite of the presence of the RESs and EVs.

B.4 Battery degradation

Deviation from the rated capacity of the charging/discharging modes may result in a significant reduction in the life cycle of EV's battery. One of the most widespread types of EV batteries is lithium-ion battery, and in the recharging periods, if the battery could not achieve 80% of its rated (original) capacity, it is the end of the battery life cycle [31]. Therefore, the investment cost of the battery should be taken into account in designing a planning strategy.

B.5 Load management

Load management concept is optimal coordination of the responsive loads and EVs as well as RESs to efficiently implement DRPs. In [32], an EMS is proposed by Internet of Things technology for smart homes considering DRPs. The goal of optimal management is to maximize profit for EV owners as well as reduct peak electrical demand by collaborating in DRPs. The work [33] presents an EMS considering EVs, smart homes, and interaction of the neighbors. Therefore, an approach based on EVs charging/discharging power scheduling is proposed for flattening the electrical power demand profile.

B.6 Voltage profile and voltage stability improvement

The presence of EVs increases the charging demand in the MG and consequently, impacts voltage profile and voltage stability. The work [34] proposed a model to assess the capacity of electrical energy supply in MG based on voltage stability indices. Moreover, the movement of EVs is taken into account to determine the load margin and the maximum number of EVs that MG can charge in critical condition. The work [35] proposes an area-based sensitivity index for determining the buses which are sensitive to voltage drop are potentially can be collapsed. Then, it determines the buses that can support voltage and increase the load margin.

B.7 Power deviation minimization

EV aggregators try to minimize power deviations in the operation of MG, this is due to the fact that deviations pertaining to EV charging uncertainty can lead to overload in the feeders of MG [36]. In an EV aggregator system, DSO centrally dispatches electric power among aggregators which dedicate electric power to EV piles. Aggregator operates in such a way that there is the minimum deviation between the power requested by the DSO and the power required by the EVs for each hour.

C. Environmental objectives

C.1 Eco-friendly products

Global policies in case of sustainable environmental pollutants and high recycling cost caused the government to propose some incentives to EVs' manufacturers to use eco-friendly materials. Two main components of EVs are battery and electric engine, which could be maintained with eco-friendly materials. As consumption of these eco-friendly materials affect infrastructure cost of EVs, in day-ahead, optimal operation of MGs should be considered.

C.2 Pollution reduction

Environmental constraints are vital indices in MG planning. Recently, international agreements on emission reduction force power system operators to satisfy some constraints about pollutants. For example, in [37] a model is presented for minimization of the pollutions in residential MGs including EVs. To preserve system reliability within peak hours, the DRP is also applied using Time of Use tariffs and curtailable loads to change the energy consumption patterns of consumers. The work [38] suggests a model for energy and reserve management of MGs including EVs to decrease environmental impacts. To effectively manage uncertainties, a robust optimization method is presented and is applied to handle the uncertain parameters.

C.3 Health and climate benefits

World Health Organization has recently reported that more than 4 million people have died each year due to health effects related to air pollution disclosure, including cardiovascular diseases and lung cancer [39]. Although researchers mostly approve that using EV has environmental profits [40], analyzing the impact of EVs on MG operation shows that detailed analysis of EV travel patterns, emissions initiated by electricity generation, and air pollutant transport is necessary.

The work [41] indicates that even it is assumed the electricity needed by EV is provided by natural gas, vehicle electrification can efficiently increase significant health benefits in the understudied MG, in comparison to the base case scenario (without EVs). If EVs are only supplied by RESs, then electrifying all passenger vehicles can increase the health and social benefits more.

Many scientific and business surveys show that emission related to carbon dioxide will lead to extensive global climate change in the future. It can be noted that the world has accepted a temperature rise of 2 °C above pre-industrial levels and it is set as the maximum boundary to sustain global economic and environmental prosperity. Many international modeling agencies are agreed that early, vital de-carbonization action is necessary to meet 2 °C targets. Postponing this vital action could increase the economic, social, and political costs of the system operation. It can be said that using low-carbon technologies such as EVs and RESs is vital to global de-carbonization [42].

According to the above-mentioned health and climate change goals, the day-ahead decision makers have to pursue such strategies that, in long term, cause a reduction in GHGs and consequently decrease in disease rate and help reach the 2 °C scenario of climate change.

8.2.2.2 Constraints

It is worth to mention that EV operation optimization problem is subjected to numerous constraints. To integrate EV to MG, constraints are introduced to bound optimal solutions with the limitations of real physical networks, EV properties, and user-defined constraints. Like the designed objectives, the constraints are also designed according to the level of planning and related tasks. The rated electric power, voltage, and current of piles are usually bounded by the chargers used in EV. The formulation of the optimization problem focuses on heavy mathematical methods. Therefore, several numerical methods have been utilized to solve this problem. DGs and ESSs are physically limited, so some numerical limitations are required for load flow and planning. Moreover, there are MG constraints on feeder capacity and feeder losses and the balance of electric demand and consumption. There are also EV charging restrictions. An EV battery has a specific charging capacity for different types of batteries and different charging needs. Manufacturers of EV implement battery specifications based on EV electric power. Therefore, there are several charge/discharge limitations for each particular EV.

8.2.3 Multi-objective optimization

EV scheduling is a complex task and includes a trade-off between operation costs and emission of pollutants, while also seeking to maximize welfare, minimize investment cost of the battery, and network support such as reducing power disruption at the aggregation level, SOC battery, and so on. Therefore, optimizing and modeling EV must actually involve multi-objective optimization to be realistic. To solve the optimization problem, there are basically two methods that are traditional methods and meta-heuristic methods. Following sections deal with the implementation of these methods in MG-EMS equipped with EVs.

8.2.3.1 Traditional mathematical optimization methods

Traditional numerical optimization methods to solve EVs integration network problems include LP [43], NP [44], Mix Integer Programming [45], DP [46], as well as some other approaches such as Game Theory (GT) [47], Queue Theory [48], and so on.

8.2.3.2 Meta-heuristic methods

One of the alternatives to optimize complicated power systems and MG programming problems is MAs. In particular, considering EVs as uncertain, controllable, and chargeable demands cause more implementation of these algorithms. MAs randomly generate initial solutions and select the optimal solutions among them. According to their optimization mechanism, MAs generally do not guarantee to harvest global optimal solutions; however, MAs are powerful algorithms to solve nonlinear, non-convex, and high-dimensional problems. Thus, they are common options for resolving objective functions and constraints that are discrete and non-differential, such as binary charge/discharge modes for a lot of EVs [49]. MAs are classified into methods based on binary values and continues values according to the variables in the formulations, or methods based on path and population with operators based on evolutionary programming (EP) [50, 51]. The most famous MAs consist of GA [52], Differential Evolution [53], PSO [54], Coordination Search [53], Evolutionary Algorithm [55], EP [56], Ant Colony Optimization [57], Covariance Matrix Adaptation Evolution Strategy [58], and Bioassay-Based Optimization [59].

8.2.4 Uncertainty modeling

The uncertainty pertaining to EV's behavior can be modeled by four different models that are scenario-based stochastic decision-making, fuzzy decision-making, robust optimization and interval-based decision-making, and information gap decision theory (IGDT) [60]. These methods are based on three different criteria: uncertain input data, uncertainty modeling technique, and decision variables and constraints. Each technique has its own advantages and disadvantages. Furthermore, the best technique is firmed based on the accessibility of the input data or severity of the uncertainty. It is observed that probabilistic approaches require accurate information about the probability density function of uncertain input data. The nature of scenario-based probabilistic optimization is the combined growth of computation burden. Fuzzy optimization requires membership function to solve multiple value problems of value-cuts and moreover is computationally costly. The interval optimization or robust optimization method requires a particular uncertainty set of input data. The interval optimization method generally requires two optimizations for each objective function. The robust optimization method regularly needs to solve bi-level optimization which is usually challenging to solve. Furthermore, IGDT method requires uncertainty set; however, compared to other methods, this uncertainty set is not required to be known precisely [61].

8.3 Real-time MG energy management

The use of *heuristic optimization algorithms* and *artificial intelligence-based methods* for RT energy management of MGs becomes a trend today, while these algorithms and methods are more flexible, easier to model the practical restrictions, and a fewer computational requirements. The authors of [62] present a sliding mode control method for MGs due to their robustness properties against a wide class of uncertainties, which is a suitable solution for mitigating uncertainties in the RES generation forecast. In [63], a two-layer optimizer is proposed to achieve a fast and accurate solution. The authors of [58] propose a new strategy based on a GA for an RT EMS for MGs to increase the integration of RES and minimize the cost of energy and emission. In [64], a multi-objective distributed RT electricity allocation scheme is proposed for large residential MGs with the aims of reducing the cost of energy for customers and improving the reliability and efficiency of the MGs. The proposed scheme then converges to an optimal Nash equilibrium with an undominated but desired solution. The authors of [65] mentioned an optimal assignment of the droop gain for RT energy management of an MG in stand-alone mode. Another method for droop control for RT energy management of MGs is presented in [66]. It is very important to mention that one of the major shortcomings of heuristic methods is the lack of guarantee for flinging the global optimal solution.

Using a *multi-agent control methodology* has been a common way in recent years to overcome the complexities of MG control. In [67], a distributed control system is introduced which is based on a multi-agent system approach for the control and RT energy management of MG. A multi-agent system for the generation planning and the management of MG with an RT digital simulator (RTDS) is proposed in [68]. The authors of [69] presented a multi-agent control system in Java for RT simulation of interactions between various components of an MG in RTDS. In [70], a reverse auction model for the RT market operation of MGs with multi-agent-based GT is addressed. The authors of [71] modeled a multi-agent-based RT optimal MG power flow using a distributed cooperative control theory approach. In [72], a self-organizing multi-agent-based system for RT control of MGs is proposed in both island and stand-alone mode.

The *reactive power dispatching* problem is one of the other interesting topics in MG RT energy management area that is typically solved with gradient-based solvers [73] and probabilistic search methods [74]. The authors of [75] present a dynamic programming optimization method for the RT distribution of reactive power in MGs.

The topic of *state estimation* is one of the most important topics in the field of MG RT energy management [76]. Although state estimation methods are well developed in both transmission and distribution levels, they are not applicable well for MGs with high RES penetration. The main reason is the failure to take into account the dynamic properties of renewables and dynamic loads. The authors of [77] suggested that the centralized state estimation method is incapable to process a large amount of data in MG, since the measurement must be sampled at high sample rates (e.g.,1 kHz) and therefore, the amount of data even for a small MG would be so large. In [78], an automatic power flow is proposed, which is based on integrated automation of the geographic information system in order to achieve an RT state estimation of the distribution network and the MGs.

Resource operational planning or *scheduling* is one of the most extensively studied topics [79–81]. The intermittent nature of renewables in MGs leads to large uncertainties in the output power, which must be taken into account in the operation and planning studies. Thus, RT planning techniques have been proposed to track unpredictable variation in the generation of MGs [82]. In [83], the optimization problem of MG's day-ahead RT operations is proposed, including the uncertainty modeling. The authors of [84] present a model for the two-layer RT dispatch of MGs, taking into account the hidden costs for the discharge option and the marginal charging of the ESS. In addition to assuming an accurate forecast for renewable energy production and loads, which is usually very difficult to achieve, another issue is that they fail to take into account the limitations of power grids. As a result, these methods cannot improve the stability of the network or lead to non-optimal economic conditions. The authors of [85] propose a new method for RT energy management of MG that takes into account the power flow and the operating limitations of the system. A chance-constrained stochastic optimization method is proposed in [86] for optimal day-ahead planning and RT energy management of MG. RT scheduling of DC MGs with multiple slack terminals is presented in [87] including the equalization of the SOC between the multiple operating distributed storage. The authors in [88] present the equalization of the SOC between the battery energy storage and the compressed air energy storage.

Demand response (DR) is an efficient solution for balancing supply and demand in RT operation [89]. The most important advantage of RT DR is the efficient provision of the capacity of the system [90]. To encourage customers to participate in DRPs, they must first actively participate in market pricing, rather than just accepting offers [91]. Thus, RT Balancing Market [92] could be a suitable solution. In [93] a decentralized demand-side management is proposed in order to compensate the RT generation of RES and storage systems with reference to the day-ahead energy market. Thus, a mixed strategy game is formulated. The authors of [48] represent a central DR approach to frequency and voltage control of MGs. The authors of [94] also present an algorithm for the RT energy management of MG using ESSs.

It can be concluded that multi-agent control algorithms with communicative RT devices are the most accepted approach for RT energy management MGs. This is mainly due to the fact that, taking into account different control parameters, i.e., the high penetration rate of DER and its uncertainty in the generation forecast, the active loads, the market requirements, technical constraints, and conflicting objects of grid and customers; multi-agent approach could perform well. The main reason is the distributed nature and potential of the multi-agent control approach to model autonomous decision-making entities in solving complex problems.

8.4 MG Energy management with EVs, seawater desalination, and RESs: a case study

With the development of human societies, new industries and needs are being formed which supplying their power and energy consumption, requires the development of equipment and infrastructure in the field of power generation and transmission. An example of these industries is seawater desalination (SWD) units. In addition, increasing attention to air pollution has led to the proliferation of EVs or the use of hydrogen as a zero-emission fuel. MGs can benefit from advantages such as optimal use of installation capacity and environmental potential by accepting to

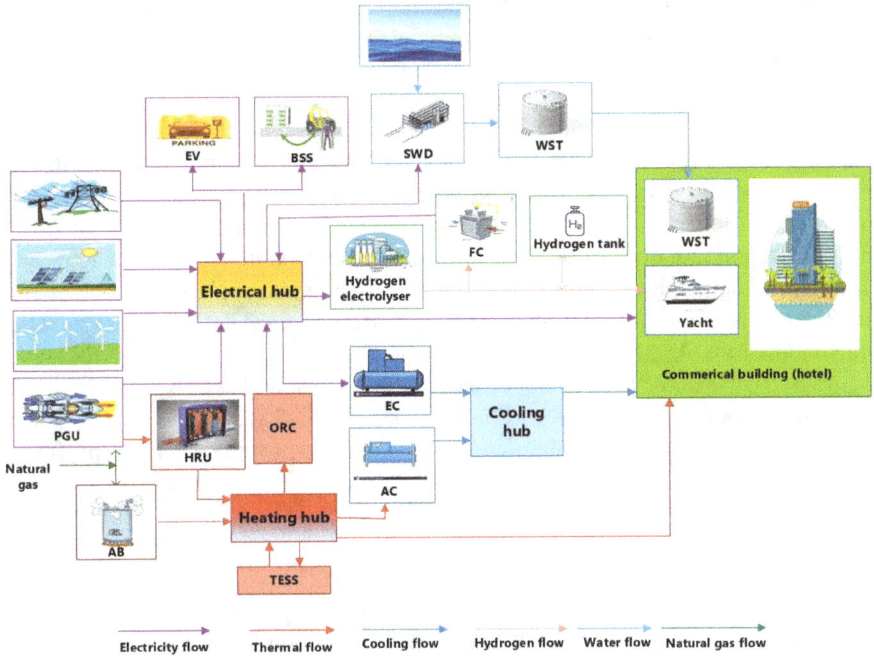

Figure 8.3 Overview of the proposed MG

provide other needs along with common requirements such as electricity, cooling, and heating. In this case, MG is able to increase its flexibility by converting energy into the desired product and storing it in different forms such as electrical or thermal energy. In this case study, MG utilizes an SWD with Reverse Osmosis (RO) technology as a flexible load and an Organic Rankine Cycle (ORC) as a new technology for generating electrical power by means of thermal power. In addition, to better interact with available EVs, one parking lot and a Battery Swapping Station (BSS) are introduced. The electrolyzer uses the electrical power to supply the required amount of hydrogen, and the hydrogen tank in addition to the fuel cell (FC) helps to balance the production and consumption of hydrogen. The proposed model has objective functions that are operational cost and emission cost subject to several technical constraints. In order to model the existing uncertainties, the scenario-based method and IGDT method have been used. The efficacy of the proposed model is evaluated on a smart commercial building in Qeshm Island, which is located a few kilometers off the southern coast of Iran.

8.4.1 Overview of the proposed MG

As shown in Figure 8.3, the proposed MG consists of power generation unit (PGU), auxiliary boiler (AB), electric chiller (EC), absorption chiller (AC), SWD with RO technology, ORC, WT, PV units, hydrogen electrolyzer, hydrogen tank, FC, EV parking lot, BSS, thermal energy storage system (TESS), and water storage tank (WST). MG inputs include natural gas and electricity purchased from the upstream grid, and its output includes the electricity, heating, cooling, hydrogen, and fresh-water demands.

PGU is a gas turbine that generates electricity by consuming natural gas. By installing a heat recovery unit (HRU), the excess heat generated can be used to prevent the losses in the part of initial energy spent in the form of dissipated heat. Another heat generator is an AB, which is powered by natural gas similar to gas turbines. Depending on the conditions, the heat generated in the MG can be directly used thermal demands, stored in TESS, supply AC to meet cooling demand, or enter to ORC to generate more electrical power. In addition to AC, an EC is also installed, which, unlike the AC, is powered by electricity.

It should be noted that in hours when heating demand is low, to store excess recycled thermal energy, there will be a need to a TESS with high capacity, while using the ORC, the extra heating power can be used to provide electrical power and enjoy its benefits such as the possibility of selling to the upstream grid.

8.4.2 Mathematical modeling and proposed algorithm

The mathematical stochastic model for optimal operation of an MG is proposed in which uncertainties are modeled using scenario-generation and the IGDT method. It should be noted that in all stochastic cases, the uncertainties of the EVs are modeled by the scenario-generation method and the uncertainties of PV, WT, electrical, heating, cooling, water, and hydrogen demands, and the gas price are modeled by the IGDT method. To evaluate the efficacy of the proposed model, the simulation results are reported for two cases: the deterministic (neutral risk) and stochastic (risk-averse). Therefore, deterministic case studies are modeled as mixed-integer programming (MIP) and are solved using CPLEX solver. However, by using information gap decision theory (IGDT) to model other uncertainties in stochastic case studies, the problem is converted to mixed-integer nonlinear programming (MINLP) and is solved using DIscrete and Continuous OPTimizer (DICOPT).

There are continuous and binary variables in the proposed mathematical model for MG optimal operation. The binary variables are presented to end the charging/discharging mode of storage devices and preventing them from running at the same time [95] and on-off status of equipment. The continuous variables include the output electrical power of combined cooling, heat & power (CCHP), the output thermal power of ABs, the exchanged electrical power with the upstream grid, the exchanged electrical power with electric batteries of EVs in the parking lot or BSS, the exchanged thermal power with TESS, the input thermal power of ORC, the input thermal power of AC, the input electrical power of EC, the output freshwater of WST, the input electrical power of SWD at each hour, the electrolyzer input electric power, the input hydrogen power to FC, and the hydrogen exchange between hydrogen tank and grid [96]. The simulation is performed on a PC with Intel Core i7, 1.8 GHz CPU with 6 GB of RAM using GAMS software.

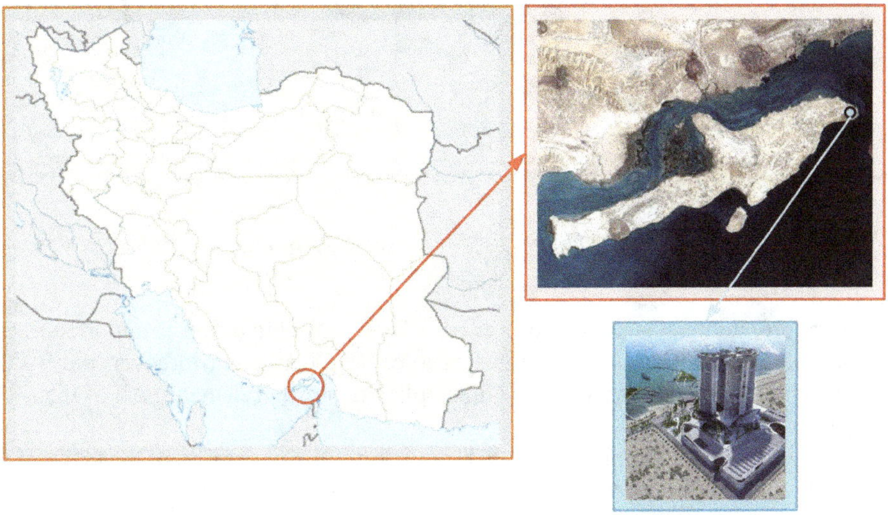

Figure 8.4 Location of the MG in the island

8.4.3 Numerical results

The proposed algorithm to solve the formulated optimization problem was applied to the MG presented in Figure 8.3. This MG is used as a test system to evaluate the efficacy of the proposed mathematical model and it belongs to a smart commercial building (hotel) in Qeshm Island, Iran, which is an arrow-shaped Iranian island in the Strait of Hormuz, separated from the mainland by the Clarence Strait/Khuran in the Persian Gulf (26°50′N 56°0′E). It is the largest island in Iran having a surface area of 1 491 square kilometers (576 square miles) and the average temperature on the island is approximately 27 °C (81 °F).

The warmest months are June through August, and the coolest from October to January. The average rainfall is 183.2 mm. The island comprises 59 towns and villages and the population was 123 774 at the 2018 Census. The local population is involved in fishing, dhow construction, trade, and services. An additional 30 000 are administrative or industrial workers or students. Figure 8.4 shows the location of the MG (hotel) in the island.

Electric hub, cooling hub, and thermal hub investigated for the MG (hotel), which not only meet electricity, cooling, and heating power needs but also provide the fresh water required by a part of the island population and the hydrogen required by the yachts of the hotel.

Electric power derived to EC, SWD, electrolyzer, and electrical demand from PGU, FC, WT, PV, and ORC are imported from the electric hub. Besides, it interacts the electric power with upstream grid, FC, EVs in a parking lot, and batteries in BSS. In addition, electrical power can be stored in other ways, such as hydrogen in a hydrogen tank, or in stagnant water in a water tank. Thermal power is imported from AB and HRU via the thermal hub while it exports to ORC, AC, and thermal demand.

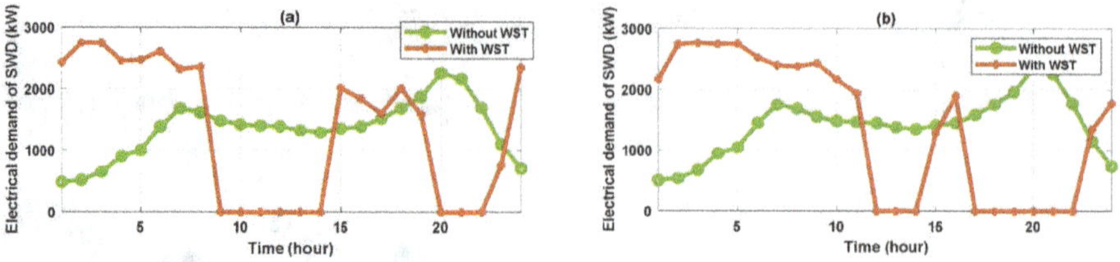

Figure 8.5 Electrical demand for SWD: (a) neutral risk and (b) risk averse

Furthermore, it interacts thermal power with TESS. Cooling power is imported from AC and EC via cooling hub and transfers to cooling demand. Moreover, the SWD supplies the WST and, consequently, it supplies freshwater demand. All of the MG data have been reported in [97].

Since the energy supply of the SWD constitutes the major part of the electricity demand to assess the effects of WST on the MG operation and the efficacy of the proposed hybrid IGDT/scenario-based model, the deterministic/stochastic energy management of MG with/without WST is carried out in following cases.

8.4.3.1 Effects of the WST on the SWD operation

Figure 8.5 displays the electrical power consumption of SWD in deterministic/stochastic energy management of EH. Regarding Figure 8.5(a), by installing a WST, SWD is turned off at peak hours between 9–14 and 20–22 while in the early hours of the day, which are off-peak hours, the activity of SWD is increased. Indeed, installing WST has added flexibility to supplying fresh water and by storing electrical power in off-peak hours in the form of fresh water, it has helped to peak load shaving. Figure 8.5(b) shows the results of the stochastic modeling. Similar to the deterministic case, WST causes displacement of the electrical demand of the SWD from peak hours 12–14 and 20–22 to off-peak hours.

8.4.3.2 Effects of the WST on the EVs operation

Another growing need of the power grid is to meet the electrical demand of EVs whose proper energy management can turn them from a threat to an opportunity. In order to better study the presence of EVs in the proposed EH, two systems have been considered for the participation of EVs including EVs parking lot and BSS. For comparison, the number, specifications, arriving time and departure time of EVs in the parking lot, and BSS are assumed equal. In addition, the initial charge status of batteries in BSS in the early hours of operation horizon is considered equal to the initial charge of arriving EVs to the parking lot.

Figures 8.6 and 8.7 illustrate the charging and discharging process of batteries in two systems. It can be seen that in the system without WST, at the early hours of the day, when the purchase price of electricity from the network is minimum, BSS is able to use the capacity of its existing batteries and charge them. As a result, the amount of power purchased to charge the batteries is reduced to zero during hours

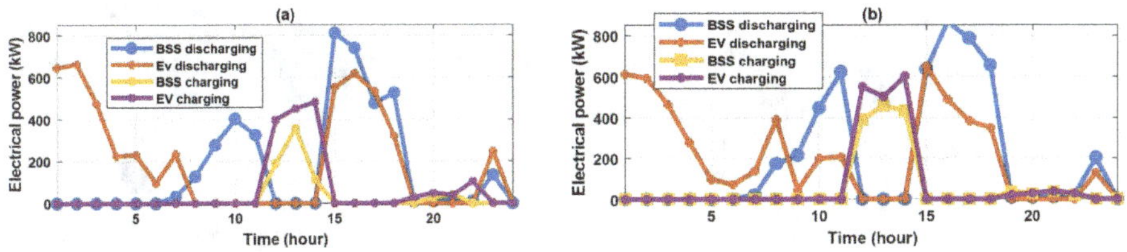

Figure 8.6 *Process of charge and discharge of EVs and BSS without WST: (a) neutral averse and (b) risk neutral*

8–15. On the other hand, it was able to sell more electrical power to the network in hours between 12–14 and 19–20, which are peak hours. Similarly, in the system with WST, the BSS used more low-price electricity in early hours, but compared to the system without WST, this exploitation was less. For this reason, neither of the two systems was able to inject a lot of electrical power into the network during peak hours. In stochastic modeling, due to the increasing demand for various carriers and especially the demand for fresh water, the possibility of supplying electrical power to BSS at the early hours of the day has also decreased compared to the deterministic modeling. However, it still has a better performance compared to the parking lot in the use of electrical power at hours with minimum cost.

8.4.3.3 Effects of the WST on the electrolyzer operation

According to Figures 8.8 and 8.9, without WST, a large part of electrolysis operation, which is responsible for supplying the required hydrogen, has been transferred to the early hours of the day (off-peak hours) in deterministic and stochastic modeling, while with WST this possibility has been denied. Figure 8.10 displays the SOC of the hydrogen storage.

8.4.4 Comparative studies

Figures 8.11 and 8.12 show electrical power balance in two case studies including without/with WST. At first glance, it can be seen that the electrical demand is reduced in some hours and is broken down into other hours, such as the early hours of the day when the price of electricity is minimum.

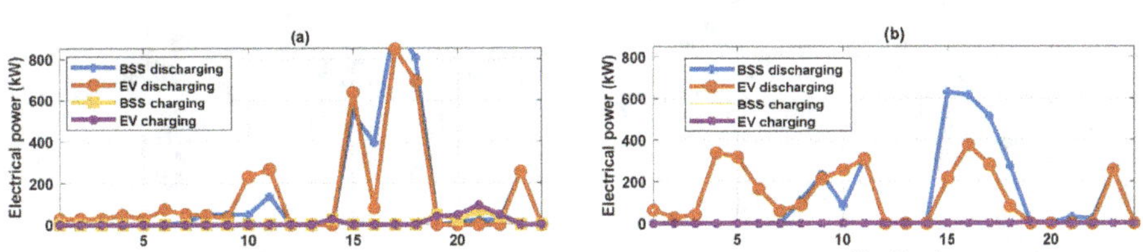

Figure 8.7 *Process of charge and discharge of EVs and BSS with WST: (a) neutral averse and (b) risk neutral*

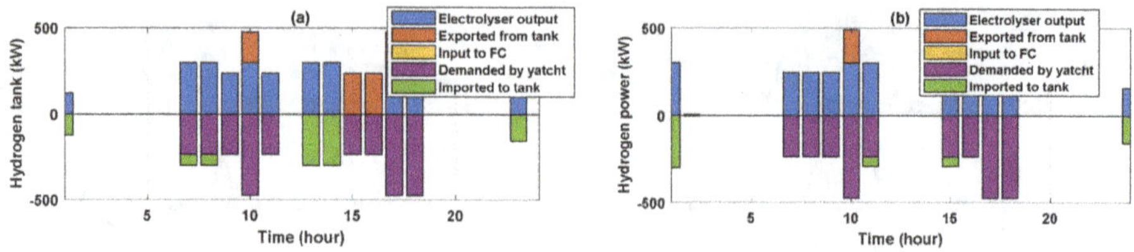

Figure 8.8 *Hydrogen power balance with WST: (a) risk averse and (b) risk neutral*

It can be noted that the presence of WST affects the operating variables as well as the charge and discharge status of other storage devices, such as BSS and hydrogen tank. Figure 8.13 compares the electrical power generated by PGU and ORC and exchanged by upstream grid together in deterministic/stochastic modeling with/without WST.

It is seen that by applying the WST, the capacity of PGU is used more efficiently and, consequently, the electrical power generated by ORC increases as well as the need for purchasing electrical power from the upstream grid decreases and at the same time, selling the electrical power to the upstream grid increases. For a closer look, the values of objective functions in deterministic/stochastic modeling with/without WST are shown in Figure 8.14.

In the deterministic modeling, it can be seen that considering WST, despite the 1.41% increase in the pollution tax, the operating cost is decreased by 7.21%, and the total cost is reduced by 6.92%. In stochastic modeling, the optimal value of the total cost is decreased by 11.08%.

Table 8.1 summarizes the results of studies in the deterministic and stochastic modeling. It is observed that using the WST reduces the costs while considering uncertainties of the demands and generations increases the related costs. Table 8.2 shows that case studies (with WST) are less affected by uncertainties. It can be noticed that uncertainties increase the total cost by 2.21% (with WST) and 6.99% (without WST). Therefore, the presence of WST has a positive role in damping the effect of uncertainties.

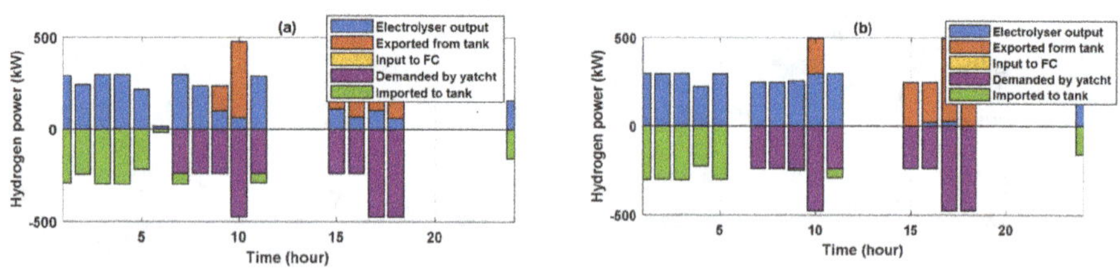

Figure 8.9 *Hydrogen power balance without WST: (a) risk averse and (b) risk neutral*

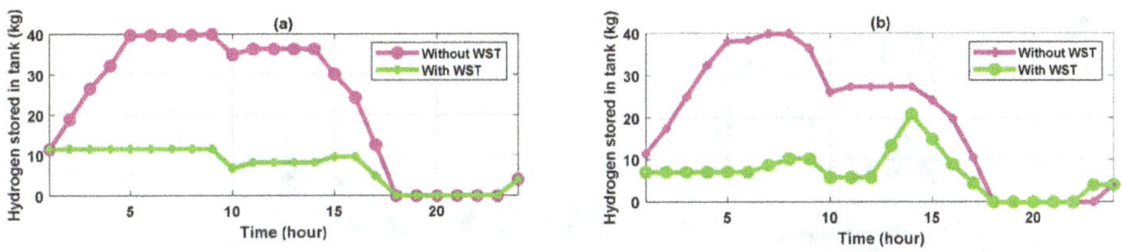

Figure 8.10 SOC of hydrogen tank: (a) risk averse and (b) risk neutral

8.5 Conclusion

This chapter analyzes the state of the art of EV scheduling and optimization approaches for integrating EV into MG. Formerly, analytical impact evaluation and charging strategies are presented. Furthermore, the scheduling methods presented as optimization problems encompassing different objective functions and limitations are investigated. It shows that conventional mathematical solution methods have some restrictions to solve complicated objective functions. This is due to the fact that these optimization problems have some constraints related to real problems that include multi-objective and high-dimension problems, i.e., probabilistic characteristic of EV charging and intermittency in RESs. In contrast, high flexibility and efficacy of MAs offer individual advantages to solve a complex objective function that involves non-smooth, non-convex, and mixed-integer problems related to the EV charging scheduling. Some MAs are effectively applied to EV charging scheduling, mostly taking into account a set of inclusive and complicated scheduling objective functions regarding the incorporation of probabilistic RESs and stochastic EV charging performances. Afterward, the energy management problem of the MG in RT applications is investigated. Finally, this chapter simulates a real case study in Iran.

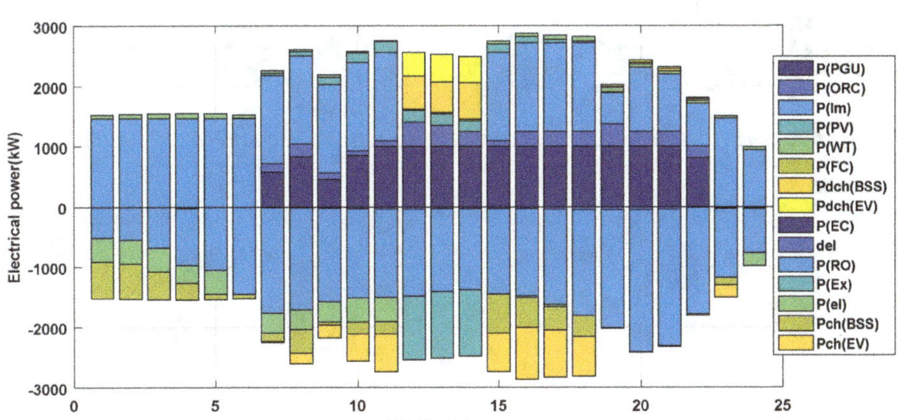

Figure 8.11 Risk averse electric power balance in the system without WST

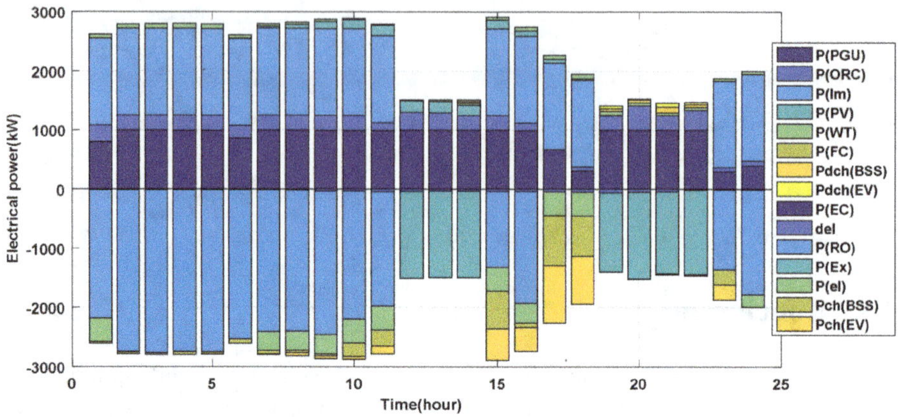

Figure 8.12 Risk neutral electric power balance in the system with WST

The simulated MG has high penetration of renewable energies, high capability in recovering waste heat, and a high capability to meet the needs of the water desalination process considering EVs. The integration of SWD, EVs, and hydrogen supply in MG also has advantages for the MG. Benefits such as the ability to sell electrical power to the upstream grid and help to make better use of the installed capacity of PGU are achievable. In the present study, by examining two systems including EVs parking lots and BSS, the advantages of them in energy management were identified. This system has more ability to adapt to the hourly price of

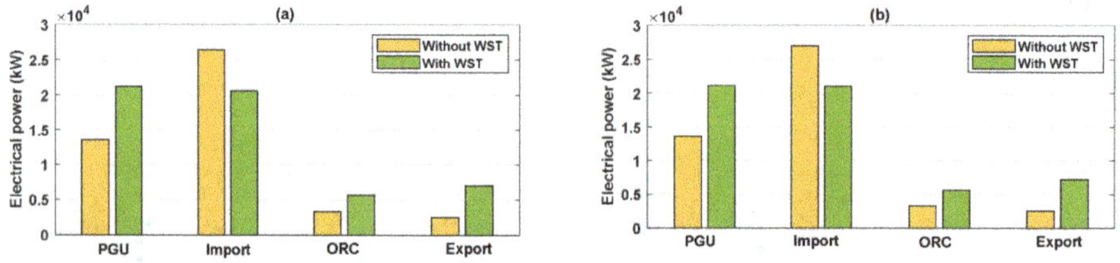

Figure 8.13 Electrical power in EH: (a) risk averse and (b) risk neutral

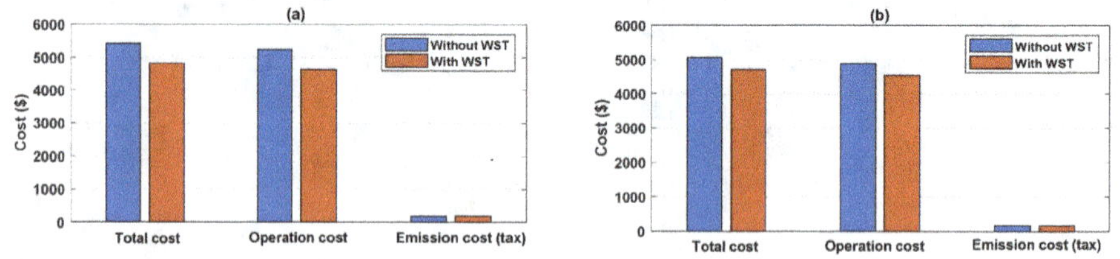

Figure 8.14 Impacts of the WST on total cost and its components: (a) risk averse and (b) risk neutral

Table 8.1 Comparison of objective functions in deterministic/stochastic modeling with/without WST

	Total cost ($)		Operation cost ($)		Emission tax ($)	
	Neutral risk	**Risk-averse**	**Neutral risk**	**Risk-averse**	**Neutral risk**	**Risk-averse**
With WST	4 716.046	4 820.544	4 543.533	4 629.295	172.512	191.248
Without WST	5 066.919	5 421.603	4 896.811	5 242.014	170.109	179.588

Table 8.2 Percentage increase of objective function values for stochastic modeling in comparison to deterministic modeling

Case	Total cost (%)	Operation cost (%)	Emission tax (%)
With WST	2.21	1.88	10.86
Without WST	6.99	7.04	5.57

electricity and also ends the problem of long charging time in parking lots. In addition, with proper energy management, the hydrogen needed for the hotel yachts is supplied. These yachts prevent environmental and noise pollution in the hotel space. In case of installing a WST inside MG and before delivering the prescribed water to water demands, flexibility is added to the freshwater supply process and considering the consumption of electrical power of SWD as a large flexible load, more benefits are available.

References

[1] Su W., Wang J. 'Energy management systems in microgrid operations'. *The Electricity Journal*. 2012, vol. 25(8), pp. 45–60.

[2] Markvart T. 'MGs: power systems for the 21st century?' *Refocus*. 2006, vol. 7(4), pp. 44–8.

[3] Núñez-Mata O., Palma-Behnke R., Valencia F., Urrutia-Molina A., Mendoza-Araya P., Jiménez-Estévez G.. ' 'Coupling an adaptive protection system with an energy management system for microgrids''. *The Electricity Journal*. 2019, vol. 32(10), pp. 106675–86.

[4] Nopmongcol U., Grant J., Knipping E. 'Air quality impacts of electrifying vehicles and equipment across the United States'. *Environmental Science & Technology*. 2017, vol. 51(5), pp. 2830–7.

[5] Gnann T., Klingler A.-L., Kühnbach M. 'The load shift potential of plug-in electric vehicles with different amounts of charging infrastructure'. *Journal of Power Sources*. 2018, vol. 390, pp. 20–9.

[6] Schneider K., Gerkensmeyer C., Kintner-Meyer M., Fletcher R. 'Impact assessment of plug-in hybrid vehicles on pacific northwest distribution systems'. 2008 IEEE Power and Energy Society General Meeting-Conversion and Delivery of Electrical Energy in the 21st Century, IEEE; 2008. pp. 1–6.

[7] Liu J., Peper J., Lin G. 'A planning strategy considering multiple factors for electric vehicle charging stations along German motorways'. *International Journal of Electrical Power & Energy Systems*. 2021, vol. 124(2), ppp. 106379–87. p.

[8] Sedighizadeh M., Shaghaghi-shahr G., Aghamohammadi M.R., Esmaili M. 'A new optimal operation framework for balanced microgrids considering reconfiguration and generation scheduling simultaneously'. *International Transactions on Electrical Energy Systems*. 2020, vol. 30(4),p. e12302.

[9] Papadopoulos P., Jenkins N., Cipcigan L.M., Grau I., Zabala E. 'Coordination of the charging of electric vehicles using a multi-agent system'. *IEEE Transactions on Smart Grid*. 2013, vol. 4(4), pp. 1802–9.

[10] Grau Unda I., Papadopoulos P., Skarvelis-Kazakos S., Cipcigan L.M., Jenkins N., Zabala E. 'Management of electric vehicle battery charging in distribution networks with multi-agent systems'. *Electric Power Systems Research*. 2014, vol. 110(1), pp. 172–9.

[11] Ramos Muñoz E., Jabbari F. 'A decentralized, non-iterative smart protocol for workplace charging of battery electric vehicles'. *Applied Energy*. 2020, vol. 272(2012), pp. 115187–96.

[12] Ivanova A., Chassin D., Aguado J., Crawford C., Djilali N. 'Techno-economic feasibility of a photovoltaic-equipped plug-in electric vehicle public parking lot with coordinated charging'. *IET Energy Systems Integration*. 2020, vol. 2(3), pp. 261–72.

[13] Alilou M., Tousi B., Shayeghi H. 'Multi-objective energy management of smart homes considering uncertainty in wind power forecasting'. *Electrical Engineering*. 2021, vol. 103(3), pp. 1367–83.

[14] Kasturi K., Bhoi S.K., Nayak M.R. 'An economic evaluation of the coordination between electric vehicle storage and photovoltaic in residential home under real-time pricing'. *Green Technology for Smart City and Society*. Springer; 2021. pp. 327–37.

[15] Shao S., Pipattanasomporn M., Rahman S. 'Challenges of PHEV penetration to the residential distribution network'. *2009 IEEE Power & Energy Society General Meeting, IEEE*; 2009. pp. 1–8.

[16] Garcia-Torres F., Vilaplana D.G., Bordons C., Roncero-Sanchez P., Ridao M.A. 'Optimal management of microgrids with external agents including battery/fuel cell electric vehicles'. *IEEE Transactions on Smart Grid*. 2018, vol. 10(4), pp. 4299–308.

[17] Vosoogh M., Rashidinejad M., Abdollahi A., Ghaseminezhad M. 'An intelligent day ahead energy management framework for networked MGS considering high penetration of electric vehicles'. *IEEE Transactions on Industrial Informatics*. 2020.

[18] Jiao Z., Ran L., Zhang Y., Ren Y. 'Robust vehicle-to-grid power dispatching operations amid sociotechnical complexities'. *Applied Energy*. 2021, vol. 281(11), pp. 115912

[19] Gayathri M.N. 'A smart bidirectional power interface between smart grid and electric vehicle'. *Intelligent Paradigms for Smart Grid and Renewable Energy Systems*. Springer; 2021. pp. 103–37.

[20] Srithapon C., Ghosh P., Siritaratiwat A., Chatthaworn R. 'Optimization of electric vehicle charging scheduling in urban village networks considering energy arbitrage and distribution cost'. *Energies*. 2020, vol. 13(2), p. 349.

[21] Gandhi O., Zhang W., Rodriguez-Gallegos C.D., Bieri M., Reindl T., Srinivasan D. 'Analytical approach to reactive power dispatch and energy arbitrage in distribution systems with DERs'. *IEEE Transactions on Power Systems*. 2018, vol. 33(6), pp. 6522–33.

[22] Hariri A.-M., Hejazi M.A., Hashemi-Dezaki H. 'Investigation of impacts of plug-in hybrid electric vehicles' stochastic characteristics modeling on smart grid reliability under different charging scenarios'. *Journal of cleaner production*. 2020, pp. 125500–11.

[23] Hariri A.-M., Hashemi-Dezaki H., A. Hejazi M. 'A novel generalized analytical reliability assessment method of smart grids including renewable and non-renewable distributed generations and plug-in hybrid electric vehicles'. *Reliability Engineering & System Safety*. 2020, vol. 196, ppp. 106746–55. p.

[24] Tian M.-W., Talebizadehsardari P. 'Energy cost and efficiency analysis of building resilience against power outage by shared parking station for electric vehicles and demand response program'. *Energy*. 2021, vol. 215(7),p.),pp. 119058–66.

[25] Alirezazadeh A., Rashidinejad M., Afzali P., Bakhshai A. 'A new flexible and resilient model for a smart grid considering joint power and reserve scheduling, vehicle-to-grid and demand response'. *Sustainable Energy Technologies and Assessments*. 2021, vol. 43(2), ppp. 100926–34. p.

[26] Holttinen H., Meibom P., Orths A. 'Impacts of large amounts of wind power on design and operation of power systems, results of IEA collaboration'. *Wind Energy*. 2011, vol. 14(2), pp. 179–92.

[27] Bhatt P., Long C., Saiyad M. 'Review of the impact of vehicle-to-grid schemes on electrical power systems'. *Advances in Electric Power and Energy Infrastructure*. Springer; 2020. pp. 199–208.

[28] Tan B., Chen H. 'Multi-objective energy management of multiple microgrids under random electric vehicle charging'. *Energy*. 2020, vol. 208(2), pp. 118360–7.

[29] Tavakoli A., Saha S., Arif M.T., Haque M.E., Mendis N., Oo A.M.T. 'Impacts of grid integration of solar PV and electric vehicle on grid stability, power quality and energy economics: a review'. *IET Energy Systems Integration*. 2020, vol. 2(3), pp. 243–60.

[30] Khan A., Memon S., Sattar T.P. 'Analyzing integrated renewable energy and smart-grid systems to improve voltage quality and harmonic distortion losses at electric-vehicle charging stations'. *IEEE Access*. 2018, vol. 6, pp. 26404–15.

[31] Spotnitz R. 'Simulation of capacity fade in lithium-ion batteries'. *Journal of Power Sources*. 2003, vol. 113(1), pp. 72–80.

[32] Zare Oskouei M., Mohammadi-Ivatloo B., Abapour M., Anvari-Moghaddam A., Mehrjerdi H. 'Practical implementation of residential load management system by considering vehicle-for-power transfer: profit analysis'. *Sustainable Cities and Society*. 2020, vol. 60(11), pp. 102144–53.

[33] Khemakhem S., Rekik M., Krichen L. 'A collaborative energy management among plug-in electric vehicle, smart homes and neighbors' interaction for residential power load profile smoothing'. *Journal of Building Engineering*. 2020, vol. 27, ppp. 100976–85. p.

[34] Lyu L., Yang X., Xiang Y., Liu J., Jawad S., Deng R. 'Exploring high-penetration electric vehicles impact on urban power grid based on voltage stability analysis'. *Energy*. 2020, vol. 198(2), pp. 117301–12.

[35] Monteiro M.R., Rodrigues Y.R., Abdelaziz M., de Souza A.C.Z., Wang L. 'New technique for area-based voltage stability support using flexible resources'. *Electric Power Systems Research*. 2020, vol. 186(2), pp. 106384–93.

[36] Heredia W.B., Chaudhari K., Meintz A., Jun M., Pless S. 'Evaluation of smart charging for electric vehicle-to-building integration: a case study'. *Applied Energy*. 2020, vol. 266(6), pp. 114803–12.

[37] Mohammadkhani N., Sedighizadeh M., Esmaili M. 'Energy and emission management of CCHPs with electric and thermal energy storage and electric vehicle'. *Thermal Science and Engineering Progress*. 2018, vol. 8, pp. 494–508.

[38] Rezaei N., Khazali A., Mazidi M., Ahmadi A. 'Economic energy and reserve management of renewable-based microgrids in the presence of electric vehicle aggregators: a robust optimization approach'. *Energy*. 2020, vol. 201(4), pp. 117629–38.

[39] World Health Organization. *Air pollution [online]*. 2021. Available from https://www.who.int/health-topics/air-pollution#tab=tab_1 [Accessed Jan 2021].

[40] Peng T., Ou X., Yan X. 'Development and application of an electric vehicles life-cycle energy consumption and greenhouse gas emissions analysis model'. *Chemical Engineering Research and Design*. 2018, vol. 131(12), pp. 699–708.

[41] Gai Y., Minet L., Posen I.D., Smargiassi A., Tétreault L.-F., Hatzopoulou M. 'Health and climate benefits of electric vehicle deployment in the greater Toronto and Hamilton area'. *Environmental Pollution*. 2020, vol. 265(6), pp. 114983–92.

[42] Ballinger B., Schmeda-Lopez D., Kefford B. 'The vulnerability of electric-vehicle and wind-turbine supply chains to the supply of rare-earth elements in a 2-degree scenario'. *Sustainable Production and Consumption*. 2020, vol. 22(1), pp. 68–76.

[43] Li D., Zouma A., Liao J.-T., Yang H.-T. 'An energy management strategy with renewable energy and energy storage system for a large electric vehicle charging station'. *eTransportation*. 2020, vol. 6(4), pp. 100076–85.

[44] Mehrjerdi H. 'Dynamic and multi-stage capacity expansion planning in microgrid integrated with electric vehicle charging station'. *Journal of Energy Storage*. 2020, vol. 29(9), pp. 101351–9.

[45] Haupt L., Schöpf M., Wederhake L., Weibelzahl M. 'The influence of electric vehicle charging strategies on the sizing of electrical energy storage systems in charging hub microgrids'. *Applied Energy*. 2020, vol. 273(1), pp. 115231

[46] Sedighizadeh M., Mohammadpour A., Alavi S.M.M. 'A daytime optimal stochastic energy management for EV commercial parking lots by using approximate dynamic programming and hybrid big bang big crunch algorithm'. *Sustainable Cities and Society*. 2019, vol. 45(December), pp. 486–98.

[47] Ma Z., Callaway D., Hiskens I. 'Decentralized charging control for large populations of plug-in electric vehicles: Application of the Nash certainty equivalence principle'. 2010 IEEE International Conference on Control Applications, IEEE; 2010. pp. 191–5.

[48] Turitsyn K., Sinitsyn N., Backhaus S., Chertkov M. 'Robust broadcast-communication control of electric vehicle charging'. 2010 First IEEE International Conference on Smart Grid Communications, IEEE; 2010. pp. 203–7.

[49] Peng M., Liu L., Jiang C. 'A review on the economic dispatch and risk management of the large-scale plug-in electric vehicles (PHEVs)-penetrated power systems'. *Renewable and Sustainable Energy Reviews*. 2012, vol. 16(3), pp. 1508–15.

[50] Yang X.-S. *Nature-inspired metaheuristic algorithms*. Luniver Press; 2010.

[51] Yang J., Yang J. 'B-spline surface construction for the complete characteristics of pump-turbine'. 2012 Asia-Pacific Power and Energy Engineering Conference; 27-29 March 2012. pp. 1–5.

[52] Goldberg D.E., Holland J.H. 'Genetic algorithms and machine learning'. *Machine Learning*. 1988, vol. 3(2), pp. 95–9.

[53] Storn R., Price K. 'Differential evolution – a simple and efficient heuristic for global optimization over continuous spaces'. *Journal of Global Optimization*. 1997, vol. 11(4), pp. 341–59.

[54] Kennedy J., Eberhart R. 'Particle swarm optimization'. Proceedings of ICNN'95-International Conference on Neural Networks, IEEE; 1995. pp. 1942–8.

[55] Kuk-Hyun Han., Jong-Hwan Kim. 'Quantum-inspired evolutionary algorithm for a class of combinatorial optimization'. *IEEE Transactions on Evolutionary Computation*. 2002, vol. 6(6), pp. 580–93.

[56] Yao X., Liu Y., Lin G. 'Evolutionary programming made faster'. *IEEE Transactions on Evolutionary Computation*. 1999, vol. 3(2), pp. 82–102.

[57] Dorigo M., Birattari M., Stutzle T. 'Ant colony optimization'. *IEEE Computational Intelligence Magazine*. 2006, vol. 1(4), pp. 28–39.

[58] Hansen N., Ostermeier A. 'Completely derandomized self-adaptation in evolution strategies'. *Evolutionary Computation*. 2001, vol. 9(2), pp. 159–95.

[59] Simon D. 'Biogeography-based optimization'. *IEEE Transactions on Evolutionary Computation*. 2008, vol. 12(6), pp. 702–13.

[60] Soroudi A., Rabiee A., Keane A. 'Information gap decision theory approach to deal with wind power uncertainty in unit commitment'. *Electric Power Systems Research*. 2017, vol. 145(1), pp. 137–48.

[61] Rezaei N., Ahmadi A., Khazali A., Aghaei J. 'Multiobjective risk-constrained optimal bidding strategy of smart microgrids: An IGDT-based normal boundary intersection approach'. *IEEE Transactions on Industrial Informatics*. 2018, vol. 15(3), pp. 1532–43.

[62] Cucuzzella M., Incremona G.P., Ferrara A. 'Decentralized sliding mode control of islanded AC microgrids with arbitrary topology'. *IEEE Transactions on Industrial Electronics*. 2017, vol. 64(8), pp. 6706–13.

[63] Tan Z., Zhang X., Xie B., Wang D., Liu B., Yu T. 'Fast learning optimiser for real-time optimal energy management of a grid-connected microgrid'. *IET Generation, Transmission & Distribution*. 2018, vol. 12(12), pp. 2977–87.

[64] Tushar M.H.K., Assi C., Maier M. 'Distributed real-time electricity allocation mechanism for large residential microgrid'. *IEEE Transactions on Smart Grid*. 2014, vol. 6(3), pp. 1353–63.

[65] Khaledian A., Ahmadian A., Aliakbar-Golkar M. 'Optimal droop gains assignment for real-time energy management in an islanding microgrid: a two-layer techno-economic approach'. *IET Generation, Transmission & Distribution*. 2017, vol. 11(9), pp. 2292–304.

[66] Kallamadi M., Sarkar V. 'Enhanced real-time power balancing of an AC microgrid through transiently coupled droop control'. *IET Generation, Transmission & Distribution*. 2017, vol. 11(8), pp. 1933–42.

[67] Jaramillo R.C., Braun J.E., Horton W.T. 'A comparative study of multi-agent control approaches for optimization of central cooling systems without significant storage'. *Science and Technology for the Built Environment*. 2020, vol. 26(8), pp. 1065–81.

[68] Kuzin A., Lukichev D., Demidova G., Anuchin A. 'Concept of hardware-in-the-loop test platform for MG with multi-agent approach'. 2020 IEEE 61th International Scientific Conference on Power and Electrical Engineering of Riga Technical University (RTUCON), IEEE; 2020. pp. 1–6.

[69] Victorio M.E.C., Kazemtabrizi B., Shahbazi M. 'Distributed real-time power management in MGs using multi-agent control with provisions for fault-tolerance'. 2020 IEEE 29th International Symposium on Industrial Electronics (ISIE), IEEE; 2020. pp. 108–13.

[70] Esfahani M.M., Hariri A., Mohammed O.A. 'Game-theory-based real-time inter-MG market design using hierarchical optimization algorithm'. 2018 IEEE Power & Energy Society General Meeting (PESGM), IEEE; 2018. pp. 1–5.

[71] Liu Y., Qu Z., Xin H., Gan D. 'Distributed real-time optimal power flow control in smart grid'. *IEEE Transactions on Power Systems*. 2016, vol. 32(5), pp. 3403–14.

[72] Tavakkoli M., Pouresmaeil E., Godina R., Vechiu I., Catalão J.P.S. 'Optimal management of an energy storage unit in a PV-based microgrid integrating uncertainty and risk'. *Applied Sciences*. 2019, vol. 9(1), p. 169.

[73] Aljohani T.M., Ebrahim A.F., Mohammed O. 'Single and multiobjective optimal reactive power dispatch based on hybrid artificial physics–particle swarm optimization'. *Energies*. 2019, vol. 12(12), p. 2333.

[74] Karmakar N., Bhattacharyya B. 'Optimal reactive power planning in power transmission system considering facts devices and implementing hybrid optimisation approach'. *IET Generation, Transmission & Distribution*. 2020, vol. 14(25) 6294–305.

[75] Levron Y., Beck Y., Katzir L., Guerrero J.M. 'Real-time reactive power distribution in microgrids by dynamic programing'. *IET Generation, Transmission & Distribution*. 2017, vol. 11(2), pp. 530–9.

[76] Shahparasti M., Mohamadian M., Teimourzadeh Baboli P., Yazdianp A. 'Toward power quality management in hybrid AC–DC MG using LTC-L utility interactive inverter: load voltage–grid current tradeoff'. *IEEE Transactions on Smart Grid*. 2015, vol. 8(2), pp. 857–67.

[77] Choi S., Meliopoulos A.P.S. 'Effective real-time operation and protection scheme of microgrids using distributed dynamic state estimation'. *IEEE Transactions on Power Delivery*. 2016, vol. 32(1), pp. 504–14.

[78] Radhakrishnan K.K., Moirangthem J., Panda S.K., Amaratunga G. 'GIS integrated automation of a near real-time power-flow service for electrical grids'. *IEEE Transactions on Industry Applications*. 2018, vol. 54(6), pp. 5661–70.

[79] Teimourzadeh Baboli P., Shahparasti M., Parsa Moghaddam M., Haghifam M.R., Mohamadian M. 'Energy management and operation modelling of hybrid AC–DC microgrid'. *IET Generation, Transmission & Distribution*. 2014, vol. 8(10), pp. 1700–11.

[80] Hashemi B., Shahabi M., Teimourzadeh-Baboli P. 'Stochastic-based optimal charging strategy for plug-in electric vehicles aggregator under incentive and regulatory policies of DSO'. *IEEE Transactions on Vehicular Technology*. 2019, vol. 68(4), pp. 3234–45.

[81] Teimourzadeh Baboli P., Bahramara S., Parsa Moghaddam M., Haghifam M.-R. 'A mixed-integer linear model for optimal operation of hybrid AC-DC microgrid considering renewable energy resources and PHEVs'. IEEE PowerTech; Eindhoven, Netherlands, *2015*; 2015. pp. 1–5.

[82] Ghanbarzadeh T., Teimourzadeh Baboli P., Rostami M., Moghaddam M.P., Sheikh-El-Eslami M.K. 'Wind farm power management by high penetration of PHEV'. 2011 IEEE Power and Energy Society General Meeting, IEEE; 2011. pp. 1–5.

[83] Aboli R., Ramezani M., Falaghi H. 'Joint optimization of day-ahead and uncertain near real-time operation of microgrids'. *International Journal of Electrical Power & Energy Systems*. 2019, vol. 107(3), pp. 34–46.

[84] Hashemi B., Teimourzadeh Baboli P., Taheri S., Wamkeue R. 'Contribution of coordinated charging of plug-in electric vehicles to urban medium voltage distribution grid'. 2019 IEEE 7th International Conference on Smart Energy Grid Engineering (SEGE), IEEE; 2019. pp. 274–8.

[85] Ji Y., Wang J., Xu J., Fang X., Zhang H. 'Real-time energy management of a microgrid using deep reinforcement learning'. *Energies*. 2019, vol. 12(12), p. 2291.

[86] Conte F., Massucco S., Saviozzi M., Silvestro F. 'A stochastic optimization method for planning and real-time control of integrated pv-storage systems: design and experimental validation'. *IEEE Transactions on Sustainable Energy*. 2017, vol. 9(3), pp. 1188–97.

[87] Abhishek A., Ranjan A., Devassy S., Kumar Verma B., Ram S.K., Dhakar A.K. 'Review of hierarchical control strategies for DC microgrid'. *IET Renewable Power Generation*. 2020, vol. 14(10), pp. 1631–40.

[88] Li Y., Miao S., Luo X., Yin B., Han J., Wang J. 'Dynamic modelling and techno-economic analysis of adiabatic compressed air energy storage for emergency back-up power in supporting microgrid'. *Applied Energy*. 2020, vol. 261(15), pp. 114448–57.

[89] Teimourzadeh Baboli P., Parsa Moghaddam M., Eghbal M. 'Present status and future trends in enabling demand response programs'. 2011 IEEE Power and Energy Society General Meeting, IEEE; 2011. pp. 1–6.

[90] Teimourzadeh Baboli P., Eghbal M., Parsa Moghaddam M., Aalami H. 'Customer behavior based demand response model'. 2012 IEEE Power and Energy Society General Meeting, IEEE; 2012. pp. 1–7.

[91] Baboli P.T., Moghaddam M.P. 'Allocation of network-driven load-management measures using multiattribute decision making'. *IEEE Transactions on Power Delivery*. 2010, vol. 25(3), pp. 1839–45.

[92] Shinde P., Hesamzadeh M.R., Date P., Bunn D.W. 'Optimal dispatch in a balancing market with intermittent renewable generation'. *IEEE Transactions on Power Systems*. 2020, vol. 36(2), pp. 865–78.

[93] Tushar M.H.K., Zeineddine A.W., Assi C. 'Demand-side management by regulating charging and discharging of the EV, ESS, and utilizing renewable energy'. *IEEE Transactions on Industrial Informatics*. 2017, vol. 14(1), pp. 117–26.

[94] Teimourzadeh Baboli P., Yazdani Damavandi M., Parsa Moghaddam M., Haghifam M. 'A mixed integer modeling of micro energy-hub system'. 2015 IEEE Power & Energy Society General Meeting, IEEE; 2015. pp. 1–5.

[95] Li Z., Ierapetritou M.G. 'A new methodology for the general multiparametric mixed-integer linear programming (MILP) problems'. *Industrial & Engineering Chemistry Research*. 2007, vol. 46(15), pp. 5141–51.

[96] Manohar M., Koley E., Ghosh S. 'Stochastic weather modeling-based protection scheme for hybrid PV–wind system with immunity against solar irradiance and wind speed'. *IEEE Systems Journal*. 2020, vol. 14(3), pp. 3430–9.

[97] Fazelpour F., Shariatzadeh M., Soltani N. 'Techno-economic feasibility of grid-connected photovoltaic power plant in Qeshm Island, Iran'. 2016 IEEE 16th International Conference on Environment and Electrical Engineering (EEEIC), IEEE; 2016. pp. 1–6.

Chapter 9

Optimal energy management strategies for integrating renewable sources and EVs into microgrids

Rahmat Khezri[1] and Amin Mahmoudi[1]

Abstract

Nowadays, microgrids are more and more tied with renewable energy sources (RESs) to improve the socio-economic advantages for sustainable developments. Due to the stochastic nature of RESs, new challenges have been raised in microgrids. Hence, effective remedies are needed to mitigate the sudden variations of available energy from RESs and to match the RES generation with load consumptions. Introduction of electric vehicles (EVs) to the grids can make a major contribution to improve the variability of microgrids considering embedded batteries in EVs. The main features of EVs in microgrids are their flexible charging/discharging capabilities, operation in vehicle-to-grid (V2G) and vehicle-to-home (V2H) modes, as well as their usual plug-in availability. By integration of RESs and EVs in microgrid, energy management system (EMS) plays an essential role for the efficient use of these components in secure, intelligent, and coordinated ways. Therefore, in this book chapter, the role of EMS in microgrids with RESs and EVs is addressed by introducing objective functions, constraints, and uncertainties. EMS strategies are introduced to solve the energy management between different components of the microgrids.

9.1 Introduction

Future energy networks should be reliable, flexible, economically efficient, and accessible to realize the targets of the smart grid initiative. Microgrid concept is a viable option to achieve these aims. Microgrids are controllable entities that are created by synchronizing a group of distributed energy resources and loads in a

[1]College of Science and Engineering, Flinders University, Adelaide, Australia

low-voltage distribution network. The main feature of microgrids is their ability to operate in both grid-connected and standalone modes [1]. Hence, microgrids are valuable for energy supply in rural and remote areas where connection to the main grid with long transmission lines is uneconomical. In microgrids, active customers are enabled to participate in control and management using real-time information [2]. Therefore, microgrids can efficiently increase the reliability and decrease the economic aspect of energy supply.

In order to improve the socio-economic advantages for sustainable development, renewable energy sources (RESs) are highly integrated into microgrids [3]. Among the integrated RESs, wind turbine (WT) and solar photovoltaic (PV) systems share the highest percentage. The stochastic character and fluctuating nature of renewable generation have added new challenges in the reliable operation of microgrids containing high penetration of wind generators and solar PV. Hence, effective remedies are needed to mitigate the sudden variations of RESs and to match the RES generation with load consumption.

By advancement of EVs, the embedded battery in EVs can make a major contribution to improve the variability of RESs in microgrids [4]. The main features of EVs in microgrids are their flexible charging/discharging capabilities, operation in vehicle-to-grid (V2G) and vehicle-to-home (V2H) modes, as well as their plug-in availability in a lot of time. By integration of RESs and EVs in microgrid, energy management system (EMS) plays an essential role for efficient use of these components in secure, intelligent, and coordinated ways. Therefore, in this book chapter, the role of EMS in microgrids with RESs and EVs is addressed by introducing objective functions, constraints, and uncertainties. The EMS strategies are introduced to solve the energy management between different components in microgrids.

This chapter is organized into four sections. Section 9.2 explains the architecture and characteristics of microgrids. Role of EVs in microgrids is discussed in Section 9.3. Section 9.4 addresses the necessity of EMS of microgrids. It also evaluates and discusses the energy management strategies for improving the power flow. Finally, the conclusions are drawn in Section 9.5.

9.2 Architecture of microgrids

A microgrid is composed of various components to supply the loads and exchange power with the main grid. Figure 9.1 demonstrates the microgrid architecture with conventional generators, RESs, energy storage systems (ESSs) and parking lots of EVs to supply commercial loads, industrial sector, and smart residential buildings. The microgrid can be connected or disconnected to the main grid at the point of common coupling.

The available components (generation and storage units as well as EVs and smart homes) are connected to the microgrid mostly using power electronic interfaces [5]. Each component is equipped with appropriate metering and protection devices and can be efficiently controlled in order to strengthen the controllability

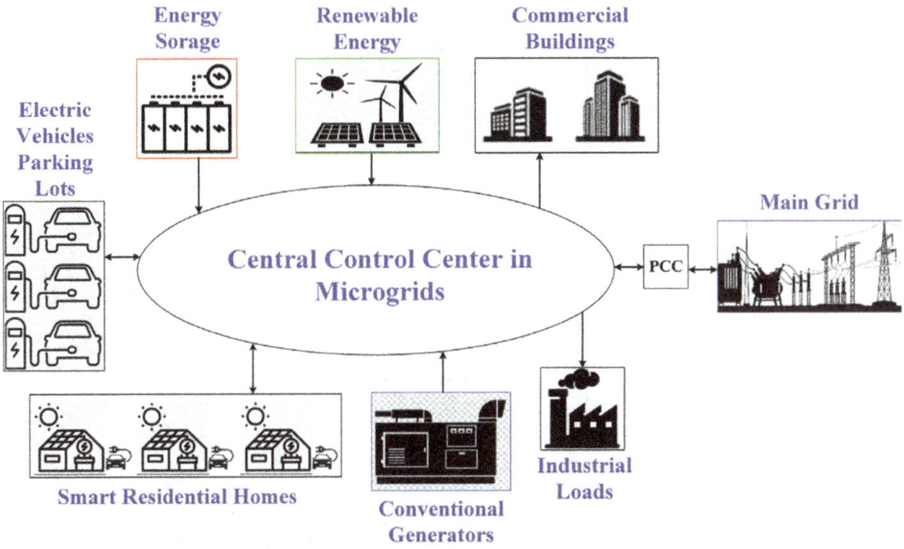

Figure 9.1 Architecture of microgrids with components and central control center

of the microgrid. The available components of the microgrid can exchange power among one another and the upstream grid.

Loads in microgrids are classified as critical, shiftable, and curtailable loads [6]. The critical loads should be supplied at any time without interruption. However, shiftable and curtailable loads are good candidates for demand response in microgrids. Demand response implies shifting or curtailment adjustment of load according to the variations in the renewable generation for enhancing the reliability of microgrids. It can also be a response to fluctuations in the market prices of electricity for increasing the microgrid profitability.

The central control of microgrids, as depicted in Figure 9.1, is responsible to control the operation of the entire microgrid. There may be local controllers too at the respective generation points, storage units, and the smart loads as well. The central control layer in microgrids coordinates with the local controllers for sudden changes in load or generation [7]. Effective coordination between the local and central controllers will result in enhanced performance of the microgrid [8].

9.2.1 Microgrid classification

Classification of microgrids can be done on the basis of control aspects, phase, power, and their operation and application, as depicted in Figure 9.2 [9]. Based on the control aspect, microgrids are categorized as centralized and decentralized. In centralized control microgrids, a central control unit is responsible for the efficient

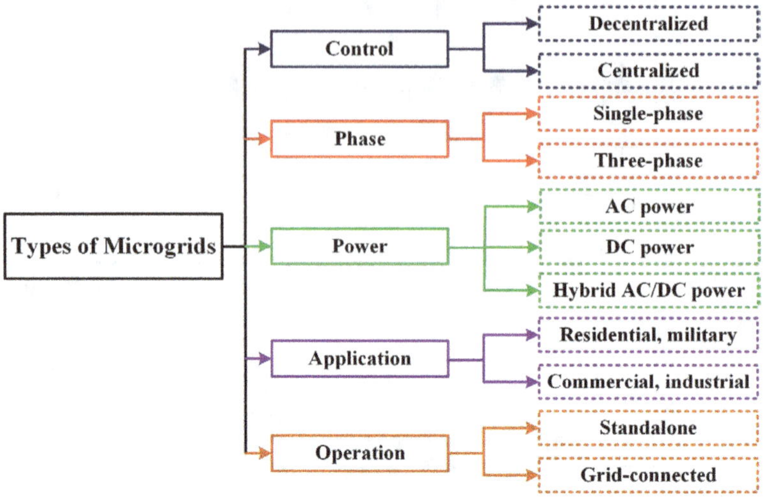

Figure 9.2 Classification of microgrids based on control, power, phase, application, and operation aspects

operation of the entire microgrid. However, in decentralized control, each part of microgrid is controlled separately based on local control.

Single-phase and three-phase types are the two classifications of microgrids based on phase. The power rating of single-phase microgrids is generally low (e.g., residential microgrids). However, three-phase microgrids are used for high-rating power applications (e.g., commercial or industrial places). Based on the type of power; the microgrids are classified as AC, DC, and hybrid AC/DC. DC and hybrid AC/DC microgrids extensively involve the use of power electronic interfaces. They also require different control schemes for proper operation. Microgrid can also be classified on the basis of their use and application. Various aims should be satisfied according to the application of microgrids. For example, reliability is the most important target for the microgrids used in military and medical applications. However, the main target for microgrids in commercial and residential areas may be the economic aspect.

The most useful classification of microgrids is based on their modes of operation. In this context, microgrids can be classified as grid-connected and standalone (isolated) operation modes. In grid-connected mode, the microgrid is consistently supported by power supply from the upstream grid. However, in low-load scenarios, the microgrids may operate in isolated mode. During the grid-connected mode, microgrid can exchange energy with the upstream grid under electricity market rates and rules. In standalone mode, all the required power of the microgrid should be supplied internally using the available generation and storage units in the microgrid. Standalone microgrids are worthy alternatives for conventional power systems in remote areas. Reliability of the microgrid is an important issue in standalone mode

Table 9.1 Components of microgrids based on the renewable generation, non-renewable generation, energy storage units, and demand response sources

Renewable generation	Fuel-based generation	Energy storage units	Demand response sources
Solar photovoltaic	Diesel generator	Battery storage	EV
Wind turbine	Gas generator	Flywheel	Smart homes
Tidal power	Coal-fired generator	Fuel cell	Lightning
Hydropower	–	Hydrogen storage	Air conditioners
Biomass	–	Thermal storage	–

especially when a major part of the energy is supplied by renewable generation units.

9.2.2 Microgrid components

Microgrids contain a wide range of generation, storage, and demand response components. Table 9.1 lists the components of microgrids based on renewable generation, non-renewable (fuel-based) generation, energy storage units, and demand response sources. There are several types of renewable generation systems: solar photovoltaic, wind turbine, tidal power, hydropower, and biomass. The application of each type of renewable generator in microgrids depends on the geographical location of the system. Hydropower is the most commonly installed renewable energy worldwide. However, remote location away from the load centers is a serious limitation of hydropower. Wind turbines and solar PV are the most matured technologies used widely, and the tidal power, however, is still in its early stages of application.

The non-renewable (fuel-based) generators were conventionally used to supply the load. Due to main disadvantages such as carbon emission, fluctuating fuel price, and high maintenance cost, the application of fuel-based generators is highly diminished. Diesel generators, gas generators, and coal-fired generators are the most applied fuel-based generation systems.

Energy storage units are increasingly integrated into the microgrids to mitigate the effect of fluctuations due to the stochastic behavior of renewable sources. There are various types of ESSs such as mechanical, chemical, and electromagnetic. Gradual advancements in battery technologies have resulted in their evolution as mature and widely used ESSs. However, the price of batteries is still a challenge for their efficient integration in microgrids.

Demand response sources are viable options for power and energy scheduling in microgrids. Demand response can be used to match the renewable energy generation with load and to schedule the microgrid based on the electricity market price of electricity. Among different sources for demand response, EVs are the best option due to the battery available in them and their availability in different levels of microgrids.

9.3 Roles of EVs in microgrids

EVs are the most promising solution for the carbon emission of the transportation system. An EV is composed of a rechargeable battery, an inverter, and the battery's controller as well as an electric motor [10]. The required energy of the electric motor is provided by the designated rechargeable battery. The inverter is responsible for converting the DC power of the battery to AC power. The provided power of the battery to the motor is managed using the controller. Lithium-ion batteries are used in most of the EVs due to high energy density, low environmental impact, and long lifetime. Because of the mobility of EVs, their designated batteries are known as mobile batteries [11]. On the other hand, the battery of EV can also be used in a bidirectional mode under V2G or V2H concepts. Thus, the battery of EV is a useful source for active and reactive power support in microgrids.

EVs can operate as an effective component of microgrids in all classification modes depicted in Figure 9.2. EVs can participate in decentralized control based on the residential customers contracts. In centralized control, the parking lots of EVs can participate in the control mechanism. The same concept can be extended for using EVs in single-phase residential microgrids and EVs parking lots in three-phase microgrids. Regarding the application modes, EVs can be used for home energy management in residential microgrids. However, in community microgrids, EVs can be used for energy scheduling. In standalone operation mode, EVs can be used to increase the reliability of the microgrids due to high variations of renewable generation and load consumption.

A few additional applications of EVs in microgrids in addition to their above-mentioned roles have been depicted in Figure 9.3. These main roles are smoothing renewable generation, spinning reserve, economic benefits, mitigating load consumption, frequency regulation, reliability improvement, peak shaving, and scheduling power exchange which all will be discussed in the subsequent sections.

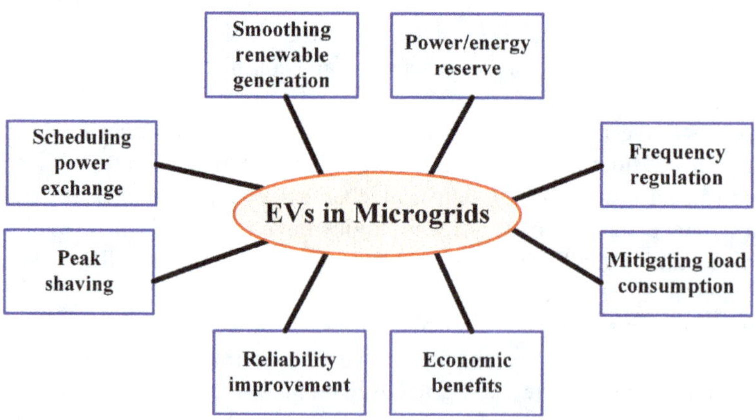

Figure 9.3 Role of EVs in microgrids

9.3.1 Smoothing renewable generation

Catering to the challenges associated with the unpredictable and intermittent nature of renewable generation had been major challenges for power system engineers. Plug-in EVs can be used effectively for smoothing these fluctuations in the output power from RESs [12]. The charging/discharging of EVs can be scheduled to track the fluctuation of renewable energy generation in microgrids. To achieve this aim, forecasting solar insolation, wind speed, ambient temperature, and any other weather data related to renewable energies is very important for EVs' scheduling. EVs can also be scheduled based on day-ahead generation of renewable energies to smooth the fluctuations.

9.3.2 Economic benefits

EVs can participate in the energy management of microgrids to improve the economic benefits. This can be achieved by scheduling the charging/discharging of EVs according to the electricity market of the upstream grid [13]. The exchange electricity (importing/exporting) rates are changed in a time-variant structure. The EVs can be charged during the low rates by importing electricity from the upstream grid. Inversely, they are discharged to export the electricity to the upstream grid during high electricity rates. In addition, the EVs can be scheduled to supply the loads during high electricity rates using the discharging state. To achieve higher economic benefits, forecasting electricity prices is efficient and should be done according to changes in the electricity market.

9.3.3 Power/energy reserve

Power and energy reserve should be maintained in microgrids to establish an equilibrium between generation and consumption at any unexpected disturbances of generation unit loss. Conventionally, fuel-based generators were used as spinning reserves in power systems. However, the penetration level of fuel-based generators has significantly decreased recently. Moreover, it is not reliable to maintain the capacity reserve by renewable energies because of their uncertainty. Hence, energy storage devices in EVs are the appropriate source for providing power and energy reserve due to their quick charging/discharging performance [14]. The reserve should be maintained by EVs parking lots due to their high capacity.

9.3.4 Mitigating load consumption

Loads in microgrids are changed consistently due to lifestyle of people as well as the working hours of commercial buildings and industry sector. The microgrids should be capable of responding to load variations in order to achieve satisfactory performance [15]. Load variations can be mitigated effectively by using battery storage in EVs. The EVs available in residential microgrids and parking lots can be managed precisely for load mitigation in stand-alone microgrids with limited generation capacity. During the peak load scenarios, the EVs can be discharged to cater the varying load requirements and power interruptions in microgrids.

9.3.5 Reliability improvement

An important issue of microgrids is the reliability and sustainability of the power supply. A reliable microgrid is a system that supplies the loads without any interruption. Any supply failure may result in customer dissatisfaction, damages of components, and high financial loss. Therefore, impressive remedies are needed to guarantee the reliability of microgrids. The available energy in batteries of EVs is an appropriate backup for microgrids to increase their reliability [11].

9.3.6 Scheduling power exchange

Microgrids exchange power with the upstream grid by importing electricity to supply its loads and exporting electricity to support the network. Limitations of the rate of exchange power may put pressure on microgrids. For example, if there is a high renewable generation for exporting and the upstream grid does not accept the extra generated power, then the microgrid should curtail the excess generation. On the other side, if the microgrid needs more power than generated, and the upstream grid could not support it, then a part of the loads in the microgrid should be shed. In such situations, EVs are a viable option for scheduling power and energy to ease the microgrid's operation due to upstream grid limitations [16].

9.3.7 Peak shaving

Microgrids have different peak loads on different days. By increasing the number of electricity consumers, the peak load is also increased every day. An increment in peak load may result in power interruption during the peak load hours on the microgrids. Hence, a major concern with microgrids is to provide uninterrupted supply to the loads during peak hours. One of the most efficient strategies for meeting peak load demand is the integration of EVs' charging stations [17]. Forecasting the peak load demands and coordinated scheduling of EVs to support the required load demands can be helpful in alleviating the load peaks. The energy surplus within the EVs can be effectively scheduled to match the load requirements during peak hours.

9.3.8 Frequency regulation using EVs

Frequency regulation is one of the most important services in microgrids where EVs can participate effectively [18]. Frequency is an important parameter for judging the balance between load consumption and power generation in a power system. Hence, it becomes necessary to achieve better frequency control in small time frames using primary frequency control. The response time of every control component participating in frequency control should match the dynamic load requirements. EVs having fast charging and discharging rates operating through controllable inverters can be used effectively for frequency regulation in microgrids. EVs can be scheduled appropriately for primary control response and can be equipped with suitable controllers for providing secondary frequency control [19].

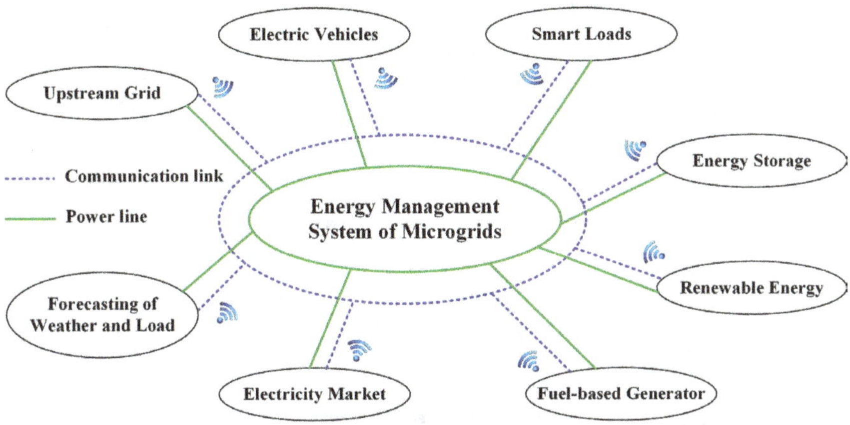

Figure 9.4 General diagram of EMS for microgrids

9.4 Energy management system of microgrids

Microgrids are rather complex systems involving generation units and loads of different nature integrated within them. Facilitating power management in microgrids is an important stage. This is implemented by EMSs. The main aim of EMS is to monitor and control the power interactions between the components in order to achieve the minimum cost with the highest reliability [20].

Figure 9.4 demonstrates a general diagram of EMS for microgrids. As illustrated, the EMS unit monitors all the components using the communication links. For example, the EMS monitors the SOC of the battery to check the remaining energy within the battery. The smart loads' data are sent to update the EMS about the available loads for demand response and load shedding. The forecasted data of renewable energies and load are sent to EMS to make the best decision to achieve a reliable and economical operation. The electricity market data and forecasted electricity rates are required to reach the minimum cost by the EMS. The EMS unit needs to be updated about the fuel rate, the capacity, and the maintenance of fuel-based generators. The associated data with EVs and charging stations are sent to EMS to check the availability and capacity of EVs for different services in the microgrid. The power flow among the components is regulated using power lines based on the data received from each one of them using the communication links.

9.4.1 Problem identification

The problem of an efficient EMS design is to find the best strategy to minimize the operation costs of the microgrids. Hence, the EMS problem, in most of the cases, is an optimization issue in which the objective functions and constraints need to be identified and discussed in detail. Figure 9.5 shows a general flowchart of EMS optimization for microgrids. The input data for EMS of microgrid should be provided at first. This contains the forecasted generation of RESs and load demand of microgrid,

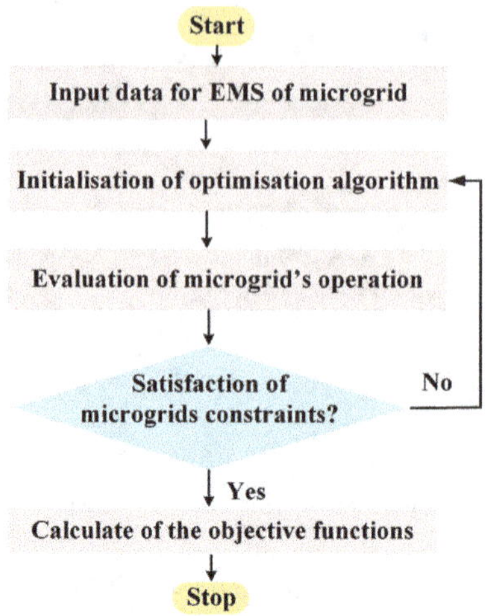

Figure 9.5 A general flowchart for optimal EMS of microgrids

the availability of EVs in the parking lots, the energy prices for the exchange with the grid, and the cost of components. The next step is the initialization of the optimization algorithm. Then, the microgrid's operation is evaluated. All the constraints of the optimization problem should be satisfied otherwise the obtained results are not acceptable. Finally, the objective function is calculated for the acceptable results and the best objective function is selected as the optimal result.

9.4.1.1 Objective functions

All the objective functions in microgrids EMS problems can be modeled as cost functions. Figure 9.6 demonstrates the used objective functions for the EMS problem. Separate objective functions can be specified for each microgrid based on its geographical location, preference of the customers, components of the microgrid, their capacity, electricity pricing programs, government policies, and the type of ESSs.

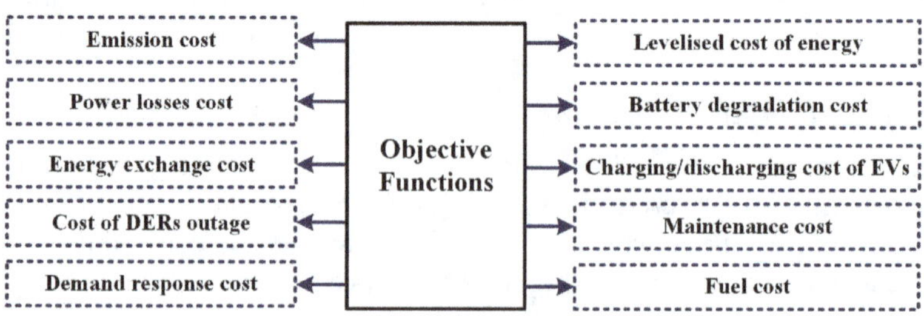

Figure 9.6 Objective functions for EMS of microgrids

The objective functions are classified on the basis of levelized cost of energy (LCOE), cost of DERs outage, as well as emission, power losses, energy exchange, demand response, battery degradation, charging/discharging, maintenance, and fuel costs [21]. All these objective functions are associated with the operational cost of microgrids containing EVs. Emission cost is incorporated due to the use of fuel-based generators producing carbon emissions. The emission cost should be minimized to achieve the lowest greenhouse gas emission. The fuel cost of generation unit i can be formulated as follows:

$$f = a_i P_{i,t}^2 + b_i P_{i,t} + c_i \tag{9.1}$$

where a_i , b_i, and c_i are the fuel cost coefficients of generation unit i, and $P_{i,t}$ is the generated power of unit i at time t.

A cost function is assigned to power losses due to the active power losses in the power lines of the microgrid. The microgrid can exchange energy with the upstream grid under designated electricity rates. The cost of energy exchange is the dominant part of EMS objective function for the grid-connected microgrids. The cost of energy exchange can be expressed by

$$f = \sum_{t=1}^{T} \left(C_{im,t}.P_{im,t} - C_{ex,t}.P_{ex,t} \right) \tag{9.2}$$

where $C_{im,t}$ and $C_{ex,t}$ are the costs for import and export energies to/from the microgrid, respectively, at time t. In addition, $P_{im,t}$ and $P_{ex,t}$ are the imported and exported powers to/from the microgrid, respectively, at time t.

When the DERs are out of service during the maintenance period, the cost of DERs outage is considered as a part of the EMS cost. However, during sudden events of load shedding, a penalty is required to be paid to the customers affected by load shedding. These costs can be formulated as follows:

$$f = C_{der,i}.t + C_{ls,t}.P_{ls,t} \tag{9.3}$$

where C_{der} is the cost of outage of DER unit i. Δt represents the time period of DER outage. $C_{ls,t}$ and C_{der} are the cost and power of load shedding, respectively, at time t.

Load shifting as a part of demand response needs some payments to the participants. LCOE is a general factor to compare energy systems like microgrids for operation and planning studies. The LCOE can be calculated based on net present costs of components and energy exchange as follows [22]:

$$LCOE = \frac{NPC_c}{E_m}.crf_c + \frac{NPC_e}{E_m}.crf_e \tag{9.4}$$

where NPC_c and NPC_e are the net present cost of components and energy exchange, respectively. E_m is the energy demand of microgrid during the operation period.

An important coefficient in calculating the LCOE is the capital recovery factor (CRF). The CRF is calculated based on the interest rate of the system and the total years of operation. crf_c and crf_e represent the CRFs of components and energy exchange in the microgrid, respectively. Due to the inflation rate, the interest rate of energy is usually higher than that of components [23]. crf_c and crf_e can be formulated as follows:

$$crf_c = \frac{r(1+r)^n}{(1+r)^n - 1}, \qquad crf_e = \frac{e(1+e)^n}{(1+e)^n - 1} \tag{9.5}$$

where r and e are the interest rates of components and energy exchange. In addition, e represents the total years of operation.

The depth of discharge and charging/discharging cycles of the battery during its operation are the factors affecting its fading capacity and consequently the battery lifetime. In order to increase the lifetime of the battery, a cost of battery degradation can be applied as an EMS cost in microgrids. The battery degradation cost can also be applied for the available EVs in the operation of the microgrid. The cost of charging/discharging of EVs in charging stations is another part of EMS's cost. The operation cost of EVs can be expressed by:

$$f = \sum_{t=1}^{T} \left(CH_{ev,t}.P_{ch,t} - DIS_{ec,t}.P_{dis,t} \right) + CD_{ev}.Cycles \tag{9.6}$$

where $CH_{ev,t}$ and n are the charging and discharging cost of EV at time t. In addition, $P_{ch,t}$ and $P_{dis,t}$ are the charging and discharging power of EV at time t. CD_{ev} is the cost of battery degradation per cycle and *Cycles* represents the total cycles of EV's battery.

Most of the components in microgrids need maintenance which is a costly part of the operation. The fuel cost is an important item for the microgrids with fuel-based generators which affects the EMS cost.

9.4.1.2 Constraints

There are various constraints in microgrids that can affect the EMS process. Ignoring the microgrid constraints may result in an incorrect EMS strategy [24]. Figure 9.7 shows the various constraints for the EMS of microgrids.

The DERs output power is the first microgrid constraint that is associated with the components. For example, the output power of fuel-based generators cannot exceed a maximum level and also cannot drop from a minimum level due to operational constraints. The energy balance of the microgrid is another constraint that

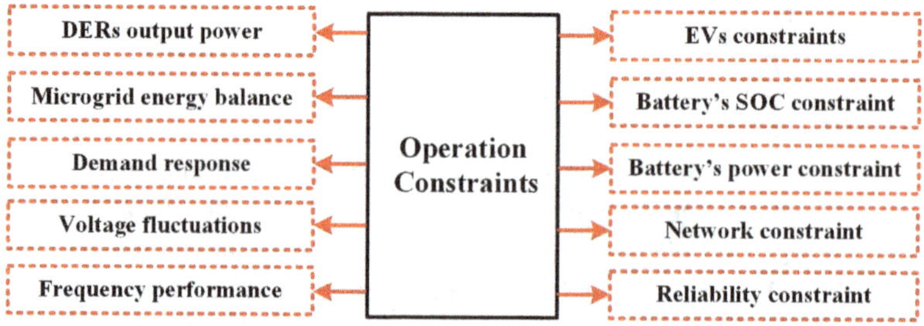

Figure 9.7 Various constraints for EMS of microgrids

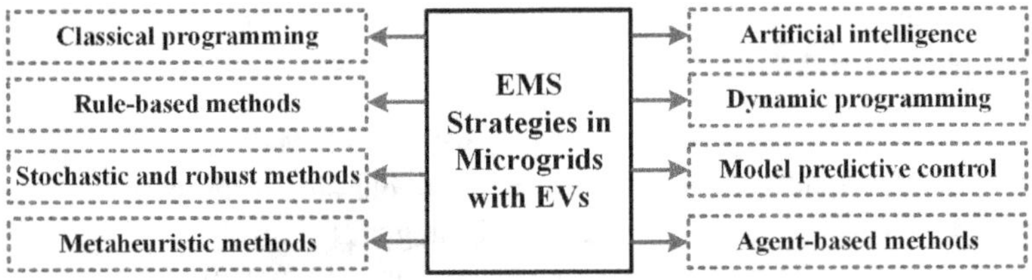

Figure 9.8 Energy management strategies in microgrids with EVs

should be preserved to reach a reliable operation within an uninterruptible power supply. If demand response is applied in the microgrid, the related constraints such as the customer satisfaction factor should also be considered. The voltage fluctuation cannot exceed maximum/minimum levels in order to maintain a stable operation of the microgrid. The frequency performance of the microgrid is limited in order to maintain the frequency within the acceptable operation range. Reliability constraints like loss of power supply probability are defined to constrain the uninterrupted loads during the operation.

Several constraints deal with EVs and ESSs in microgrids. The availability of EVs is an important constraint in the EMS problem. The availability is clarified by arriving and departure time as well as the initial SOC of EV in the microgrid. This depends on the travel distance of the users. The SOC of battery in EVs and ESSs should be limited during the operation to achieve the lowest degradation and hence the highest lifetime. The battery's power is limited due to its size and hence the ESSs or EVs cannot support power beyond their capacity.

After assigning proper objective functions and microgrid constraints, appropriate EMS strategy should be used to achieve the most efficient operation of microgrids with EVs.

9.4.2 EMS strategies for microgrids with EVs

Various strategies are discussed to overcome the EMS problem of microgrids with EVs. Figure 9.8 illustrates the energy management strategies in microgrids with EVs. These approaches are classical, rule-based, stochastic, and robust, metaheuristic, artificial intelligence (AI), dynamic programming, model predictive control, and agent-based methods. The illustrated methods in Figure 9.8 are discussed in the subsequent sections.

9.4.2.1 Classical programming for EMS

The first and most common strategy is to solve the EMS challenge by classical programming. This method is also known as mathematical optimization. The EMS problem is mathematically modeled by specifying the objective function and design constraints [25]. Since the microgrid is integrated with EVs, the constraints of EVs should also be mathematically modeled. The deployed EMS strategies by classical methods can be solved as a linear programming (LP) or non-linear programming (NLP) problem. In the LP problem, all objective functions and microgrid constraints should be linearly modeled and if there is any non-linear parameter in the problem, it should be linearized using the appropriate method. If there is any non-linear parameter in objective function or constraints, and the EMS should be solved within the non-linearities, the NLP method should be used. The EMS problem may contain some integer decision variables. In this state, the classical programming will solve the problem as a mixed-integer LP or mixed-integer NLP problem. The EMS problems with classical programming can be solved using any one of MATLAB, GAMS, CPLEX, or PAYTON software.

9.4.2.2 Rule-based EMS methods

These EMS strategies are designed based on appropriate rules which are defined by the designers. The rules are designed in a way to minimize or maximize the designated objective function by considering the microgrid's constraints. The main advantages of rule-based EMS strategies are their user-friendliness, simple structure, and easy understanding of operation [26]. In these methods, all the assigned rules are explicitly understandable for both the customers and designers, and hence, the rules can be easily updated if the operating condition of the microgrid is changed. Despite all the advantages of rule-based strategies, these methods cannot be easily used if the system dimension increases. For example, if there are several components (ESS, PV, WT, EV, etc.) in the microgrid, it is difficult to design a rule-based EMS by considering all possible operating conditions.

9.4.2.3 Stochastic and robust EMS methods

The EMS of microgrids is generally impacted by uncertainties of load consumption, renewable generation, and the electricity market. If the stochastic behavior of these parameters is not considered in the strategy, the EMS cannot work properly for all operating conditions of the microgrid. Another important uncertainty for the microgrids integrated with EVs is the travel distance, SOC, departure time, and arrival time of EVs. Hence, efficient strategies are needed to overcome these uncertainties within the EMS of microgrids with EVs.

Stochastic and robust methods are the most efficient approaches for solving the EMS of microgrids by considering uncertainties. In stochastic methods, a number of scenarios are generated based on the uncertain parameters of the microgrid (e.g., EVs uncertainties) [27]. The scenarios can be generated based on random numbers using Monte Carlo simulation. It is very important to generate random scenarios approximate to real conditions in stochastic methods. For example, in most of the cases, the uncertainty of the arrival time of EVs cannot be between 11 p.m. and 5 a.m. [28]. This should be considered in the uncertain parameter of the EVs model by limiting the arrival time. The main deficiency with stochastic methods is the high number of scenarios, which needs high computational time. As an example, if there are PV, WT, and EVs in the microgrid and each component will be modeled by five scenarios, 125 scenarios are needed to create all the possible combination of components. To decrease the computational complexity and the number of scenarios, an efficient approach is a robust optimization. Robust methods for EMS optimization are emerging areas in microgrids. In robust optimization, the uncertain parameters of microgrid are modeled within bounded distribution and range of variation [29]. Then, the worst-case scenario is extracted, and the optimization problem is solved by minimizing or maximizing the objective function.

9.4.2.4 Metaheuristic EMS methods

The metaheuristic methods are efficient algorithms broadly applied for microgrid's EMS optimization problems. The main advantages of using metaheuristic methods for the EMS problem of microgrids with EVs are flexibility, easy handling of system's non-linearity, and capability of solving highly complex problems [30]. However, these methods cannot guarantee that the global optima solution will be achieved during the solution of the problem.

Metaheuristic methods, also known as evolutionary algorithms, are population-based or nature-inspired approaches. For this reason, several metaheuristic methods are extracted which can be used for solving EMS problem. Among different approaches, particle swarm optimization (PSO) [31] and genetic algorithm (GA) [32] are two efficient and widely used metaheuristic methods.

Another advantage of using metaheuristic methods is their capability to be used in single-objective and multi-objective EMS optimization problems. By applying multi-objective evolutionary algorithms for microgrids EMS, the problem can be solved by minimizing two objective functions. The algorithm generates appropriate Pareto-fronts as the results to show the compromise between the objective functions.

9.4.2.5 Artificial intelligence for EMS

AI is an impressive strategy for EMS in microgrids. The AI techniques are based on human intelligence to implement the way of thinking and decision-making. Thus, these methods are expert based and higher knowledge of the expert about the microgrid operation will assist to design a more efficient EMS strategy. Fuzzy logic and artificial neural network (ANN) are two well-known AI methods. In fuzzy logic, the imprecise and vagueness information of microgrid can be handled. This makes fuzzy logic desirable for EMS in microgrids due to various imprecise and uncertain parameters [33]. Fuzzy logic can be deployed to model the imprecise information of EVs like the travel distance and SOC at the arrival time. In ANN, different layers connect neurons to analyze the data for making the best decision. Due to accurate learning and decision-making, ANN is worthy for EMS in microgrids when there are imprecise parameters in the model [34].

9.4.2.6 Dynamic programming for EMS

Dynamic programming can be used for complex EMS problems where there are several decision variables and components in the microgrid [35]. The dynamic programming method breaks the problem into several simple sub-problems. Then, it solves each sub-problem and stores all the solutions of the sub-problems. In microgrid's EMS, one sub-problem can be solved by finding the most appropriate charging/discharging of EVs to increase the lifetime of EV's battery, and the other sub-problem can be solved to increase the profitability of the microgrid.

9.4.2.7 Model predictive control EMS

Model predictive control (MPC) can predict the future behavior of microgrids over a limited time interval. Then, an optimal control input is computed to ensure that the microgrid constraints are satisfied, and the objective function is minimized [36]. The MPC is a worthy option for the EMS of microgrids because of the high number of uncertainties. The stochastic methods cannot update the produced uncertain parameters during the operation. This is while, some uncertainties like load consumption are updated any time depending on the behavior of customers. In such cases, the MPC method is more suitable since it can update the prediction based on the system behavior.

9.4.2.8 Agent-based EMS methods

Due to the high complexity of microgrids, distributed methods are gaining more attention for energy management. Lower cost as well as higher reliability and security can be achieved through distributed management. An efficient distributed management system is agent-based EMS. In this method, the microgrid's EMS is divided into multiple agents that interact with each other [37]. Each agent may follow a defined target. Such methods are much better for EMS problems where an individual management system cannot solve the problem. The interaction between agents is so important in agent-based EMS. This can be achieved using intelligent methods.

9.5 Conclusions

This chapter addressed the EMS of microgrids in presence of EVs and renewable energy sources. Different aspects of the integration of renewable energy sources and EVs in microgrids were discussed including the architecture, classifications, and generation-storage components. The roles of EVs in microgrids were explained considering aspects like smoothing renewable generation, frequency control, and load management. The fundamental of EMS of microgrids in the presence of EVs and RESs was discussed by categorizing objective functions and design constraints. Different strategies to implement the EMSs in microgrids with EVs were explained. It was found that energy management is solved using different methods. However, the high penetration of renewable energies and EVs in microgrids causes several problems due to uncertainties of renewable generation, travel distance, arrival time, and departure time of EVs. It is expected that the EMS strategies not only manage the charging/discharging of the EVs but also be able to consider all uncertainties to achieve robust results.

References

[1] Olivares D.E., Mehrizi-Sani A., Etemadi A.H., *et al.* 'Trends in Microgrid control'. *IEEE Transactions on Smart Grid.* 2014, vol. 5(4), pp. 1905–19.

[2] Chowdhury S., Chowdhury S.P., Crossley P. *Microgrids and active distribution networks.* London, U.K: The Institution of Engineering and Technology; 2009.

[3] Zhang Y., Gatsis N., Giannakis G.B. 'Robust energy management for microgrids with high-penetration renewables'. *IEEE Transactions on Sustainable Energy.* 2013, vol. 4(4), pp. 944–53.

[4] Oshnoei A., Khezri R., Muyeen S.M., Oshnoei S., Blaabjerg F. 'Automatic generation control incorporating electric vehicles'. *Electric Power Components and Systems.* 2019, vol. 47(8), pp. 720–32.

[5] Rajesh K.S., Dash S.S., Rajagopal R., Sridhar R. 'A review on control of Ac microgrid'. *Renewable and Sustainable Energy Reviews.* 2017, vol. 71(5), pp. 814–9.

[6] Siano P. 'Demand response and smart grids—A survey'. *Renewable and Sustainable Energy Reviews.* 2014, vol. 30(3), pp. 461–78.

[7] Bevrani H., Habibi F., Babahajyani P., Watanabe M., Mitani Y. 'Intelligent frequency control in an AC microgrid: Online PSO-based fuzzy tuning approach'. *IEEE Transactions on Smart Grid.* 2012, vol. 3(4), pp. 1935–44.

[8] Bevrani H., Franï B., Ise T. *Microgrid dynamics and control.* John Wiley & Sons; 2017.

[9] Zia M.F., Elbouchikhi E., Benbouzid M. 'Microgrids energy management systems: a critical review on methods, solutions, and prospects'. *Applied Energy.* 2018, vol. 222(6), pp. 1033–55.

[10] Liu L., Kong F., Liu X., Peng Y., Wang Q. 'A review on electric vehicles interacting with renewable energy in smart grid'. *Renewable and Sustainable Energy Reviews.* 2015, vol. 51, pp. 648–61.

[11] Mortaz E., Valenzuela J. 'Microgrid energy scheduling using storage from electric vehicles'. *Electric Power Systems Research.* 2017, vol. 143(4), pp. 554–62.

[12] Latifi M., Khalili A., Rastegarnia A., Sanei S. 'A Bayesian real-time electric vehicle charging strategy for mitigating renewable energy fluctuations'. *IEEE Transactions on Industrial Informatics.* 2018, vol. 15(5), pp. 2555–68.

[13] Luo F., Ranzi G., Kong W., Dong Z.Y., Wang F. 'Coordinated residential energy resource scheduling with vehicle-to-home and high photovoltaic penetrations'. *IET Renewable Power Generation.* 2018, vol. 12(6), pp. 625–32.

[14] Mohan V., Singh J.G., Ongsakul W. 'Sortino ratio based portfolio optimization considering EVs and renewable energy in microgrid power market'. *IEEE Transactions on Sustainable Energy*. 2016, vol. 8(1), pp. 219–29.

[15] Yang H., Pan H., Luo F., *et al.* 'Operational planning of electric vehicles for balancing wind power and load fluctuations in a microgrid'. *IEEE Transactions on Sustainable Energy*. 2016, vol. 8(2), pp. 592–604.

[16] Ahmad F., Alam M.S., Shariff S.M., Krishnamurthy M. 'A cost-efficient approach to EV charging station integrated community microgrid: a case study of Indian power market'. *IEEE Transactions on Transportation Electrification*. 2019, vol. 5(1), pp. 200–14.

[17] Uddin M., Romlie M.F., Abdullah M.F., Abd Halim S., Abu Bakar A.H., Chia Kwang T. 'A review on peak load shaving strategies'. *Renewable and Sustainable Energy Reviews*. 2018, vol. 82, pp. 3323–32.

[18] Khezri R., Oshnoei A., Tarafdar Hagh M., Muyeen S.M. 'Coordination of heat pumps, electric vehicles and AGC for efficient LFC in a smart hybrid power system via SCA-based optimized FOPID controllers'. *Energies*. 2018, vol. 11(2), p. 420.

[19] Xia S., Bu S.Q., Luo X., Chan K.W., Lu X. 'An autonomous real-time charging strategy for plug-in electric vehicles to regulate frequency of distribution system with fluctuating wind generation'. *IEEE Transactions on Sustainable Energy*. 2017, vol. 9(2), pp. 511–24.

[20] Li Y., Nejabatkhah F. 'Overview of control, integration and energy management of microgrids'. *Journal of Modern Power Systems and Clean Energy*. 2014, vol. 2(3), pp. 212–22.

[21] Ahmad Khan A., Naeem M., Iqbal M., Qaisar S., Anpalagan A. 'A compendium of optimization objectives, constraints, tools and algorithms for energy management in microgrids'. *Renewable and Sustainable Energy Reviews*. 2016, vol. 58(1), pp. 1664–83.

[22] Khezri R., Mahmoudi A., Aki H. 'Optimal planning of renewable energy resources and battery storage system for an educational campus in South Australia'. 2020 IEEE Energy Conversion Congress and Exposition (ECCE), Oct 11; 2020. pp. 3677–83.

[23] Combe M., Mahmoudi A., Haque M.H., Khezri R. 'Cost-effective sizing of an AC mini-grid hybrid power system for a remote area in South Australia'. *IET Generation, Transmission & Distribution*. 2019, vol. 13(2), pp. 277–87.

[24] Khezri R., Mahmoudi A. 'Review on the state-of-the-art multi-objective optimisation of hybrid standalone/grid-connected energy systems'. *IET Generation, Transmission & Distribution*. 2020, vol. 14(20), pp. 4285–300.

[25] Corchero C., Cruz-Zambrano M., Heredia F.J. 'Optimal energy management for a residential microgrid including a vehicle-to-grid system'. *IEEE transactions on smart grid*. 2014, vol. 5(4), pp. 2163–72.

[26] Khezri R., Mahmoudi A., Haque M.H. 'Optimal capacity of solar PV and battery storage for Australian grid-connected households'. *IEEE Transactions on Industry Applications*. 2020, vol. 56(5), pp. 5319–29.

[27] Su W., Wang J., Roh J. 'Stochastic energy scheduling in microgrids with intermittent renewable energy resources'. *IEEE Transactions on Smart Grid.* 2013, vol. 5(4), pp. 1876–83.

[28] Tushar M.H.K., Zeineddine A.W., Assi C. 'Demand-side management by regulating charging and discharging of the EV, ESS, and utilizing renewable energy'. *IEEE Transactions on Industrial Informatics.* 2017, vol. 14(1), pp. 117–26.

[29] Hu W., Wang P., Gooi H.B. 'Toward optimal energy management of microgrids via robust two-stage optimization'. *IEEE Transactions on Smart Grid.* 2016, vol. 9(2), pp. 1161–74.

[30] Meng L., Sanseverino E.R., Luna A., Dragicevic T., Vasquez J.C., Guerrero J.M. 'Microgrid supervisory controllers and energy management systems: a literature review'. *Renewable and Sustainable Energy Reviews.* 2016, vol. 60(4), pp. 1263–73.

[31] AlRashidi M.R., El-Hawary M.E. 'A survey of particle swarm optimization applications in electric power systems'. *IEEE Transactions on Evolutionary Computation.* 2008, vol. 13(4), pp. 913–18.

[32] Batool M., Shahnia F., Islam S.M. 'Impact of scaled fitness functions on a floating-point genetic algorithm to optimise the operation of standalone microgrids'. *IET Renewable Power Generation.* 2019, vol. 13(8), pp. 1280–90.

[33] Chaouachi A., Kamel R.M., Andoulsi R., Nagasaka K. 'Multiobjective intelligent energy management for a microgrid'. *IEEE Transactions on Industrial Electronics.* 2012, vol. 60(4), pp. 1688–99.

[34] Rosato A., Panella M., Araneo R., Andreotti A. 'A neural network based prediction system of distributed generation for the management of microgrids'. *IEEE Transactions on Industry Applications.* 2019, vol. 55(6), pp. 7092–102.

[35] Zeng P., Li H., He H., Li S. 'Dynamic energy management of a microgrid using approximate dynamic programming and deep recurrent neural network learning'. *IEEE Transactions on Smart Grid.* 2018, vol. 10(4), pp. 4435–45.

[36] Parisio A., Wiezorek C., Kyntäjä T., Elo J., Strunz K., Johansson K.H. 'Cooperative MPC-based energy management for networked microgrids'. *IEEE Transactions on Smart Grid.* 2017, vol. 8(6), pp. 3066–74.

[37] Mao M., Jin P., Hatziargyriou N.D., Chang L. 'Multiagent-based hybrid energy management system for microgrids'. *IEEE Transactions on Sustainable Energy.* 2014, vol. 5(3), pp. 1–46.

Chapter 10

Charging infrastructure layout and planning for plug-in electric vehicles

P. Sivaraman[1], C. Sharmeela[2], and S. Logeshkumar[3]

Abstract

This chapter contributes to the technologies and various standards associated with the latest Electric Vehicles (EVs) and Electric Vehicle Service (supply/charging) Equipment (EVSE) on par with the standard along with associated infrastructure for their erection. A broad review of infrastructure along with various, highway and vehicle safety standards are adjourned in this chapter. The chapter is written in such a way that it covers the infrastructure development and also assesses the barriers, problems and challenges of deploying an extended network of EV charging stations. The chapter also emphasizes some recommendations to help a few standardizations in charging facilities and expedite EVSE infrastructure deployment to support the continuous growth of EVs to achieve a green environment through lighter exposure of carbon settlements on environments, which in turn encapsulates the globe by slowing down the sea level rise and ozone depletion.

This chapter focuses on EVSE and the infrastructure for charging the Plug-In Hybrid Electric Vehicles (PHEVs) and Battery-Electric Vehicles (BEVs), collectively known as Plug-In Electric Vehicles (PEVs). The results are restricted to the International Electrotechnical Commission (IEC) standards, regulations, and deployment of EVSE across the globe. The standards followed in this study are applicable to any capacity two wheelers/Light Commercial Vehicles (LCV)/ Medium Commercial Vehicles (MCVs).

10.1 Introduction

An Electric Vehicle Charging Station, also known as Electric Vehicle Supply Equipment (EVSE), supplies electrical power to the Plug-in Electric Vehicle (PEV)

[1]Vestas Technology R&D Chennai Pvt Ltd, India
[2]Department of Electrical and Electronics Engineering, Anna University, India
[3]Department of Electronics and Communication Engineering, Saveetha Engineering College, India

and a few other Electric Vehicle (EV) batteries. Charging stations are classified according to the charging current, charging speed and location of the charging stations [1, 2].

The EVSEs are categorized into alternating current (AC) charging and direct current (DC) charging, based on the nature of the charging current. Batteries can only be charged with DC electricity. The EVSEs are graded as AC charging and DC charging based on where this AC-DC conversion takes place. When AC-DC conversion takes place ON board (inside the vehicle) then it is called AC charging and if it takes place fOFF-board (inside EVSE) it is called DC charging [3].

The charging station are classified as slow charging (C/5 – C/20) and fast charging (C/5 – 2C) based on the speed of charging or time taken for reaching 100% state of charge (SOC) of the battery [4]. Slow charging means charging overnight, or charging that takes about six to eight hours or more to charge an EV battery pack [1, 5]. Fast charging systems are generally described as high-power chargers that can charge a vehicle in as little as 15 minutes to 80%. Such charging points are currently common and are within the 20–50 kW range. Charging speeds are based on the EVSE power rating and battery capacity. Hovewer, the charger is usually referred to as an ultra-fast/super-fast/rapid charger when charging time is less than 30 minutes [6].

Charging stations are also classified as home/residential charging and public charging station (PCS) based on where the charging process is carried out. Home charging is commonly practiced for two wheelers where the wattage of the charger is low, such that done at household supplies [7]. The residential charging has its own advantages in terms of economy and ease of work but has a few disadvantages such as improper protocols are followed, improper wiring slot between the source to converter and converter to the battery [8]. The residential charging usually involves slow chargers because of the flexibility in available time for charging. Electric two-wheeler owners prefer to charge their vehicles overnight in their home because of less electricity tariffs when compared to commercial or industrial electricity tariffs on which they are billed in the public charging infrastructure [9–11]. EV owners roughly perform 80% of the charging in their home charging station. PCSs are the charging facility for EVs installed and operated by the utility and other private companies licensed by the utility [12–14]. Charging facility at home/operational roads/highways are a few places where either own charging facility or distributed charging facility with enough variations for charging were interrogated. A survey on 2019 reveals there will be 30% growth in the EV in total vehicle user in 2025. If this is asserted as charging infrastructure development at multiple places for charging and battery swapping[15–18]. A battery-swapping facility is also provided in some charging stations where the discharged batteries are swapped with fully charged battery, thereby eliminating the waiting time at the charging stations. In addition to this, a large number of research and development is ongoing in implementing commercial wireless charging for EVs [16].

Standardization has been a key consideration for increasing the uptake of EVs in global markets. International Electrotechnical Commission (IEC) [19] and similar bodies such as the International Organization for Standardization

(ISO), Institute of Electrical and Electronics Engineers (IEEE) [20], Underwriters Laboratories (UL), Automotive Research Association of India (ARAI), Japan Industrial Standard (JIS), National Technical Committee for Automotive Standardization (NTCAS). The National Grid (ESO) have been identified key global standards regarding the overall safety for charging station, AC or DC charging connector, EV–EVSE communication protocols and Vehicle-to-Grid (V2G) [17, 18, 21–23].

In regards to the overall Standard, IEC 61851-1 covers the general requirements for an EVSE with a rated supply voltage up to 1000 V AC or up to 1500 V DC. Standard IEC 61851-21-1 covers the EV ON-board charger requirements for conductive connection to AC/DC supply. IEC 61851-23 covers the requirements of the EV DC charging station for conductive connection to the vehicle with rated supply voltage up to 1000 V AC or up to 1500 V DC. IEC 61851-24 specifies the standards and requirements for digital communication between charging stations and EV . Hovewer, for wireless charging applications the IEC 61980-1 is used for Inductive Charging from Grid to EV . IEC 60364-7-722:2018 is used for the supply of energy to EV and feeding back Electricity from EV.

IEC 61140 covers the common aspects of installation and equipment and IEC 61000-6-2 covers the EVSE's immunity standards for the industrial environment for EVSE safety IEC 61000-6-3 covers the EVSE's emission standards for residential, commercial, and light-industrial environments.

In regards to charging connectors, IEC 62196-2, IEC 60309-1, IEC 60309-2, SAE J 1772, and GB/T 20234.2-2015, AIS 138-8 are applicable for AC charging and its connectors. IEC 62196-3, GB/T 20234.3-2015, and AIS 138-2 are applicable for DC charging and its connectors.

International Organization for Standardization (ISO) has framed the ISO 15118 for Vehicle to Grid communication interface. ISO 15118-1 covers general information and definitions, ISO 15118-2 covers network and application protocol requirements, ISO15118-3 covers physical and data link layer requirements, ISO 15118-4 covers network and application protocol conformance tests.

IEC 61851-3-1 and IEC 61581-3-2 cover requirements for protection that rely on double or reinforced insulation for AC/DC conductive power supply systems. IEC 61851-3-3 covers requirements for Light Electric Vehicle Battery Swap Systems. IEC 63110-1, IEC 63110-2, and IEC 63110-3 cover the Protocol for Management of Electric Vehicles Charging and Discharging Infrastructures.

This chapter discusses the various types of charging stations, specifications and guidelines given by various international standards organization for EV charging infrastructure.

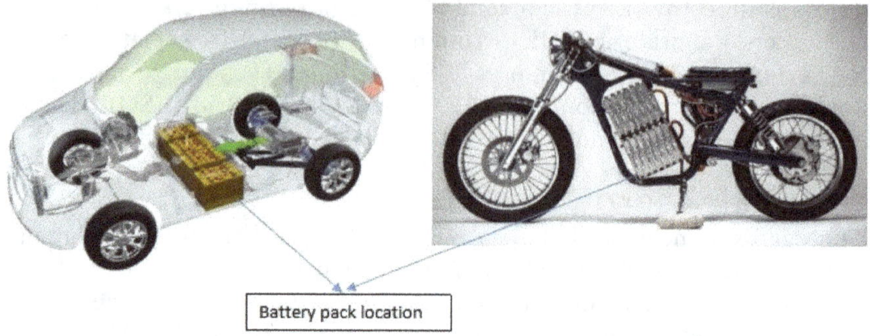

Figure 10.1 *Battery position in electric two wheeler and electric light vehicle*

10.2 Electric vehicle supply equipment technology

An EV charging station or infrastructure is also called EVSE, electronic charging station, an EV charging station, charging point, or electric charging point [3]. It provides the electrical energy to recharge the batteries of BEVs and PEVs [24].

The basic principle behind the EVSE operation: the electrical energy from the source of electricity (normally 240 V or 110 V) is used to recharge the battery of a PEV [16, 25–30, 31]. The charging stations are commonly referred to as EVSE units [32].

Figure 10.1 illustrates how compatibly the battery pack placement on EVs (four-wheeler and two-wheeler) both for the newly developed and retrofit models is shown for clarity.

10.3 Basic EVSE components

10.3.1 EVSE

The EVSE is an electrical power source, and this equipment can provide the power supply in the form of either AC or DC supply to the EVs for battery charging. EVSE charging capacities are selected based on how fast batteries are being charged [3, 33, 34]. Nevertheless, the PEV on-board adapter must be able to reach the maximum performance of the EVSE in order to know the fastest charging times.

10.3.2 Electric vehicle connector

The device/equipment connected e between EVSE and the EV is called connector. The connectors are of two types: slow/fast chargers and rapid chargers. Generally, there are three prevalent EV connectors are used in EVs [35, 36]. They are SAE J1772, CHAdeMO connector, and Supercharger Connector Type 2. The different types of connectors and their power ratings are given in Table 10.1.

Table 10.1 Power/charging rate of different connector types

S. No.	DC connector type	Typical power ratings(kW)
1	CCS - SAE J1772 connector	50 150 350
2	CHAdeMO connector	50
3	Type 2 connector	130

10.3.3 Electric vehicle inlet

The inlet of EVs is the equipment that provides the physical connection between the connector and EVSE. More than one inlet ports are available in many EVs at different locations. Three common types of inlets for AC charging are: (1) Type 2 inlet-IEC 62196, for which electric power can be provided as single-phase or three-phase AC or DC; (2) SAE J3068 is a three-phase AC connector that included inlet; and (3) the Guobiao standard GB/T 20234.2-2015 adjuncted inlet.

10.4 PEV battery systems

PEV battery system is basically classified into two types:

1. A smaller system
2. A larger system.

The smaller system is a 12-V conventional battery and both lead-acid and lithium-ion stacks are available in this configuration [37]. Latter the larger system of over and above 12 V is called as the traction batteries which are 24 V, 36 V, 48 V, 60 V, 72 V, and many others. For instance, to create a lead-acid battery with a capacity of 60 V, 5 smaller batteries of capacity 12 V are added in series connection, which are made with an instinct on the capacity and model of the batteries chosen. In lithium-ion batteries, 10S4P or 12S6P configurations are some aspects of the physical connection of the batteries used to get the general voltage output. Even though smaller battery stacks with lead-acid could be made to deliver the required voltage, that system is used as an auxiliary power for the on-board systems. The larger batteries provide propulsion to the vehicle. The traction batteries come in a wide range of power ratings and are configured to suit the unique needs of the PEV model. The more advanced classification of rechargeable energy storage system is now being popular for traction batteries [38, 39].

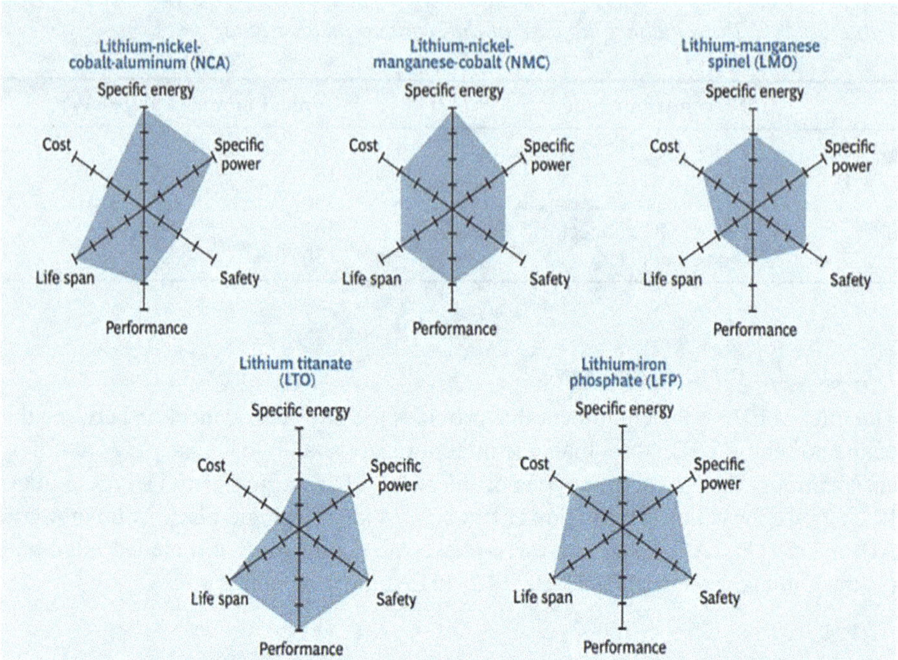

Figure 10.2 Lithium-ion battery comparison using spider graph (Source: Battery University)

10.4.1 Battery technology—a power unit of EV

NiMH batteries have higher specific energy than lead-acid, prototype EVs deliver up to 200 km of range. But the lithium-ion battery-equipped EVs provide 320–480 km of range per charge [40]. Lithium is also less expensive than nickel. Nickel-zinc batteries are cheaper and lighter than Nickel-cadmium batteries. They are also cheaper than (but not as light as) lithium-ion batteries. There exist different configurations of Li-ion batteries listed as

1. Lithium nickel-cobalt-aluminum (NCA)
2. Lithium nickel-manganese (NMC)
3. Lithium manganese-oxide spinel (LMO)
4. Lithium titanate-oxide (LTO)
5. Lithium iron phosphate.

The spider graph in Figure 10.2 [41] shows the potential capacity in terms of six parameters cost, specific energy, specific power, safety, performance, and life span of conglomerating all the factors, NMC configuration suits to be the best in terms of performance and energy density. Stacks of these batteries will be made depending on the requirement.

Table 10.2 Charging comparison in EVs

S. No.	Conductive charging	Inductive charging	Battery switching
1	Connection-based charging	Contactless charging	Physically swapping the battery at the charging station
2	The connector used between the charger (power source) and the vehicle battery	Wireless charging is used to transfer from a transmitting pad laid beneath the roadway to the receiving pad which is connected to the vehicle underneath	Depleted EV battery (discharged battery) pack is swapped with a fully charged battery pack
3	This type of charging is more suitable for residential areas	This type of charging is utilized in city road	This type of charging is utilized in highway roads

10.5 Charging system

A charging system for EVs is developed to increase the capacity of the battery system associated with it. The charging system can be

1. Conductive charging
2. Inductive charging
3. Battery swapping

The comparison of conductive charging, inductive charging, and battery swapping is listed in Table 10.2.

10.5.1 Options for electric vehicle supply equipment

a. Charging Infrastructure Options for Short-Distance Commuters

- Level 1 (Slow charging)
- Level 2 (Medium charging)

b. Charging Infrastructure Options for Longer-Distance Drivers

- Level 3 (Fast charging)
- Battery-swapping Charging Infrastructure

c. Advanced Charging Scenarios

- V2G
- Inductive Electricity Charging (Wireless charging)

d. EVSE Deployment Options for Local Governments

- Develop an informed strategy
- Public Charging Infrastructure: maximize visibility and assuage consumer range anxiety
- Make chargers available for multi-unit dwellings without garages
- Adapt urban planning codes
- Use clean electricity for municipal chargers
- Educate all EV stakeholders about charging options and operations
- Public–Private Partnerships: establishing viable charging infrastructure markets
- Lobby policymakers on the state and federal level
- Establish finance mechanisms for chargers
- Streamline the infrastructure permitting processes

Even though multiple stacks for charging are discussed above, a few standards are discussed below on their positive substantiation.

10.6 Battery charger

The chargers or battery chargers are used to charge the EV batteries. For the increased lifetime of the battery, a C/5 or C/2 rating is preferred for charging and this will take more time to charge the battery. Taking this into accounting, charging at the rate of 2C (2 times of charging rate), 5C (5 times of charging rate), and 10C is possible in EV. Even nowadays swappable battery trend is more preferred, which reduces the waiting time to charge the vehicle at the charging station.

DC Fast Chargers (DCFC) is used to supply high current to charge the PEV batteries. As discussed earlier, DC chargers are OFF-board chargers and the EVSE is placed outside the EV. It is significant to note while purchasing the PEV that easy charging of the PEV is available by AC or ON-board chargers because of easy availability of AC power supply for charging. Figure 10.3 shows the typical charging arrangement of ON-board charger.

10.7 EVSE charger classifications

EVSE is basically graded as Level 1, Level 2, or DC fast charge. In general terms, the EVSE classification relates to the power level given by the equipment for recharging the batteries of a PEV. The use of higher charging levels will substantially reduce the time taken for the batteries to recharge. Level 1, Level 2 and DCFC are the most commonly used charger levels.

An AC charging station, as the name implies, provides AC power from the grid to the EV, which is then converted to DC using the ON-board charger to charge the

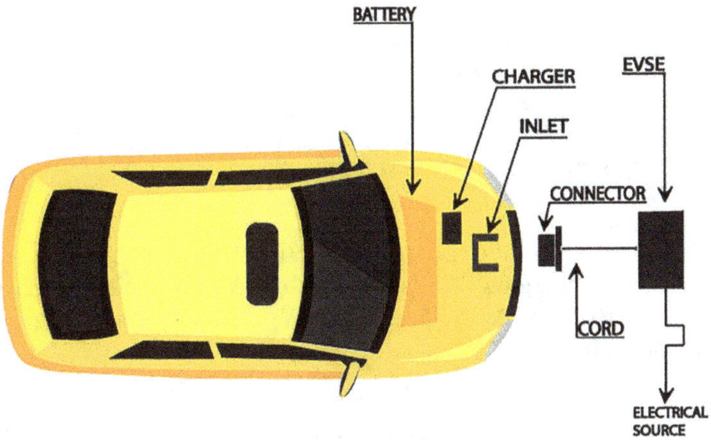

Figure 10.3 EV supply/charging technology portfit

vehicle. These chargers are also called Level 1 and Level 2 chargers that are used in residential and commercial places. The advantage of an AC charging station is that the ON-board charger will regulate the voltage and current as required for the EV hence it is not mandatory for the charging station to communicate to the EV. The disadvantage is low output power, which increases the charging time. A typical AC charging system is shown in Figure 10.4. As we can see, the AC from the grid is supplied directly to ON-board charger through EVSE and the ON-board charger then converts it to DC and charges the battery through the BMS. The pilot wire is used to sense the type of charger connected to the EV and set the required input current for the ON-board charger. We will discuss more on this later.

A DC charging station gets AC power from the grid and converts it to DC voltage and uses it to charge the battery pack directly bypassing the ON-board charger. These chargers normally output high voltage of up to 600 V and current up to 400 A, which enables the EV to be charged in less than 30 minutes as compared with 8–16 hours on an AC charger. These are also called Level 3 chargers and commonly known as DCFC or superchargers.

The advantage of this type of charger is its fast charging time while the disadvantage is its complex design and in engineering aspects, where it needs to have handshaking communication with EV to charge it efficiently and safely. The EVSE is arranged in stacks to provide high current a single stack will not be able to provide high current due to power switch limitations [42–45].

Level 1 charger is meant for residential use, Level 2 charger is provided as an update for Level 1 and Level 3 charger or Supercharger is meant for PCSs alone, Level 2 and Level 3 chargers are considered to be more efficient than the Level 1 charger. The typical EVSE for Level 1 and Level 2 AC charging station is shown in Figure 10.4 and the Level 3 AC charging station is shown in Figure 10.5.

The comparison of Level 1, Level 2, and Level 3 chargers is listed in Table 10.3.

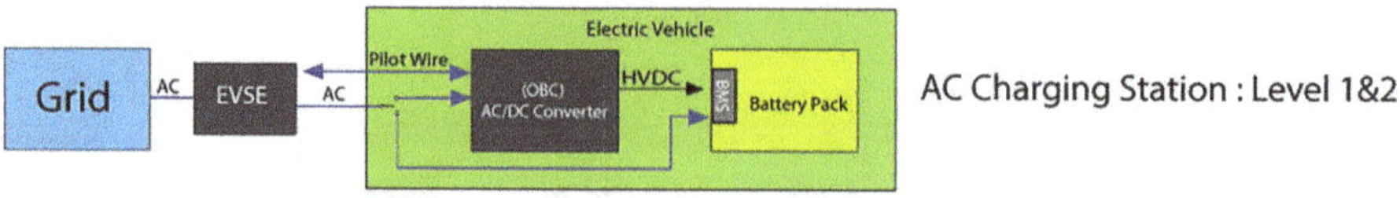

Figure 10.4 AC charging station, Levels 1 and 2

The typical connector of CHAdeMO-DC-50 kW rapid charger type 2 is shown in Figure 10.6, and Charging Standard-DC or AC-50 kW rapid charger type 2 is shown in Figure 10.7, and Mennekes-DC-7 kW to 22 kW rapid charger type 2 is shown in Figure 10.8.

10.8 EVSE signaling and communications

EVSE and PEV interaction during the battery recharging process can be an interactive and dynamic process that requires communications between both elements. Multiple, ongoing communications exchanges occur during charging; one of the primary purposes of these communications is to regulate the amount of current provided to charge the vehicle. The EVSE informs the vehicle of the maximum current available, allowing the PEV to manage current flow within the EVSEs service breaker capacity. Additional primary communications and interactions take place that monitor the SoC of the batteries and also allow the PEV to bypass the ON-board charger and use the EVSE charger if a DCFC station is being used.

10.9 Vehicle-to-grid

V2G technology would allow EVs to both draw electricity from and return electricity to the grid. This innovative technology capitalizes on the fact that electrified vehicle batteries store the same type of energy used to power homes and offices in demand on futuristic cases (60 Hz AC). In fact, an electrified vehicle battery is capable of producing enough power to satisfy the electricity demand of 10 US homes (which needs 110 V and 60 Hz supply). Ideally, vehicle owners would recharge their vehicles during OFF-peak hours (overnight, for example) when electricity is the least expensive. During peak hours, when electricity prices are highest, electricity stored in the battery would be sent back to the grid.

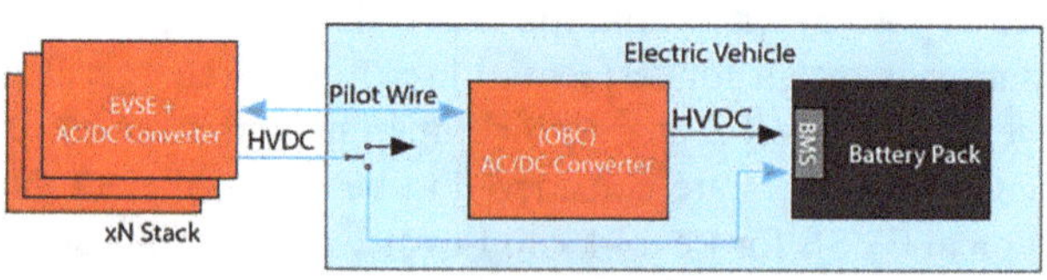

Figure 10.5 AC charging station, Level 3

Table 10.3 Charger comparison on the basis of input and the time taken for charging

Charging station type	Charger level	AC supply voltage and current	Charger power(kW)	Time to charge a 24 kWH battery pack
AC charging station	Level 1 – Residential	Single-phase – 120/230 V and ~12 to 16 A	~1.44 to 1.92	~17 hours
AC charging station	Level 2 – Commercial	Split-phase – 208/240 V and ~15 to 80 A	~3.1 to 19.2	~8 hours
DC charging station	Level 3 – Supercharger	Single-phase – 300/600 V and ~400 A	~120 to 240	~30 minutes

Figure 10.6 CHAdeMO-DC-50 kW rapid charger type 2 connector

V2G provides a wide range of benefits to vehicle owners, utility companies and the general public. Vehicle owners may save more money by recharging when electricity is inexpensive and returning electricity to the grid when it is more expensive. Utility companies may benefit because V2G would reinforce grid stability by reducing electricity production requirements during peak daytime hours when electricity demand is at its highest. Furthermore, V2G may allow utility companies to increase their use of intermittent renewable energy such as solar and wind. Increased reliance on renewable energy would decrease greenhouse gas emissions and air pollution from the electricity sector. Currently forecasts on the integration of V2G and stationary batteries to store surplus wind energy.

10.10 Wireless charging

The typical schematic diagram of wireless charging of an EV is shown in Figure 10.9.

10.10.1 Inductive and resonant technologies

The term inductive charging, also called Wireless Power Transfer, is an egressing field that allows PEV recharging without a virtual connection through a wire or cabled connection. The most common prevailing application uses a charging pad installed on the car and a receiving pad installed on the pavement. This inductive coupling utilizes the lenz law for the flux linkage and the EMF produced due to the rate of change of flux. The electrical current flows through the pavement pad, which

Figure 10.7 Combined Charging Standard-DC or AC-50 kW rapid charger type 2 connector

Figure 10.8 Mennekes-DC-7 kW to 22 kW rapid charger type 2 connector

generates an inductive magnetic field detected by the charging pad to power vehicle batteries for the cars.

Induction chargers produce an alternating electric current from the interior of a charging base plate and a second induction coil from the portable device (i.e, PEV) takes power out of the electromagnetic field and by the converse effect of lenz law converts it back to electric current for charging the battery. The position of the two induction coils combines to form an electric transformer.

Although the inductive charging device uses resonant inductive coupling, greater distances can be achieved between sender and receiver coils. Recent enhancements to this resonant system include the use of a handheld transmitting coil, with the use of silver-plated copper or aluminum components for the receiver coil.

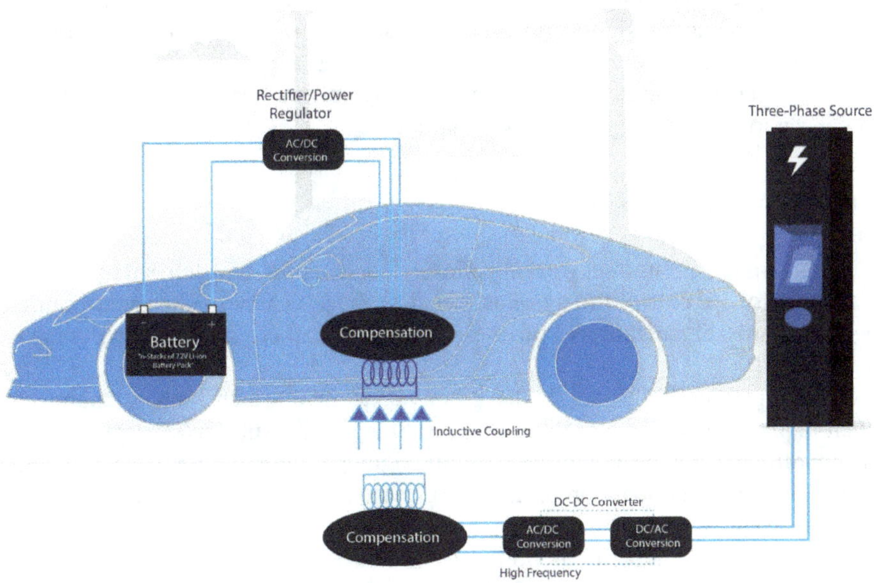

Figure 10.9 Illustrating the inductive coupling between the charging and receiving pads

Figure 10.10 EV, plug-in heavy EV safety standard

10.10.2 Research on wireless charging

The scientific community, academic fraternity, governmental and private and pivotal industries are making significant efforts in research and development to help realize the promise of untethered charging of PEV batteries.

The MIT, Utah State University and leading universities are continuously developing smarter technologies for the strangling deployment.

10.11 Vehicle design

Several organizations are working closely together to develop safety standards and tests for PEVs and the systems found in those vehicles. The most prominent US standards organizations and the interaction of their standards are shown in Figure 10.10.

Requirements differ so all the groups concerned are seeking to "harmonize" their requirements in order to prevent severe disputes. A substantial degree of EVSE standardization has been achieved but an ongoing controversy exists in the United States on the standardization of the DCFC system used to connect the charging station to the car.

Some PEV manufacturers prefer the standard CHAdeMO connector and some other manufacturers have selected the standard SAE J1772 Combo T2. Several automobiles are capable of handling both types of connectors for recharging. CHAdeMO connectors have been the most widely used fast charging connector; the new 1 772

Combo has the flexibility to use for both Level 2 and DCFC connections. Both connectors follow all safety requirements and are somewhat similar in the way they are used and the selection of individual vehicle supplier connectors is largely based on technical preferences. The most striking aspects are the CHAdeMO is considerably larger and limited to 62.5 kW while the smaller 1 772 Combo is rated at 100 kW.

10.11.1 Society of automotive engineers

SAE International, initially founded as the Automotive Engineers Society, is a US-based professional association and organization for the development of standards to the technical practitioners in various industries. The principal interest of SAE is in the transport sector, such as the automotive, aerospace and commercial vehicles. SAE coordinates the development of technical standards based on the suggestion of SAE stack holder committees and task forces SAE is involved in developing standards used in the US automotive industry and by extension, those applied to PEV production and supporting infrastructure.

10.12 Innovative charging solutions

This innovative charging solutions can increase deployment and accessibility

- Off-grid solar and storage
- Mobile storage
- DC fast charge plazas
- Streetside lamp/power pole
- Shared mobility fleets
- Modular automated parking garages.

10.12.1 Solar charging

Very recently, EVs have been rapidly gaining popularity as they are an eco-friendly alternative to gasoline-driven cars. According to a report by International Energy Agency, the global EV ownership has shown a rapid growth of 71% from 2019 to 2020. Studies also suggest that EVs, as compared to fossil fuel cars, have considerably lower greenhouse gas emissions that can even be brought down to zero, provided green electricity is used for charging the EVs. Access to green electricity is also becoming a reality due to the paradigm shift in the global energy generation landscape. On the world stage, a lot of innovations and investments are happening in this space and specifically in the solar power generation sector [46]. The global advisory firm DNV GL in its Energy Transition Outlook 2018 has reported that 80% of global electricity production will be powered by renewable energy sources (solar energy and wind power). Figure 10.11 shows the typical solar-based EV charging station.

Figure 10.11 Alternative EV charging—solar charging platform

10.12.2 Development hindrances in EVSE infrastructure expansion

The PEV sales across the globe will grow rapidly with a 30% growth overall in the next 2 years. Limited access to recharging facilities is one of the most commonly cited barriers to the continued adoption of PEVs and while direct equality of infrastructure growth to PEV sales is not required, the continued accelerated deployment of PEVs must be supported by accessible and reliable recharge facilities. So now common infrastructure that matches all EVs for charging is almost developed.

10.12.3 Governmental awareness

Although the extension of the EVSE system has many major obstacles, the primary obstacle is the insufficient alignment of the needs and specifications for developing an easily accessible PEV recharge network. The department of Energy clean cities program identifies fewer barriers in charging station installation. All countries across the globe face major air pollution due to 25% of the emissions from light-duty vehicles, 68% of NOx is produced by trucks, buses and OFF-road vehicles. To maintain the air quality standards, every country must electrify transportation to meet its climate and air quality goals.

10.12.4 Financial surprises

EVSE technology is an upcoming transport technology and there is a lot of work being done to bring standardization to all elements of the infrastructure. Deployment of readily available charging infrastructure will be delayed until considerable work on standards such as PEV/EVSE/grid communications and management is completed.

Earlier the biggest barrier is the lack of financial incentives to support the deployment of a readily accessible charging infrastructure to support independence for the environment and national energy. But nowadays government support for this egressing field is achieving optimum attention. Developing network standards is going to be very useful due to the government's monetary support to install charging stations. Ideations on the development of charging stations at parking lots, petrol stations and separate charger units installed and swappable batteries for two wheelers are current solutions to coir on the current situation.

10.12.5 Standards

PEV owners do not reap the advantages of centralized refueling facilities that are common to conventionally powered vehicle owners. Existing PEVs and recharging facilities are secure and effective, requiring a substantial increase in dedication, effort and standardization to achieve the necessary level of EVSE availability and convenience.

Creation of standards to enable expansion of the EVSE network has been sluggish and sometimes interrupted. Combining the lack of standardization of the physical layer and reliability with the proprietary nature of the existing PEV recharge networks can present the average motorist with a real challenge in considering and converting to an EV. Culminating all the barriers is due to the lack of coherent strategy for developing EVSE infrastructure. The efforts of national strategic goals on forecasting the EVSE infrastructure are to be considered in equal on par with other consideration.

10.13 Site visit, evaluation and selection

The selection of charging station is affected by various factors and many literature are available that may reveal the attributes of economic and land factors that are paid with great attention in the site selection. There are few other factors for serious consideration for convenience and security. For service availability, the charging station should be located at a place that comprises traffic conditions and huge demand for service. Therefore, the attributes considered for site selection of charging stations are: (1) economic factors, (2) engineering feasibility, (3) service availability, (4) social factors, (5) environmental factors, and (6) land factors. The economic factors include construction cost, operation and maintenance cost, investment payoff period. The engineering feasibility includes distance from the substation, influence on the power system and availability of resources. The service availability includes the convenience of transportation, service capability and service radius.

10.14 Planning and selection of charging station

Figure 10.12 shows the charging station site planning and construction. The location is selected in such a way that what sort of EV is going to get charged, whether two wheelers, three wheelers, four wheelers, heavy wheelers called as

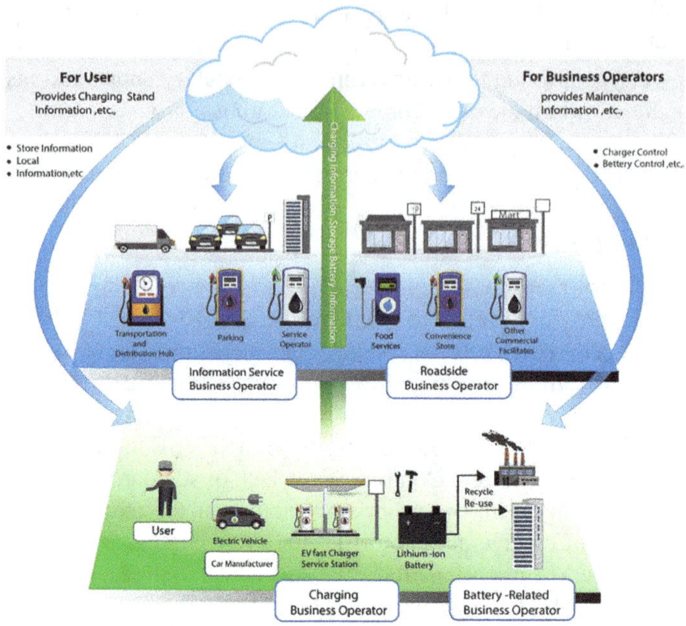

Figure 10.12 *Charging station site planning and construction*

trucks or a multi-optional exclusive charging station that can be used to charge all types. Once the necessary information about the type of charging that is going to be performed, the location can be planned. Once the necessary information is gathered and appointments are coordinated, the contractor will visit the site. The first site visit will answer any additional questions and resolve any decisions yet to be made. The charging station selected will influence the site selection and vice versa. If the customer has selected both the charging station and the site in advance, it will be important to check that NEC is adhered to and potential problems are avoided.

10.15 A few initiatives and recommendation for accelerating the development of EVSE infrastructure

The following recommendations are made in an effort to focus attention on some of the key elements needed to stabilize and accelerate the establishment of a nationwide PEV recharging infrastructure.

1. Policy and decision-maker involvement and support for the expansion of EVSE infrastructure includes:

a. Overall plan for EVSE network expansion
b. Government incentives linked to the deployment and ROI of the EVSE infrastructure to support a business model for deployment
c. Specific attention is needed to develop the PEV recharge infrastructure for urban and multi-family environments
d. Requirements for the creation of an EVSE network in the federal highway system. Adoption of infrastructure reliability and performance standards similar to those required of conventional refueling stations
e. State and local strategies to deploy EVSE in primary travel corridors, airports and major public meeting places
f. Finding and developing wireless charging EVs
g. Increased V2G research and development
h. PEV and EVSE awareness campaigns for the general public
i. Standardized methodology for setting schedules for OFF-peak electric power rates for PEV owners
j. Standardized accounting methodology enabling all public EVSE operators in all countries to charge kilowatt-hours
k. Incentives to recognize PEV environmental contributions allowing power companies to set tariff levels for publicly available EVSE
l. Restoration of tax credits for public and private EVSE installations.

2. Industry support work groups and stakeholders:

a. Strategy for EVSE network expansion
b. PEV and EVSE public education campaigns
c. Strategies to deploy EVSE in primary travel corridors, airports and major public meeting places
d. Recharge station location capabilities based on non-Internet for PEV owners lacking Internet access or smartphones
e. Models for EVSE and ROI strategic deployment
f. Improved coordination among stakeholder groups in promoting the expansion of EVSE infrastructure.

3. EVSE/network/auto industry standards—uninterrupted contribution to standards creation and adoption for:

a. Payment systems and applications and back-office correspondence
b. Public relations network-to-network EVSE
c. Communications via EVSE V2G
d. Reliability and quality of the station performance measurements.

10.16 Feasibility of accelerating EVSE installation

While local governments do not have the authority or resources required to implement the entire process model for accelerated EVSE installation recommended here unilaterally, the above action areas emphasize the importance of local governments in ensuring timely EVSE deployment. Recognize that the organizational culture and inertia of the relevant local government institutions may prevent the process reforms recommended above from being implemented. However, given the societal imperatives associated with the deployment of electrified vehicle technologies, Bay Area municipalities should seek to overcome institutional hurdles and set up an EVSE installation process that contributes to environmental protection and energy independence by accelerating consumer acceptance of EV technologies.

The overall development of the charging infrastructure is explained with utmost care, taking all the facts into account.

10.17 Conclusion and recommendations

The chapter then focuses on the deployment of EVSE infrastructure—describing available options, highlighting existing possible support from the government and outlining the steps that local governments can take to address the challenge of deployment at EVSE.

This chapter makes the case that EVs will play a significant role in reducing transport sector air pollution and GHG emissions. It identifies obstacles to the penetration of the EV market and argues that to reduce these barriers, the public sector, including local governments, should act.

10.17.1 Key recommendations

1. PCSs
2. Curbside charging station deployment for multi-unit dwellings
3. Adaptation of urban planning codes
4. Stakeholder education
5. Public–private partnerships
6. Establishment of a financing mechanism for private charging stations
7. Streamlining the charging station installation process.

Finally, this chapter discusses a simplified, consumer-friendly deployment process for charging stations and the process improvements required to achieve this. The EV development sector provide these guidelines in the expectation of customers interest and will be able to set up a practical, efficient and duplicable regional EVSE implementation model that represents the complexity and urgency associated with the current environmental and energy challenges faced across the EV interested areas.

References

[1] Sivaraman P., Sharmeela C. 'Power quality problems associated with electric vehicle charging infrastructure' in Padmanaban S., Sharmeela C., Holm-Nielsen J.B. (eds.). *Power Quality in Modern Power Systems*. Academic Press; 2020.

[2] Sharmeela C., Sanjeevikumar C., Holm-Nielsen C. (eds.). *Microgrid Technologies*. Scrivener Publishing; 2021. Available from https://onlinelibrary.wiley.com/doi/book/10.1002/9781119710905.

[3] Sivaraman P., Sharmeela C. 'Power system harmonics' in Padmanaban S., Sharmeela C., Holm-Nielsen J.B. (eds.). *Power Quality in Modern Power Systems*. Academic Press; 2020.

[4] Sivaraman P., Sharmeela C. 'IoT based battery management system for hybrid electric vehicle' in Chitra A., Padmanaban S., Holm-Nielsen J.B., Himavathi S. (eds.). *Artificial Intelligent Techniques for Electric and Hybrid Electric Vehicles*. Scrivener Publishing; 2020.

[5] Sivaraman P., Sharmeela C. 'Introduction to electric distribution system' in Baseem K., Hassan H.A., Ghassan. (eds.). *Handbook of Research on New Solutions and Technologies in Electrical Distribution Networks*. Hershey: PA: IGI Global; 2020. pp. 1–31.

[6] Steen D., Tuan L.A. 'Impacts of fast charging of electric buses on electrical distribution systems'. CIRED 24th International Conference on Electricity Distribution, Paper 0802; Glasgow, 12–15 June; 2017.

[7] Dubey A., Santoso S. 'Electric vehicle charging on residential distribution systems: impacts and mitigations'. *IEEE Access*. 2015, vol. 3, pp. 1871–93.

[8] Anable J., Darnton A., Pangbourne K., Lane B., Banks N., Henry N. 'Evidence base review of business travel behaviour to inform development of a segmentation of businesses'. *Government social research*. 2015.

[9] Plug-in electric vehicle handbook [online], August 2013. Clean cities publications. Available from http://www.afdc.energy.gov/uploads/publication/pev_workplace_charging_hosts.pdf [Accessed 13-08-2021].

[10] SAE International. *Communication Between Plug-in Vehicles and Off-Board DC Chargers* [online]. Available from http://standards.sae.org/wip/j2847/2/ [Accessed 13-08-2021].

[11] Wikipedia. *Inductive charging* [online]. Available from http://en.wikipedia.org/wiki/Inductive_charging [Accessed 13-08-2021].

[12] Center for Sustainable Energy. California air resources board clean vehicle rebate project, EV consumer survey dashboard [online]. 2015. Available from http://energycenter.org/clean-vehicle-rebateproject/survey-dashboard [Accessed 15 Apr 2020].

[13] Alternative Fuels Data Center: Laws and Incentives. Electric vehicle infrastructure definitions [online]. Available from http://www.afdc.energy.gov/laws/6534 [Accessed 13-08-2021].

[14] Alternative Fuels Data Center. *Codes and standards basics* [online]. Available from http://www.afdc.energy.gov/codes_standards_basics.html [Accessed 13-08-2021].

[15] Argonne national laboratory. *Codes and standards support for vehicle electrification, theodore bohn* [online]. 2013. Available from http://energy.gov/sites/prod/files/2014/03/f13/vss053_bohn_2013_o.pdf [Accessed 13-08-2021].

[16] Bloomberg New Energy Finance. Electric vehicle outlook 2019 [online]. 2019. Available from https://about.bnef.com/electric-vehicle-outlook/ [Accessed 13-08-2021].

[17] International organization for standards. Available from http://www.iso.org/iso/home/about.htm [Accessed 13-08-2021].

[18] International Electrotechnical Commission [online]. Available from http://www.iec.ch/ [Accessed 13-08-2021].

[19] Sivaraman P., Sharmeela C., Thaiyal Nayagi A., Mahendran R. *Basic Electrical and Instrumentation Engineering*. USA: John Wiley; 20219781119765141. Available from https://onlinelibrary.wiley.com/doi/book/10.1002/9781119765141.

[20] Sanjeevikumar P., Sharmeela C., Holm-Nielsen J.B., Sivaraman P. *Power Quality in Modern Power Systems*. USA: Academic Press; 20209780128233467. Available from https://www.elsevier.com/books/power-quality-in-modern-power-systems/sanjeevikumar/978-0-12-823346-7.

[21] University of Delaware. Vehicle to grid technology [online]. 2010. Available from http://www.udel.edu/V2G/ [Accessed 13-08-2021].

[22] Sivaraman P., Sharmeela C., Ghassan. 'Existing issues associated with electric distribution system' in Kahn B., Alhelou H.H. (eds.). *Handbook of Research on New Solutions and Technologies in Electrical Distribution Networks*. Hershey, PA: IGI Global; 2020. pp. 1–31.

[23] Sivaraman P., Sharmeela C. 'Power quality and its characteristics' in Padmanaban S., Sharmeela C., Holm-Nielsen J.B. (eds.). *Power Quality in Modern Power Systems*. Academic Press; 2020.

[24] Jolly J. '2020 set to be year of the electric car, say industry analysts'. *Guardian*. 25 Dec 2019.

[25] UTILITIES: LEAD THE CHARGE IN eMOBILITY. 2019. Available from https://www.accenture.com/_acnmedia/pdf-97/accenture-utilities-lead-charge-emobility.pdf [Accessed 13-08-2021].

[26] A sustainable future preparing for electric vehicles. Emu analytics; 2018. Available from https://www.emu-analytics.com/whitepapers/electric_vehicles [Accessed 13-08-2021].

[27] Uptake of ultra low emission vehicles in the UK. BROOKLYNDHURST; 2015. Available from https://assets.publishing.service.gov.uk/government/uploads/system/uploads/attachment_data/file/464763/uptake-of-ulev-uk.pdf#:~:text=This%20report%20presents%20the%20findings%20from%20a%20rapid,the%20uptake%20of%20ULEVs%20from%202015%20to%202020. [Accessed 13-08-2021].

[28] Electric car charging points. Available from https://www.edfenergy.com/electric-cars/charging-points [Accessed 13-08-2021].

[29] Anable J., Kinnear N., Hutchins R., Delmonte E., Skippon S. 'Plug-in vehicles infrastructure project. consumers and vehicles: consumer segmentation and demographic patterns'. WP 1.3.5 Project Report. University of Aberdeen and Transport Research Laboratory; 2011.
[30] Anable J., Schuitema G., Stannard J. Consumer responses to electric vehicles literature review. Report for Energy Technology Institute Plug-in vehicles infrastructure project (June 2010). Transport Research Laboratory Published Project Report PPR728; 2014.
[31] Sharmeela C., Sivaraman P., Balaji S. 'Design of hybrid DC mini grid for educational institution: case study'. *Lecture Notes in Electrical Engineering*. 2020, vol. 580, pp. 125–34.
[32] 'Diesel faces global crash as electric cars shine'. *Financial Times*. 11 December 2016.
[33] Schorske R. Idaho national labs: observations from the EV project in Q4 2013 [online]. 2009. Available from http://avt.inel.gov/pdf/EVProj/EVProjectSummaryReportQ42013FINAL.pdf [Accessed Feb, 2014].
[34] Alternative Fuels Data Center. Electric Vehicle Charging Station Locations. Available from https://afdc.energy.gov/fuels/electricity_locations.html#/find/nearest?fuel=ELEC [Accessed 13-08-2021].
[35] Inside EVs, Monthly Plug-In Sales Scorecard [online]. Available from http://insideevs.com/monthly-plug-in-sales-scorecard/ [Accessed 13-08-2021].
[36] California Legislative Information, Assembly Bill 2565 [online]. Available from http://leginfo.legislature.ca.gov/faces/billNavClient.xhtml?bill_id=201320140AB2565 [Accessed 13-08-2021].
[37] CalCars. CalCars frequently asked questions. 2010. Available from http://www.calcars.org/faq.html [Accessed 13-08-2021].
[38] Alternative Fuels Data Center. Signage for plug-in electric vehicle charging stations [online]. Available from http://www.afdc.energy.gov/fuels/electricity_charging_station_signage.html [Accessed 13-08-2021].
[39] ADA.gov, Information and Technical Assistance On The Americans With Disabilities Act, Standards For Accessible Design. Available from http://www.ada.gov/2010ADAstandards_index.htm [Accessed 13-08-2021].
[40] Logeshkumar S., Manoharan R. 'Influence of some nanostructured materials additives on the performance of lead acid battery negative electrodes'. *Electrochimica Acta*. 2014, vol. 144, pp. 147–53.
[41] Available from https://batteryuniversity.com/ [Accessed Jul 17, 2021].
[42] Cornell University Law School, Legal Information Institute. U.S. Code title 26, alternative fuel vehicle refuelling property credit [online]. Available from http://www.law.cornell.edu/uscode/text/26/30C [Accessed 13-08-2021].
[43] Open Charge Alliance [online]. Available from http://www.openchargealliance.org/ [Accessed 13-08-2021].
[44] NissanNews.com [online]. Available from http://nissannews.com/en-US/nissan/usa/releases/nissan-launches-programs-to-make-leaf-charging-free-and-ez [Accessed Apr 16, 2014].

[45] A Guide to the Lessons Learned from the Clean Cities Community Electric Vehicle Readiness Projects [online]. Available from http://www.afdc.energy.gov/uploads/publication/guide_ev_projects.pdf [Accessed 13-08-2021].

[46] Sivaraman P., Sharmeela C. 'Solar micro-inverter' in Zbitou J., Pruncu C., Errkik A. (eds.). *Handbook of Research on Recent Developmentsin Electrical and Mechanical Engineering*. Hershey, PA: IGI Global; 2020. pp. 283–303.

Chapter 11

Power loss and thermal modeling of charger circuit for reliability enhancement of EV charging systems

Md. Mazharul Islam[1] and Md. Rabiul Islam[2]

The fast growth of vehicles is significantly increasing the energy demand in the transportation sector in the world. The excessive use of natural gas and crude oil as the main fuel in the transportation sector is creating extra pressure on the fossil fuel system. The burning of fossil fuel not only diminishes the reserve but also increases the amount of carbon dioxide (CO_2) in our atmosphere, which traps heat, increases temperature, and also creates significant impacts on our health, environment, and climate. To mitigate this accelerating crisis, advanced electric and hybrid electric vehicles (EVs) can be considered as an alternative solution for transportation. EVs have a fast-growing mass market where reliable technologies are essentially required in a cost-effective manner. Power electronic devices act as a vital component in the EVs' traction drive. Failure of any component may lead to a sudden breakdown of the whole power conversion system. Most of the failure mechanisms of power electronic converter-based power conversion are triggered by abnormal temperature conditions. The electro-thermal stress introduced by the massive power loss within power semiconductor devices reduces the lifetime of the power semiconductor modules as well as costly power electronic converters. Therefore, a reliable power conversion system for EVs' application is necessary, which can be guaranteed by a reduced power loss vs. operating temperature profile of semiconductor devices and traction converters. Since the electrical characteristics of power electronic devices are temperature dependent, a fast and accurate electro-thermal model is required to calculate the junction temperature and temperature distribution within power semiconductor modules accurately. This chapter presents an analytical model to calculate temperature-dependent intrinsic power losses of power semiconductor devices for EVs' converters. Advance modulation techniques for power converters and an accurate fast electro-thermal model for reliability analysis of power modules

[1]Department of Electrical and Electronic Engineering, Z. H. Sikder University of Science and Technology, Shariatpur, Bangladesh
[2]Faculty of Engineering and Information Sciences, University of Wollongong, Australia

are also considered in this chapter. This chapter includes (1) an analytical model of power losses with advanced modulation techniques used in motor drives and corresponding numerical conduction and switching power losses of power electronic semiconductor modules; (2) the impact of the heat spreading effect, thermal coupling, and temperature-dependent thermal properties of materials are comprehensively investigated in the ANSYS/Icepak simulation environment (compared to the temperature-dependent thermal properties, the thermal coupling and heat spreading effect have a significant impact on reliability study); and (3) a 3D thermal model, which considers both the thermal network properties and the required accuracy for the calculation of temperature distribution.

11.1 Introduction

Worldwide continuously rising hydrocarbon-based fuel energy consumption and greenhouse gas emission have stimulated the change in vehicular technology. Industries are working towards advance technologies for future emission-free vehicles. To reduce the emission of pollutants in the air and to prevent our fuel reservoir from being run out, the electrification of propulsion has become a challenging issue in vehicle design. In European countries, United States, China, and Japan, the CO_2 emission per traveled km was roughly 200 g in 2000, which is very high compared to the CO_2 emission of 100 g/km in 2020 [1].

The development of reliable power electronic converters, highly efficient electric machine, and energy storage systems has accelerated the revolution of future electric vehicular technology. The development of reliable power electronic converter units is one of the most critical and important issues for efficient battery charging and drive trains, which results in size reduction, fast speed/torque dynamic, and better utilization of battery power. However, reliable power conversion can be achieved via the thermal management of power electronic semiconductor components, because those elements introduce additional heat sources to the system. Figure 11.1 shows a photograph of the power converter section of a hybrid electric vehicle (HEV). To ensure the reliability of converters and sufficient performance of the electric vehicles, thermal management of power semiconductor components acts as the main role. Power electronic modules are liable for approximately 50% of the power converter costs [2]. The component size and cost can be reduced with an efficient, effective, and optimized thermal model and cooling system of the power converters.

Insulated-gate bipolar transistor (IGBT) power modules act as a vital component in battery charging and electric vehicle traction drives, their failures could lead to a sudden shutdown of the power converters. The massive amount of energy losses in power semiconductor components generates significant thermal stress, which reduces the lifetime of the power IGBT modules. Moreover, both internal combustion engine (ICE) and one or more electric motors are used in HEVs. The use of ICE already increases the temperature of

Figure 11.1 A photograph of the power converter section of a HEV

the compartments of the engine and the surrounding environment at a sufficiently higher level and at some locations, it can be more than 100 °C. DC–DC converters and inverters operate at this high-temperature environment to drive the electric motors. High thermal loading conditions may initiate bond-wire lift-off and thermo-mechanical stresses, which initiate continuous decay of the lifetime of the device by introducing fatigues inside the device package or solder cracking [3]. Therefore, efficient thermal management is essentially required for power converters, as the silicon-based power semiconductor components have a limited maximum operating temperature range and it is typically around 125 °C–150 °C [4].

Reliability analysis of power converters for EV applications has been performed in the past and presented in several literature [5, 6]. A model for the reliability prediction of inverters in HEVs was presented in [5]. This model mainly focuses on the impact of temperature and its fluctuation on the life of the main components, including the power semiconductor devices and the DC-link capacitor. Here, a software-based simulation model was developed to determine the component stresses particularly for complex mission profiles. Reliability prediction based on complex mission profiles for semiconductors is not well established and trusted. This method emphasized the Coffin-Manson and Arrhenius equations to predict the thermal stress of the most important types of components. Though this method appears to be simple, it requires data on failure rates for life-time prediction and reliability analysis of power semiconductors.

A fast electro-thermal model of traction inverters was studied in [6]. In this method, the average power dissipation of IGBT modules at rated current was determined using the curves and the corresponding information from the data sheet. The rise in temperature of the heated chips was measured using a (1D) lumped RC thermal impedance structure. Such models have the limitations to represent temperature at different layers. Because the conventional Foster network does not represent actual thermal property at any layer, and it has equivalent junction to case response for the chips only. However, this model lacks the temperature distribution at different positions in each layer of the power electronic modules.

A better and systematic thermal modeling and reliability model for high-power semiconductor modules used in HEVs have been developed here. This method provides a better coupling effect and precision for transient temperature measurement between junctions of several underlying layers of power IGBT modules. The heat spreading and thermal coupling depend on temperature-dependent thermal conductivity (λ) and specific heat capacity (c_p) of materials. Silicon (Si), Aluminum (Al), and Copper (Cu) are the widely used materials for the fabrication of power IGBT modules. A constant approximation of λ and c_p of those materials for power modules' thermal behavior analysis is not sufficiently accurate. Therefore, temperature-dependent thermal conductivity and specific heat capacity of materials are considered here with the structural geometry of power modules. This model besides considering the thermal couplings among the chips also takes into account the couplings between the layers beneath the silicon chips. Moreover, to obtain the thermal distribution at a certain layer, several equidistant monitoring points are also developed in this model.

11.2 Power electronic converters in EVs

The electric motor provides the required propulsion to the EV, which is powered by several energy sources with the associated power converters. The automotive companies are broadly focusing on four major types of vehicular technologies such as EVs, HEVs, plug-in hybrid electric vehicles (PHEVs), and fuel-cell electric vehicles. Desirable optimum power can be utilized in HEVs by the combination of electric motor and an ICE to save energy. Hybrid vehicles (HVs) are generally classified as series HVs and parallel hybrid vehicles [7]. A PHEVs can be also series or parallel hybrid. A generalized block diagram of PHEV is depicted in Figure 11.2. In PHEVs, a high-power battery is installed with ICE like as HEVs and the battery can

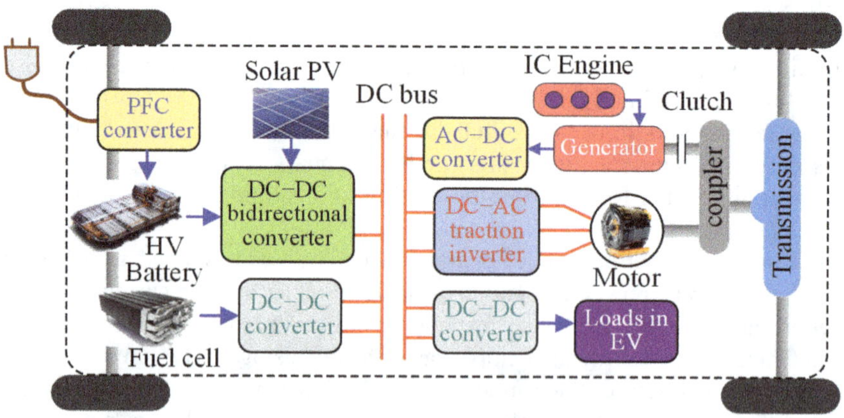

Figure 11.2 General block diagram of HEV [7]

Figure 11.3 A photograph of battery pack installed in a HEV

be restored to full charge by connecting it through a charger to an external electric power source as in an EV. Therefore, the PHEVs provide a greater range of performance compared to other types of HEVs.

Usually the battery power is utilized via a power electronic DC−AC inverter to run the electric motor, whereas an onboard AC−DC battery charger circuit is installed to charge the EV battery. In HEVs, two or more energy sources, e.g., battery and flywheel, and/or two or more power sources, e.g., fuel cell, battery, and ultra-capacitor are installed onboard [8]. Figure 11.3 shows a photograph of battery bank installed at the back side of a HEV.

In a HEV, different types of power electronic converters are used for various applications. An AC–DC converter is used for charging the high-power batteries using the power from the grid while a DC–AC inverter is used to feed power to the propulsion motor. Besides these, separate bidirectional DC–DC converters are used to transfer power between the HV battery and traction system and an auxiliary battery for feeding other loads. Moreover, there are other converters required for running air-conditioning and control systems [9–11].

The topology for DC fast charging station is shown in Figure 11.4, where multiple identical modules are connected in parallel to increase the output power handling capacity, such as Tesla supercharger, which contains 12 paralleled modules [13]. Each module has two power electronic conversion stages to convert the three-phase AC voltage up to 480 V to the required DC voltage. The first stage contains an AC–DC rectifier with a power factor correction (PFC) topology.

The three-phase AC input voltage can be converted to an intermediate DC voltage by a unidirectional three-phase boost-type PFC rectifier, which also contains the configuration of T-type Vienna Rectifier [12].

In the second stage, a DC–DC converter with high-frequency (HF) transformer is used to convert the intermediate DC voltage into the required DC voltage to charge the HV battery. The HF transformer provides galvanic isolation between the AC grid and the DC battery. However, if bidirectional power flow between DC bus and battery is required, a dual active bridge (DAB) converter can be used to regulate the charging and discharging operations of the battery, which is shown in Figure 11.5. Though the DAB converter produces a significant amount of switching losses, the

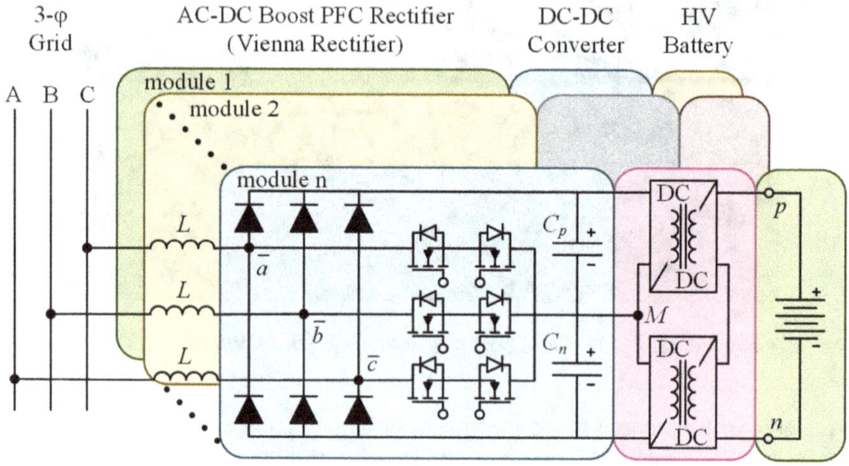

Figure 11.4 Multiple paralleled modules of three-phase PFC rectifier with series-connected isolated DC–DC converters [12]

development of silicon carbide (SiC) metal-oxide semiconductor field-effect transistor (SiC-MOSFET) as switching device and amorphous nano-crystalline soft magnetic materials as HF magnetic link has improved the higher efficiency and power density of the converter [14].

Nowadays renewable energy sources (RESs) with smart micro-grids and energy storage units-based EV charging stations have become more and more popular around the world. Using RES for charging facilitates environmental cleanliness can be guaranteed due to zero gas emissions [15]. In order to cater the charging demands during peak hours besides improving the quality of electricity and minimizing the effect of load shedding, several other distributed energy sources can be integrated into the smart micro-grids for feeding EV chargers. Smart grids fed with RESs and charging facilities powered through fast charging AC–DC converters have facilitated charging of EVs even in the parking lots, thereby alleviating the range anxiety of EV drivers [16].

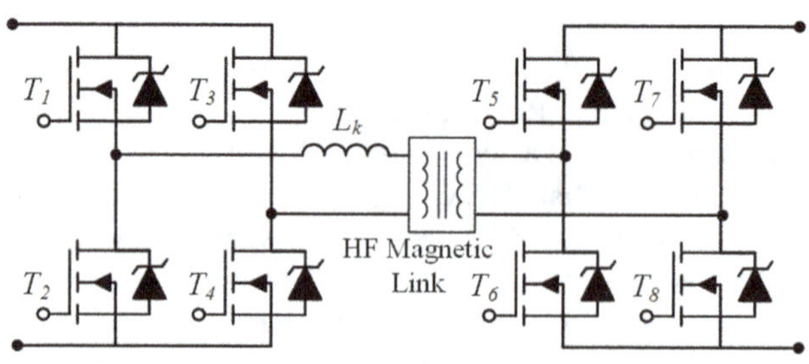

Figure 11.5 Bidirectional full-bridge DAB converter [14]

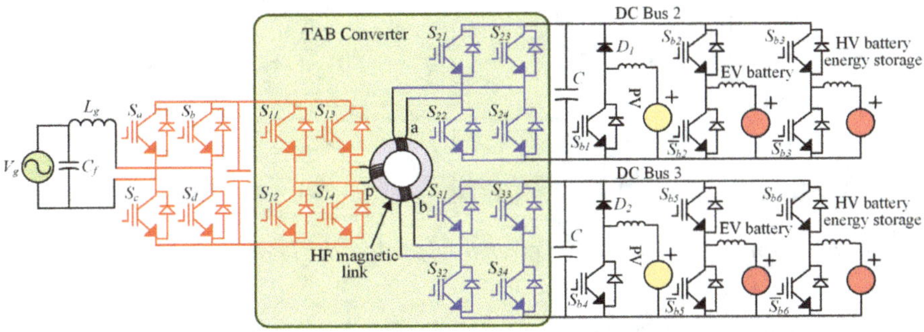

Figure 11.6 TAB converter with multidirectional power flow capability [17]

Three or multi-port converters containing HF magnetic link is the best possible solution for interfacing EVs, solar PV into the smart grids. These converters facilitate galvanic isolation besides reducing the stages for power conversion and component counts. Hybrid multi-port converters with intelligent controls can also be used for providing balance of power among RESs, EVs, and grids [17]. Multi-port converters provide multi-directional power flow capabilities and possess bidirectional AC–DC converter, bidirectional DC–DC converter, and triple active bridge converter, as depicted in Figure 11.6.

Continuous operations of power semiconductor modules lead to a significant amount of power loss in the form of heat. The electro-thermal stresses introduced due to power loss significantly reduce the lifetime of the power modules. Thermal stresses increase the junction temperature consequently the failure rate of power modules increases exponentially by altering their operating characteristics.

Hence, efficient and reliable thermal management systems are required to be developed for effectively managing the heating and cooling of power train components in EVs. Therefore, designing thermally reliable power conversion systems for hybrid and EVs has become a challenging and focus of research.

11.3 Modulation and analytical power loss model of power electronic converters

Modular power inverters containing IGBT modules have emerged as reliable systems for the propulsion of electric motors in EVs depicted in Figure 11.7. They are, however, manifested with a considerable amount of power loss in the form of heat due to hard switching. The switching and conduction power losses result in higher junction temperature and lower thermal stability [18]. Moreover, the bidirectional DC–DC converter has two H-bridge circuits interfaced via an HF magnetic link. The intrinsic properties of IGBTs and anti-parallel diodes (APDs) result in a significant amount of power loss due to forward voltage drop, on-state resistance, and switch on-off along with diode reverse recovery energy losses. Hence, the HF inverters, HF magnetic links, and rectifiers in bidirectional converters generate power losses and

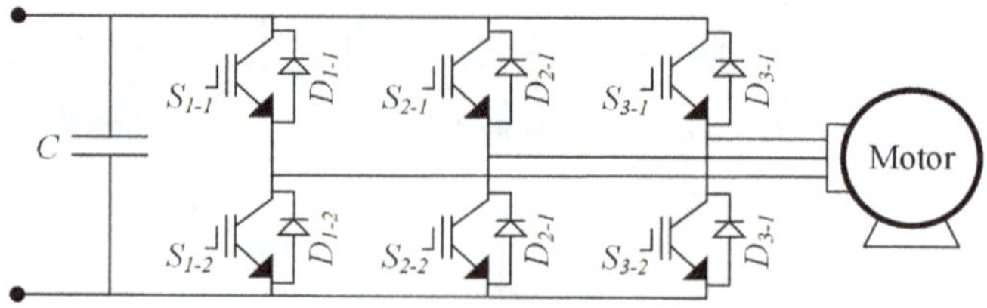

Figure 11.7 Traction inverter in electric vehicle application [18]

consequently the system already produces notable joule heating and high junction temperature. Finally, the hard switching operations at high load currents in traction inverter using pulse width modulation (PWM) techniques generate an enormous amount of power losses in the stacked IGBT modules. Therefore, the continuous higher junction temperature reduces the reliability of converter operations by initiating their failure and sometimes causing permanent breakdown. Hence, it becomes necessary to reduce, analyze, and accurately model the power losses within the power electronic converters to avert the associated severe damages due to failure and high cost of replacement.

The symmetrically designed three-level traction inverter drive for EVs using discontinuous pulse width modulation (DPWM) provides lower switching losses for the full power factor (PF) range and harmonic distortion [19, 20]. The carrier-based DPWM schemes are represented in Figure 11.8, where the value of a reference signal is constant (100%) for the duration of 120° within a complete cycle. Therefore, no switching operations will occur due to reference-carrier comparison scheme within this period as well as no switching losses. Another advantage of DPWM is that the effect of lower-order harmonic components within the line voltage is significantly little.

Typical line to line output voltage and corresponding total harmonic distortion (THD) of traction inverter with DPWM are shown in Figures 11.9 and 11.10, respectively.

However, the development of the analytical power loss model is a complex and challenging issue due to the discontinuous switching mode of IGBTs for bus

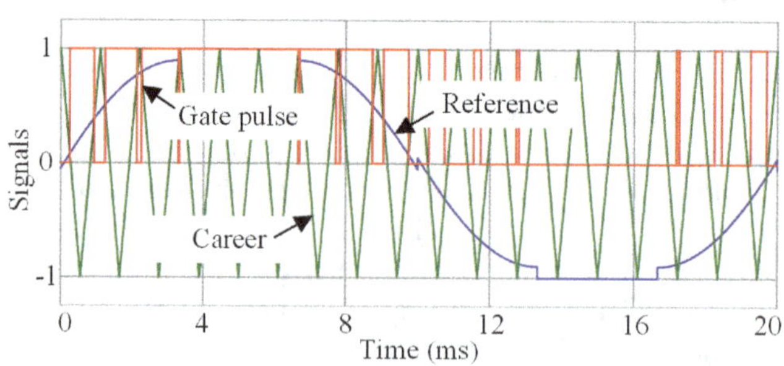

Figure 11.8 Modulation schemes of traction inverter with DPWM [21]

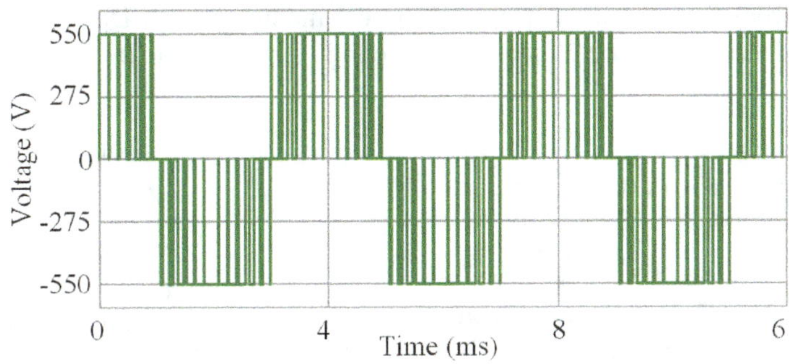

Figure 11.9 *Typical output line to line voltage of traction inverter with DPWM (DC-link voltage = 550 V, modulation index = 1, RMS voltage = 245.5 V) [22]*

clamping PWM in traction inverter [23, 24]. IEC 60747-9 only provides a general device under test (DUT) method for measuring switching losses (P_{SW}) and junction to case thermal impedance (Z_{j-c}) for multichip IGBT modules, which is not sufficient for calculating power losses and thermal impedance with DPWM.

Moreover, no international standards have been provided to calculate the switching and conduction losses of traction inverter for DPWM. Overall energy losses of the power modules are obtained as a combined effect of switching and conduction losses. Moreover, both of these losses depend on load current, PF, and the applied modulation techniques for a particular converter [21]. In this regard, analytical equations for calculating the average and RMS currents of IGBTs and APDs have been discussed in this section.

11.3.1 Conduction power losses in traction inverters

The conduction power loss of IGBTs and APDs for a fixed temperature is the product of device on-state voltage drop and load current. But the relation between them can

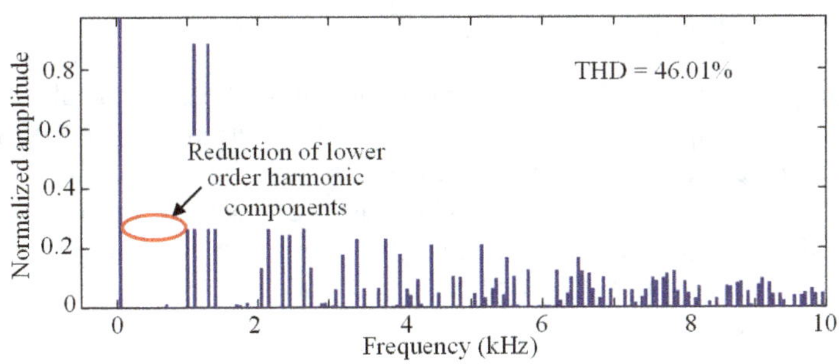

Figure 11.10 *Frequency spectrum of output voltage of traction inverter (where modulation index =1 and ratio of the frequency of carrier signal to reference signal = 21) [22]*

be found from the data sheet, provided by the manufacturers. During steady-state conduction mode IGBT collector-emitter voltage drop (v_{ce}) can be represented as

$$v_{ce} = v_{ceo} + R_c \cdot i_c \tag{11.1}$$

where v_{ceo}, R_c, and i_c are IGBT on-state zero current voltage drop, on-state resistance, and collector current, respectively. Similarly, diode forward voltage drop (v_F) can be expressed as

$$v_F = v_{Fo} + R_F \cdot i_F \tag{11.2}$$

where v_{Fo}, R_F, and i_F are diode zero current forward voltage drop, on-state resistance, and forward current, respectively. Hence, the instantaneous $P_{ct}(t)$ and average P_{ct} conduction losses of IGBT for a fixed temperature can be calculated by

$$P_{ct}(t) = v_{ce}(t) \cdot i_c(t) = v_{ceo} \cdot i_c(t) + R_c \cdot i_c^2(t) \tag{11.3}$$

$$P_{ct} = \frac{1}{2\pi} \int_0^{2\pi} \left[P_{ct}(t)\right] d(\omega t) = v_{ceo} \cdot i_{c-avg} + R_c \cdot i_{c-rms}^2. \tag{11.4}$$

where i_{c-avg} and i_{c-rms} are the IGBT on-state average and RMS conduction current, ω is the angular frequency, and t is the instantaneous time. Similarly, average diode conduction loss (P_{cd}) can be calculated as

$$P_{cd} = v_{Fo} \cdot i_{F-avg} + R_F \cdot i_{F-rms}^2 \tag{11.5}$$

where i_{F-avg} and i_{F-rms} are diode on-state average and RMS current. The dependency of on-state voltage drops (v_{ce}, v_F) on operating temperature can be found in the data sheet of power IGBT modules. Therefore, the effect of the change of temperature on zero-current forward voltage drops (v_{ceo}, v_{Fo}) and conduction resistance (R_c, R_F) has a significant outcome on conduction power loss calculation. However, the non-linear relationship of v_{ceo}, v_{Fo}, and R_c, R_F with operating temperature is depicted in Figure 11.11, which was extracted from the data sheet of INFINEON PrimePACK™3 IGBT module. Hence, the average conduction power losses for IGBT and diode at a specific temperature 'T' are represented by (11.6) and (11.7), respectively.

$$P_{ct}(T) = v_{ceo}(T) \cdot i_{c-avg} + R_c(T) \cdot i_{c-rms}^2 \tag{11.6}$$

$$P_{cd}(T) = v_F(T) \cdot i_{F-avg} + R_F(T) \cdot i_{F-rms}^2 \tag{11.7}$$

An accurate power loss profile of the semiconductor modules can be obtained using (11.6) and (11.7) expressing the average and RMS currents flowing through the semiconductor devices. Therefore, the analytical equations representing the average and RMS currents for IGBT and APD have been derived in the next section. The effect of parameters such as change in load current, PF, and modulation index in traction converters with DPWM has also been included for deriving the model.

11.3.1.1 Average and RMS currents in traction inverter

The discontinuous switching operation in DPWM makes the calculation of RMS and average currents in the switching devices of traction inverters a tricky issue [21]. The two basic modes of inverter operation are switching mode and continuous

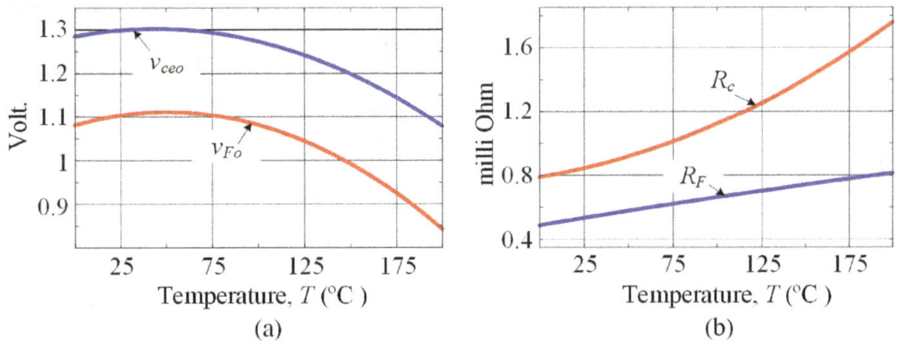

Figure 11.11 On-state temperature-dependent properties: (a) IGBT and diode zero current voltage drop characteristics and (b) conduction resistances of IGBT and diode [22]

conduction mode. Each half cycle contains a continuous conduction mode (for 60°) and two switching mode (for 120°) of operation. An inverse switching mechanism exists for vertically opposite IGBT-diode pairs in each leg. The duty cycle of odd-numbered IGBT in any leg can be represented as

$$D(t) = \frac{1}{2}\left[1 + \frac{2M}{\sqrt{3}}\sin(\omega t) + (1 - \mathrm{M})(-1)^{\mathrm{m}} + \frac{\mathrm{M}}{3\sqrt{3}}\sum_{\mathrm{i=0}}^{\mathrm{i=2}} \mathrm{a_i}\sin(6\mathrm{i} + 3)(\omega t)\right] \tag{11.8}$$

where M is modulation index and m is region number shown in Figure 11.12 as r_1, $r_2, \cdots, r_6$. For inductive load in traction motors, the phase current lags the voltage by phase angle φ. Hence, the instantaneous value of phase current for leg-1 is given by (11.9).

$$i(t) = I_p \sin(\omega t + \varphi) \tag{11.9}$$

where I_p is the peak value of load current. Mutually scalable modulation and structurally modular nature of both traction motor and inverter make the same current flowing through each leg. Moreover, the average and RMS currents of vertically opposite IGBTs and APDs at every leg are equal.

Duty cycle of odd-numbered ($S_{1\text{-}1}/D_{1\text{-}1}$) and even-numbered ($S_{1\text{-}2}/D_{1\text{-}2}$) IGBTs and diodes for the first leg are depicted in Figure 11.12. The continuous conduction (*CC*) mode (from $\pi/3$ to $2\pi/3$ and from $4\pi/3$ to $5\pi/3$) is only limited for IGBTs. Same effective average duty cycle exists for odd-numbered ($S_{1\text{-}1}$) top IGBTs, which is for ωt from φ to $(\pi+\varphi)$. Similarly, the average duty cycle of even-numbered IGBT ($S_{1\text{-}2}$) will be equal for the ωt from $(\pi+\varphi)$ to $(2\pi+\varphi)$. Hence, the average current of odd-numbered IGBTs can be given by (11.10).

$$i_{c-avg@S_{1-1}} = \frac{1}{2\pi}\int_{\varphi}^{\pi+\varphi} [i(t) \cdot D(t)]\, d(\omega t) \tag{11.10}$$

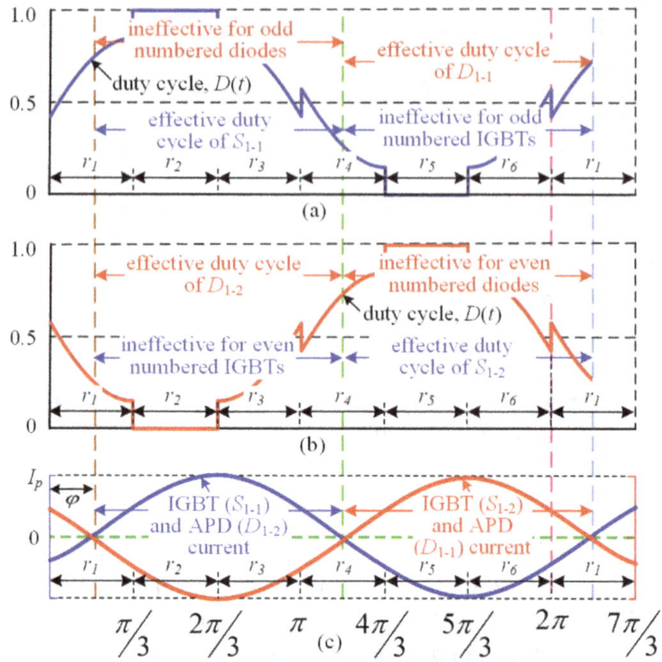

Figure 11.12 Duty cycle and load current of IGBTs and APDs in traction inverter with DPWM: (a) effective and ineffective duty cycle of odd-numbered IGBTs and APDs, (b) duty cycle of even-numbered IGBTs and APDs, and (c) forward current of IGBTs and APDs

The discontinuous switching within a full cycle can be divided into six regions (r_1, r_2, $\cdots$, r_6), as depicted in Figure 11.12. Regions ($r_1 + r_3 + r_4$) and ($r_4 + r_6 + r_7$) are the effective total duty cycle of switched conduction (*SC*) mode for the corresponding IGBTs of S_{1-1} and S_{1-2}. Although no *CC* mode is allocated for the APDs. Hence, the average current $\left(i_{c\_avg\_S_{1-1}}\right)$ of odd-numbered IGBTs (S_{1-1}) can be rewritten as

$$i_{c-avg\_S_{1-1}} = i_{c-avg\_S_{1-1}\ @\ SC} + i_{c-avg\_S_{1-1}\ @\ CC} \tag{11.11}$$

where $i_{c-avg\_S_{1-1}@CC}$ represents average current at *CC* mode and $i_{c-avg\_S_{1-1}@SC}$ stands for average currents of S_{1-1} at *SC* mode. In the *SC* mode, the equation of duty cycle represented in (11.8) will be accompanied by both S_{1-1} and D_{1-1} at regions ($r_1 + r_3 + r_4$) and ($r_4 + r_6 + r_7$), respectively. However, the average current of S_{1-1} for *SC* mode can be written as

$$\begin{aligned} & i_{c-avg\_S_{1-1}\ @\ SC} \\ & = \frac{1}{2\pi} \int_{\varphi}^{\pi+\varphi} \left[i(t) \cdot D(t) \right] d(\omega t) \quad \begin{matrix} \text{when } \omega t \leq \frac{\pi}{3} \\ \text{and } \omega t \geq \frac{2\pi}{3} \end{matrix} \\ & = \frac{1}{2\pi} \int_{\varphi}^{\pi/3} \left[i(t) \cdot D(t) \right] d(\omega t) + \frac{1}{2\pi} \int_{2\pi/3}^{\pi+\varphi} \left[i(t) \cdot D(t) \right] d(\omega t) \end{aligned}$$

$$
\begin{aligned}
&= \frac{MI_P}{4\pi\sqrt{3}} \cdot \cos\varphi \cdot \left(\frac{2\pi}{3} - \frac{\sqrt{3}}{2}\right) + \frac{MI_p}{4\pi}\left\{\cos 2\varphi - \cos\left(\frac{\pi}{3} + \varphi\right)\right\} \\
&+ \left[\begin{array}{l} \frac{MI_P}{12\pi\sqrt{3}} \sum_{i=0}^{2} \frac{a_i}{2}\left[\frac{6i+3}{18i^2+18i+4}\left\{\cos\left(\frac{\pi}{6} - \varphi\right) + \sin\varphi\right\}\right] \\ + \frac{MI_P}{12\pi\sqrt{3}}\left\{\cos\left(\frac{\pi}{6} + \varphi\right) - \sin\varphi\right\}\left\{\frac{3a_o}{8} + \frac{9a_1}{80} + \frac{15a_2}{224}\right\} \end{array}\right] \\
&\qquad + \frac{MI_P}{4\pi}\left\{\cos\varphi - \sin\left(\frac{\pi}{6} + \varphi\right)\right\}
\end{aligned}
\tag{11.12}
$$

$$
\begin{aligned}
&i_{c-avg\_S_{k-1}\,@\,SC} = i_{c-avg} \\
&= \frac{MI_P}{4\pi\sqrt{3}} \cdot \cos\varphi \cdot \left(\frac{2\pi}{3} - \frac{\sqrt{3}}{2}\right) \\
&+ \frac{MI_P}{2\pi\sqrt{3}} \cdot \cos\frac{\pi}{6} \cdot \cos\varphi \cdot \left(\frac{a_o}{8} + \frac{3a_1}{80} + \frac{5a_2}{224}\right) \\
&+ \frac{1}{2\pi}\left(\frac{MI_P}{2}\cos\varphi - I_p\cos\varphi + I_P\cos 2\varphi\right)
\end{aligned}
\tag{11.13}
$$

During CC mode, no switching will occur and full load current will flow through both S_{1-1} and S_{1-2} for the duration of $\omega t = \pi/3$ in regions r_2 and r_5. The average load current can be written as

$$
i_{c-avg\_S_{1-2}\,@\,CC} = i_{c-avg\_S_{1-1}\,@\,CC} = 2I_P\cos\varphi \cdot \cos\frac{\pi}{3} \tag{11.14}
$$

Similarly for the APDs (D_{1-1}), the average current can be expressed as

$$
i_{F-avg\_D_{1-1}} = i_{c-avg\_S_{1-1}\,@\,SC} + \frac{1}{2\pi}\left(I_p\cos\varphi - 2I_p\cos 2\varphi\right) \tag{11.15}
$$

In regions r_4, r_6, and r_7, the effective duty cycle of the IGBTs (S_{1-2}) for SC mode will follow the identical duty cycle of the APDs (D_{1-2}) in the same region. However, due to the repetitive structure of sinusoidal functions, the average currents of S_{1-2} for the effective duty cycle of regions r_4, r_6, and r_7 are equivalent to the average currents of S_{1-1} for regions r_1, r_3, and r_4. The average currents of even and odd-numbered IGBTs and APDs can be written as

$$
i_{c-avg\_S_{1-2}} = i_{c-avg\_S_{1-1}} = i_{c-avg} \tag{11.16}
$$

$$
i_{F-avg\_D_{1-2}} = i_{F-avg\_D_{1-1}} = i_{F-avg} \tag{11.17}
$$

The RMS current of odd-numbered IGBTs can be expressed as

$$
i_{c-rms\_S_{1-1}} = i_{c-rms\_S_{1-1}@SC} + i_{c-rms\_S_{1-1}@CC} \tag{11.18}
$$

$$
i_{c-rms\_S_{k-1}@SC} = \sqrt{\frac{1}{2\pi}\left[\int_{\varphi}^{\pi+\varphi} i^2(t) \cdot D(t)\, d(\omega t)\right]} \quad \text{for} \quad \begin{array}{l} \omega t \le \frac{\pi}{3} \\ \omega t \ge \frac{2\pi}{3} \end{array} \tag{11.19}
$$

Similarly, the RMS current of S_{1-2} can be represented as

$$
i_{c-rms\_S_{1-2}} = i_{c-rms\_S_{1-2}@SC} + i_{c-rms\_S_{1-2}@CC} \tag{11.20}
$$

$$
\begin{aligned}
&i_{c-rms\_S_{1-1}@SC} \\
&= \sqrt{\frac{1}{2\pi}\left[\int_{\varphi}^{\pi+\varphi} i^2(t) \cdot D(t)\, d(\omega t)\right]} \quad \text{for} \quad \begin{array}{l} \omega t \le \frac{\pi}{3} \\ \omega t \ge \frac{2\pi}{3} \end{array}
\end{aligned}
$$

$$
= \sqrt{\begin{array}{l}
\int_{\varphi}^{\pi+\varphi} \frac{MI_p^2}{4\pi\sqrt{3}} \sin(\omega t) \cdot \left[1 - \cos\{(2\omega t) + 2\varphi\}\right] + \\
\int_{\varphi}^{\pi+\varphi} \frac{MI_p^2}{24\pi\sqrt{3}} \left[\begin{array}{c} \left[\sum_{i=0}^{2} a_n \sin(6i+3)(\omega t)\right] \cdot \\ \left[1 - \cos\{(2\omega t) + 2\varphi\}\right] \end{array}\right] + \quad \text{for} \quad \begin{array}{l} \omega t \leq \frac{\pi}{3} \\ \omega t \geq \frac{2\pi}{3} \end{array} \\
\int_{\varphi}^{\pi+\varphi} \frac{I_p^2}{8\pi} \left[\begin{array}{c} \left[1 - \cos\{(2\omega t) + 2\varphi\}\right] \cdot \\ \sum_{m=1}^{4} (1-M)(-1)^m \end{array}\right]
\end{array}}
$$

$$
= \sqrt{\begin{array}{l}
\frac{MI_p^2}{4\pi\sqrt{3}} \left[-3 + 2\cos\varphi - \frac{5}{6}\cos 2\varphi + \cos 3\varphi - \frac{1}{3}\cos 5\varphi\right] + \\
\frac{MI_p^2}{24\pi\sqrt{3}} \sum_{i=0}^{2} \left[a_i \left\{ \begin{array}{c} \frac{2}{6i+3}\left[1 + \cos(6i+3)\varphi\right] - \frac{1}{6i+1}\cos(i^2+3i+1)\varphi \\ -\frac{1}{6i+5}\cos(6i+7)\varphi + \frac{6i+3}{36i^2+36i+5}\cos 2\varphi \end{array} \right\} \right] + \\
\left[-\frac{MI_p^2}{4\pi}\left(\varphi - \cos 3\varphi \sin\varphi + \frac{\sqrt{3}}{4}\cos 2\varphi - \frac{\pi}{3}\right) + \frac{I_p^2}{4\pi}\left(\varphi - \cos 3\varphi \sin\varphi\right) \right]
\end{array}} \tag{11.21}
$$

Because of the similar structure of each leg, at the *SC* regions, the RMS currents of similar type switching devices (S_{k-1}/S_{k-2} and D_{k-1}/D_{k-2}) are equal, which can be summarized as

$$
i_{c-rms\_S_{1-1}} = i_{c-rms\_S_{1-2}} = i_{c-rms} \tag{11.22}
$$

$$
\begin{array}{l}
i_{F-rms\_D_{1-1}} = i_{F-rms\_D_{1-2}} = i_{F-rms} \\
= \frac{I_p^2}{4}\left(\pi - \sqrt{3}\cos 2\varphi + \frac{M\pi}{6} + \frac{4M}{\sqrt{3}}\right) - i_{c-rms\_S_{1-1}@SC}
\end{array} \tag{11.23}
$$

During the *CC* mode at regions r_2 and r_5, the RMS currents of IGBTs can be written as

$$
\begin{array}{l}
i_{c-rms\_S_{1-1}@CC} = i_{c-rms\_S_{1-2}@CC} \\
= \frac{I_p^2}{2}\left(\frac{\pi}{3} + \frac{\sqrt{3}}{2}\cos 2\varphi\right)
\end{array} \tag{11.24}
$$

Average conduction power losses of IGBTs and APDs at each leg are equal due to the symmetrical structure of the converter and balanced three-phase load currents. The average conduction power losses of IGBTs and APDs at any leg for a particular temperature "T" are represented by (11.25) and (11.26), respectively.

$$
P_{ct-avg\_k-1,k-2}(T) = P_{ct-avg}(T) = v_{ceo}(T) \cdot i_{c-avg} + R_c(T) \cdot i_{c-rms}^2 \tag{11.25}
$$

$$
P_{cd-avg\_k-1,k-2}(T) = P_{cd-avg}(T) = v_{Fo}(T) \cdot i_{F-avg} + R_F(T) \cdot i_{F-rms}^2 \tag{11.26}
$$

11.3.2 Analytical model of switching power losses

Power losses due to turn-on and turn-off at precise instants during hard switching operation are experienced in traction inverters. Switching energies are the function of load current at specific conditions, such as operating temperature, gate resistance, and blocking voltage of the devices. However, all the devices operate under same switching frequency, same gate resistance, and equal DC cell voltage in the traction

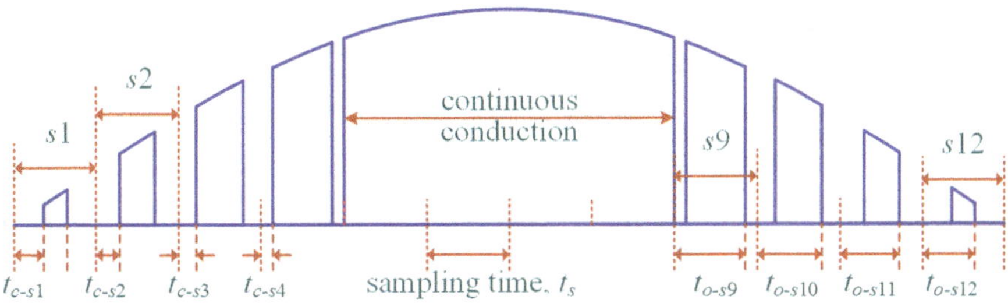

Figure 11.13 Device turn-on and turn-off lag-times at each sample

inverter. Therefore, the alterations of switching losses depend on the conduction current at the operating switch transition points. The change of device on-state load current at switching transition points causes the variation of switching energy losses.

To calculate the switching transition times analytically, it is required to know the gate pulse turn-on and turn-off times of the PWM signal and corresponding load current at those instants. Based on the sampling time, a discrete equation is required to precisely represent the analytical expression of the pulse transition instants for DPWM. Overall variation of switch turn-on and turn-off transition points at each sampling time is shown in Figure 11.13. In every sampling time, the switches do not turn on at the moment of starting time of that particular sample.

For the switching frequency (f_s) of 1.2 kHz, there will be 24 samples at each fundamental time period (T_m) of 20 ms. To remove the fractional number of samples in a fundamental period, the switching frequency should be the multiple of six times of fundamental frequency (f_m = 50 Hz). Hence, the number of sampling (N) in each *SC* region is represented by (11.27).

$$N = \frac{f_s}{6f_m} \tag{11.27}$$

Device turn-on and turn-off lag-times (t_c and t_o) from the starting point of that corresponding sample are shown in Figure 11.13. During all the three *SC* regions in a fundamental period, the lag-times change their values. Both continuous and discrete representations for turn-on lag-times of IGBT $S_{1\text{-}1}$ for different sample numbers ($s1$, $s2$, …, $s12$) are shown in Figure 11.14. However, IGBT $S_{1\text{-}2}$ is turned off at the same time when IGBT $S_{1\text{-}1}$ is turned on to avoid a direct short circuit between the terminals of each leg. Therefore, t_c of $S_{1\text{-}1}$ and t_o of $S_{1\text{-}2}$, both are equal at any sample. However, the turn-on and turn-off lag-times of switch $S_{k\text{-}1}$ and $S_{k\text{-}2}$ differ from one leg to other. A continuous representation of sample-based discrete lag-times ($t_{c@sn}$ and $t_{o@sn}$) for the switches $S_{1\text{-}1}$ and $S_{1\text{-}2}$ is depicted in Figure 11.14. The discrete form of the equation for $t_{c\_1\text{-}1@sn}$ and $t_{o\_1\text{-}2@sn}$ at any sample (sn) can be written as

$$t_{c\_11@sn}/t_{o\_12@sn} = \frac{1}{4f_s}\left[-D_{SH}\left[n\right]\right] \tag{11.28}$$

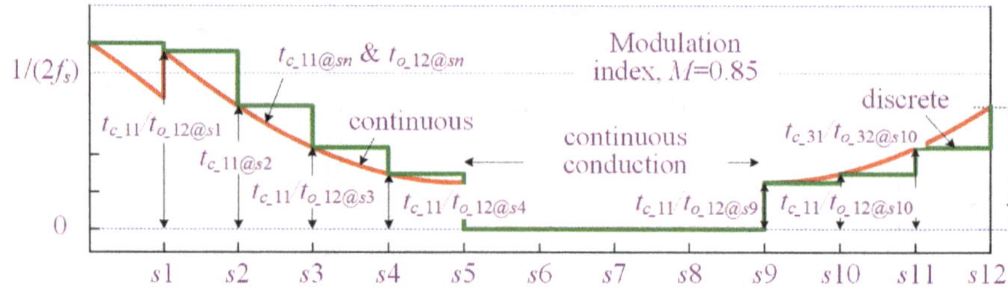

Figure 11.14 Continuous and discrete representation of turn-on and turn-off lag-times of S_{1-1} *and* S_{1-2} *[23]*

$$D_{SH}[n] = \frac{M}{3\sqrt{3}}\sum_{i=0}^{2} a_i \sin(6i+3)\left(\omega n T_S - \frac{13\pi}{42N}\right) + \frac{2M}{\sqrt{3}}\sin\left(\omega n T_S - \frac{13\pi}{42N}\right) + (1-M)(-1)^m \tag{11.29}$$

where $D_{SH}[n]$ is a discrete type of equation with a phase shift angle, which can be obtained from (11.1), T_S represents the time period of each sample, and $n = 1, 2, 3, \cdots, 6N$. Therefore, the turn-on transition time (t_{on}) of S_{1-1} and turn-off transition time (t_{off}) of S_{1-2} can be represented as

$$t_{on\_1-1@sn}/t_{off\_1-2@sn} = (n-1)\,T_S + t_{c\_11@sn}/t_{o\_12@sn} \tag{11.30}$$

$$t_{c\_1-2@\,sn}/\,t_{o\_1-1@\,sn} = \frac{1}{4f_s}[D_{UN}[n] + 3] \tag{11.31}$$

Similarly, switch turn-off lag-time of S_{1-1} and turn-on lag-time of S_{1-2} are shown in Figure 11.15, which is only limited for leg-1. Therefore, the equation for switching lag-times can be represented by (11.31) and (11.32).

$$D_{UN}[n] = \frac{2M}{\sqrt{3}}\sin(\omega n T_S) + (1-M)(-1)^m + \frac{M}{3\sqrt{3}}\sum_{i=0}^{2} a_i \sin\left[(6i+3)(\omega n T_S)\right] \tag{11.32}$$

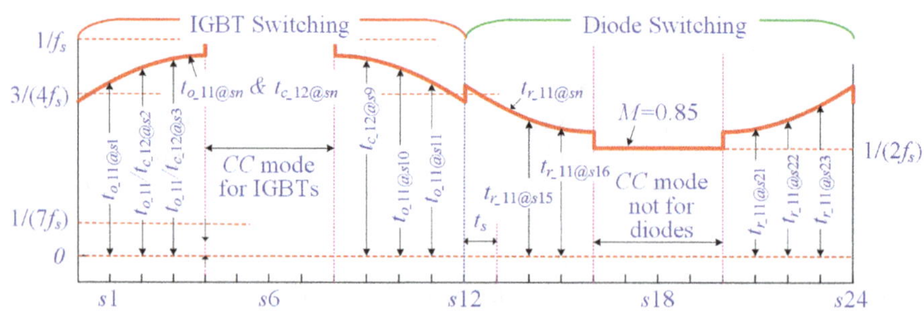

Figure 11.15 Turn-off, turn-on, and reverse recovery lag-time of S_{1-1}, S_{1-2}, *and* D_{1-1}

where $D_{UN}[n]$ is a discrete type of equation without any phase shift angle, which can be derived from (11.1). The APDs will conduct only for the negative half cycle (from $\pi+\varphi$ to $2\pi+\varphi$ for D_{1-1} and from φ to $\pi+\varphi$ for D_{1-2}) of the load current. But the diode turn-on losses being less than 1% in modern IGBT power modules can be neglected for the analysis being carried out here while the reverse recovery losses are considered. However, reverse recovery losses of the APDs are considered. So, for the negative half cycle of the load current, the diode turn-off lag times at any cell are represented by (11.33).

$$t_{r\_1-1,\,1-2,@sn} = t_{o\_1-1,\,1-2,@sn} \text{ for } \mathrm{D}_{1-1,\,1-2} \tag{11.33}$$

Also, the turn-off transition time for APDs at every sample is represented by (11.34).

$$t_{rr\_1-1,\,1-2@sn} = (n-1)\,T_S + t_{r\_1-1,\,1-2@sn} \text{ for } D_{1-1,\,1-2} \tag{11.34}$$

Separate methods for estimation of energy loss due to switching in IGBT module can be represented in terms of switch turn-on (E_{on}), switch turn-off (E_{off}), and diode reverse recovery (E_{rr}) energy losses. These switching losses are function of load currents defined at the instants of switch transition as extracted from the manufacturer data sheet. Hence, the load currents of IGBTs at the device turn-on and turn-off points are given by (11.35) and (11.36), respectively.

$$I_{S\_on\_1-1,1-2@sn} = I_P \sin\left[\omega\left(\mathrm{t_{on\_1-1,1-2@sn}}\right) + \varphi\right] \tag{11.35}$$

$$I_{S\_off\_1-1,1-2@sn} = I_P \sin\left[\omega\left(\mathrm{t_{off\_1-1,1-2@sn}}\right) + \varphi\right] \tag{11.36}$$

$$\left\lceil \frac{3N\varphi}{\pi} \right\rceil \le n \le \left\lceil 3N\left(1+\frac{\varphi}{\pi}\right)\right\rceil \tag{11.37}$$

Similarly, the forward currents of APDs at the points of reverse recovery are given by (11.38) and (11.39).

$$I_{D\_rr\_1-1,1-2@sn} = I_P \sin\left[\omega\left(\mathrm{t_{rr\_1-1,1-2@sn}}\right) + \varphi\right] \tag{11.38}$$

$$\left\lceil 3N\left(1+\frac{\varphi}{\pi}\right)\right\rceil \le n \le \left\lceil 3N\left(2+\frac{\varphi}{\pi}\right)\right\rceil \tag{11.39}$$

The equations representing average switching power losses of IGBTs and APDs ($P_{SW\_1-1,1-2}$ and $P_{RRD\_1-1,1-2}$) are given by (11.40) and (11.41).

$$\mathrm{P}_{\_SW\_1-1,1-2} = \frac{1}{T_m}\frac{V_{DC(min)}}{V_{DC(\mathrm{nom})}} \sum_{n=\left\lceil \frac{3N\varphi}{\pi}\right\rceil}^{\left\lceil 3N\left(1+\frac{\varphi}{\pi}\right)\right\rceil} \left\{ \begin{array}{c} E_{on}\left(I_{S\_on\_1-1,1-2@\mathrm{sn}}\right) + \\ E_{off}\left(I_{S\_off\_1-1,1-2@sn}\right) \end{array} \right\} \tag{11.40}$$

$$P_{RRD\_1-1,1-2} = \frac{1}{T_m}\frac{V_{DC(min)}}{V_{DC(\mathrm{nom})}} \sum_{n=\left\lceil 3N\left(2+\frac{\varphi}{\pi}\right)\right\rceil}^{\left\lceil 3N\left(1+\frac{\varphi}{\pi}\right)\right\rceil} \left\{E_{on}\left(I_{D\_rr\_1-1,1-2@\mathrm{sn}}\right)\right\} \tag{11.41}$$

where $V_{DC(\mathrm{min})}$ and $V_{DC(\mathrm{nom})}$ are the minimum and nominal DC bus voltages, respectively, and T_m is the fundamental time period. Moreover, the switching times for each leg differ from one another. But the average switching power losses of IGBTs and APDs at each leg are equal due to the balanced sinusoid load currents, symmetrical structure of converter, and similar modulation technique with $2\pi/3$ phase shifting. Hence, the average switching power losses for IGBTs (S_{k-1}/S_{k-2}) and APDs (D_{k-1}/D_{k-2}) are given by (11.42) and (11.43), respectively.

$$P_{SW\_k-1/k-1} = P_{SW\_1-1/1-2} \tag{11.42}$$

$$P_{RRD\_k-1/k-1} = P_{RRD\_1-1/1-2} \tag{11.43}$$

11.3.3 Power loss profile in traction inverter

In this section the analytical model obtained has been analyzed by calculating the average and RMS currents for IGBTs and APD. The results so obtained have been compared with the results obtained from simulation. For investigating the conduction power losses in the inverter operating with reduced switching method, the analytical equations for average and RMS currents of IGBTs and diode are used.

Here, a three-phase traction inverter with ratings 165 kVA, 225 V (line-to-neutral for maximum modulation index) operating with DPWM is designed. The parameters of commercially available half-cell PrimePACK™3 IGBT module of 1.7 kV, 900 A from INFINEON are considered for simulating the inverter; 600 V is taken as the DC-link voltage, which is lower than the collector-emitter nominal voltage (900 V) for the IGBT. As depicted in Figure 11.10, the sampling frequency of 1 200 Hz is used here for obtaining minimum harmonic distortion in the lower-order region. Moreover, the conduction losses also rely on the device on-state zero current voltage drops. Temperature-dependent voltage drop of IGBT and diode can be found from the data sheet of power module provided by the manufacturer, which is shown in Figure 11.11.

Average and RMS currents are analytically calculated for the load phase current of 300 A and are compared with simulation results. A comparison of errors between the analytical and the simulated results for average and RMS currents of IGBTs and APDs in the inverter is depicted in Figure 11.16. However, some error may occur between the analytical calculation and simulation result, because during circuit-level simulation the load current cannot keep at a fixed value as compared to analytical calculation.

It's worth mentioning that the comparison between analytical and simulated average/RMS currents is performed for different PF angles and modulation index (M) maintaining a constant load current. The IGBTs transition times at each sample are calculated analytically for the *SC* mode of the first leg. Numerically calculated switch transition times are then compared with switching times of simulated circuit, as depicted in Figure 11.17(a) and (b), respectively.

The comparative study of device turn-on and turn-off transition times represents that there is almost zero error between analytical and simulation results. During continuous conduction mode no switching occurs as represented by a flat line in Figure 11.17.

The transition times of IGBTs exist for the first half of the total number of samples (6N) in a fundamental time period. Similarly, diode turn-off points exist for the remaining second half of the total samples. To investigate the conduction losses the PF angle is varied in the range from 0° to 90° for different modulation indexes. The average conduction power loss of IGBT and APD is depicted in Figure 11.18. IGBT

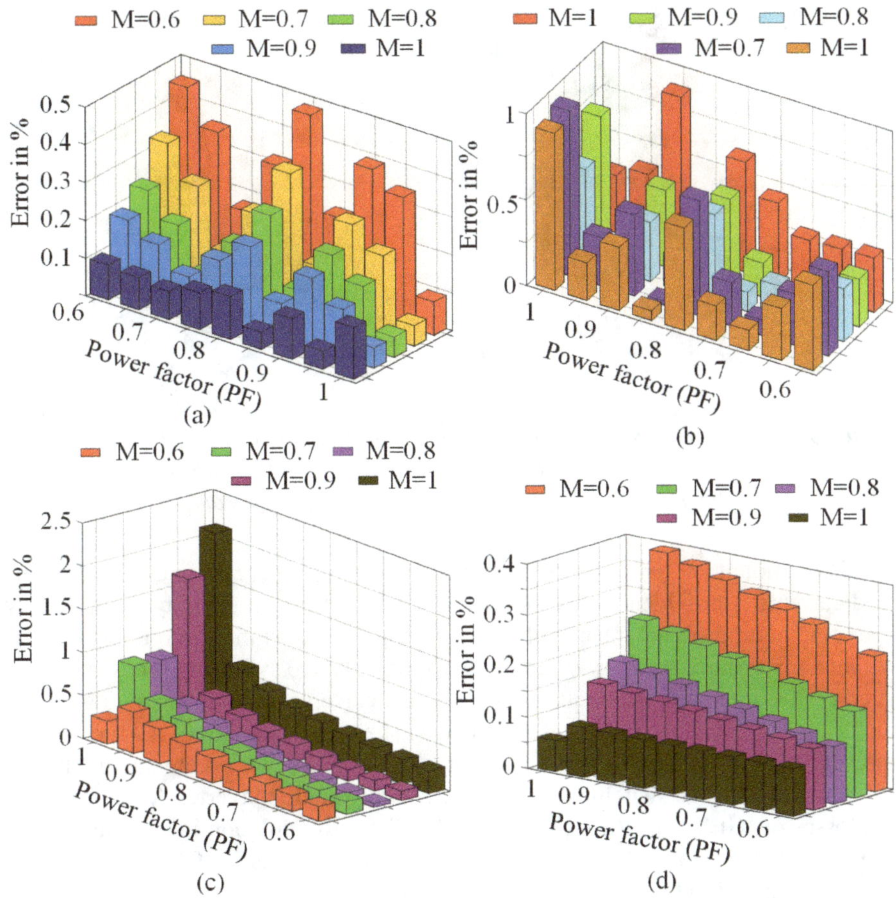

Figure 11.16 Error between analytical and simulated average and RMS currents of IGBT and APD for different PF angles and MIs: error for (a) IGBT average currents, (b) IGBT RMS currents, (c) diode average currents, and (d) diode RMS currents

conduction loss is maximum at $M = 1$, whereas diode conduction loss is minimum for the highest modulation index at PF = 0.

Due to the intrinsic turn-on and turn-off delay of semiconductor devices during on-load switching operation, a significant amount of power loss is generated in the multilevel converter. Though the switching mechanism lasts only for a few μs, it occurs repetitively several hundred times within a second. Therefore, switching losses are directly proportional to the sampling frequency. Moreover, the switching energy losses are also proportional to off-state blocking voltage and the function of device load currents at switching transition instants, which are represented by (11.40) and (11.41).

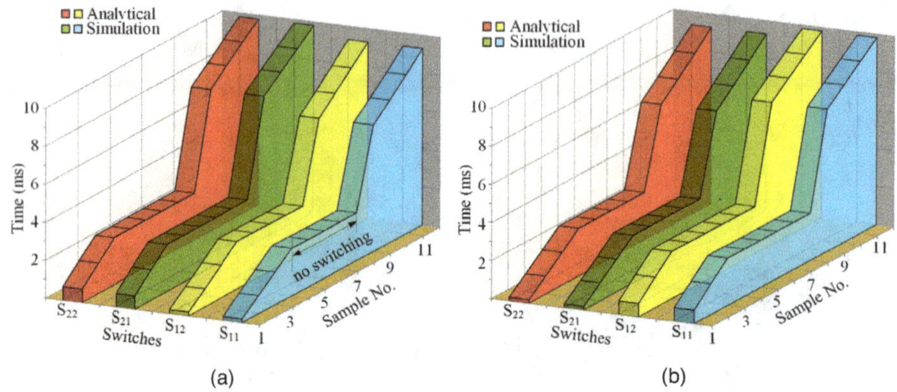

Figure 11.17 Comparative study between analytical and circuit-level simulation: (a) turn-on transition times of IGBTs at different samples, and (b) turn-off transition times of IGBTs at different samples

The amount of temperature-dependent switching energy losses for different load currents is provided by the manufacturers. The thermal stability at the highest operating point can guarantee a reliable and long-term operation of traction inverter without any failure. At every sampling instant the device switching times are calculated by discrete analytical equations. The load currents and corresponding energy losses at those instants are calculated via data sheet approximation. All turn-on and turn-off energies of IGBTs and APDs for a fundamental period are summed up to calculate the average switching power losses in the traction inverter. Overall switching losses of IGBTs and APDs for leg-1 at different modulation indexes are shown in Figure 11.19. Here, the total amounts of switching losses of IGBTs are much higher compared to the turn-off losses of diodes.

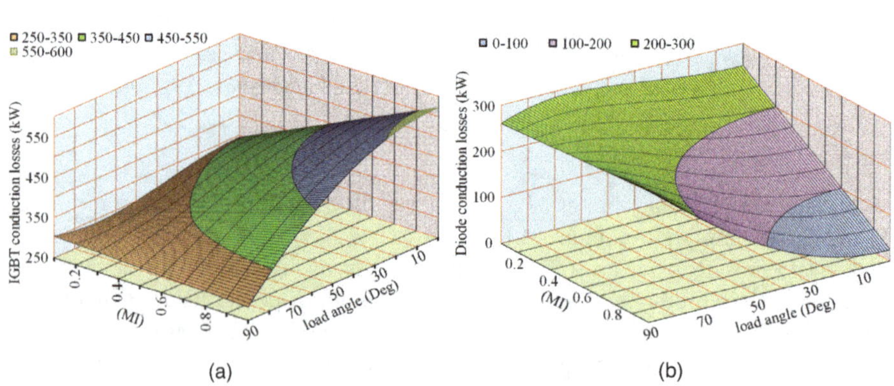

Figure 11.18 Conduction loss property for different PF angles and modulation indexes: (a) IGBT conduction power losses and (b) diode conduction power losses

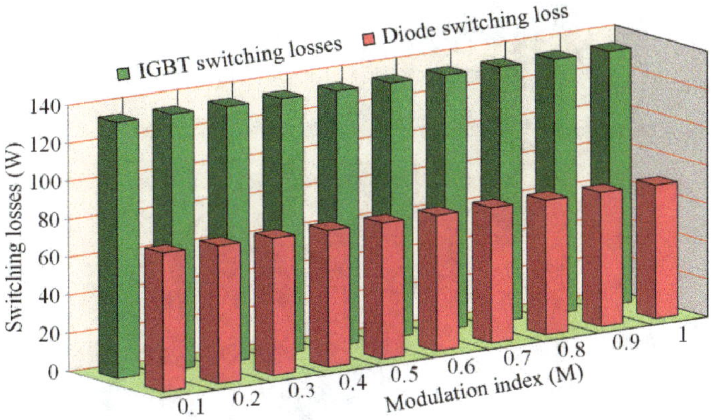

Figure 11.19 *Total switching power losses of IGBTs and APDs at leg-1 for different modulation indices*

Due to the symmetrical structure of converter, balanced load currents, and symmetrical modulation technique, the average and RMS currents as well as switching and conduction losses of IGBTs and APDs of every leg are equal in the traction converter. As a result, equal thermal stress exists for every half-bridge IGBT module, which guarantees a reliable and longer life of traction converter.

11.4 Thermal reliability of power converters

The adverse thermal stress at high-power range may initiate the reliability problems of IGBT power modules, which terminate in catastrophic failure owing to higher junction temperature. To obtain a high confidence level in converter design a precise thermal model is essentially required to predict the chip temperatures of IGBT modules. However, the precise estimation of junction temperatures of silicon chips is a challenging task. Values of Foster network-based lumped RC parameters are generally mentioned for the first four terms in the data sheet of manufacturers. Generally the rise of temperature of the heated chips is calculated analytically using 1D lumped RC Foster thermal impedance model. Such model has the limitation to represent temperature at different layers, because each RC element of the Foster network no longer represents the layer sequence. The Foster network nodes do not have any physical significance, even it does not provide actual thermal property at any layer and it represents equivalent junction to case response only. Moreover, this model does not show the temperature distribution at different positions in every layer. Hence it is desired that a thermal model should contain the influence of operating junction temperature, thermal resistance, and boundary condition on the physical parameters of the semiconductor devices.

Figure 11.20 High-power PrimePACK™ 3 half-bridge dual IGBT module from INFINEON

11.4.1 Electro-thermal behavior of power IGBT modules

The medium and high power converters are made up of semiconductor IGBT modules as depicted in Figure 11.20. The cross-sectional view of a standardized wire-bonded 16-chip high-power IGBT module is depicted in Figure 11.21. The complete structure is mounted on a copper plate and isolated through a direct copper-bonded (DCB) ceramic substrate (Al_2O_3). The layer of copper on both sides of DCB acts as a bonding material for holding the semiconductor chips on the same layer and keeps the whole structure firmly on the base plate. The soldering material (SnAgCu) is used to attach the silicon chips with DCB and also for mounting the full DCB

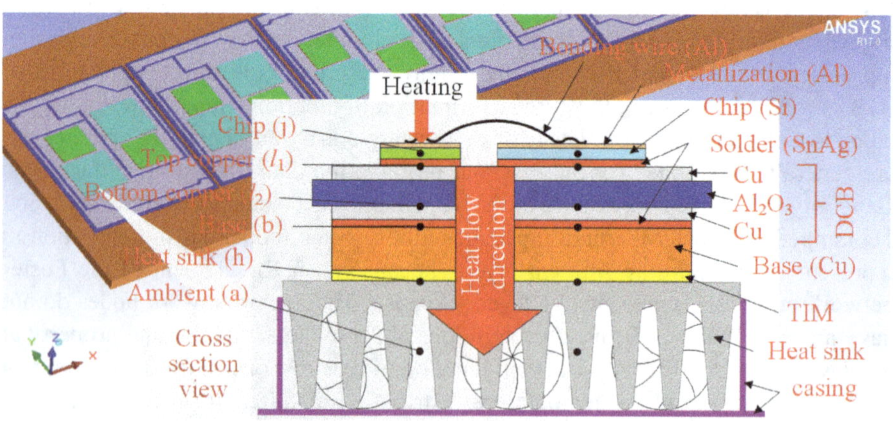

Figure 11.21 Standardized cross section view and heat flow direction of high-power IGBT module

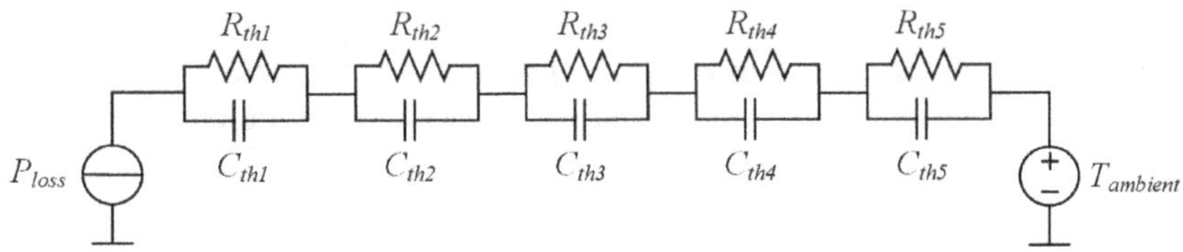

Figure 11.22 Foster network equivalent thermal model

structure on the Cu base plate. The complete structure from the chip-to-base plate is also often termed as stacked multilayer structure.

Bond wires are commonly used to connect the upper side of silicon chips to substrates and to connect substrates with terminals. The IGBT modules are mounted on the heat sink by applying thermal interfacing material (TIM) to provide maximum long-term thermal stability by optimizing metal-to-metal heat conducting area. Generally the heat transfer between layers takes place by conduction, convection, and radiation. But in IGBT modules, as the transfer is mostly by conduction and convention, the heat transfer through radiation is neglected.

The thermal resistor-capacitor (RC) network is the most widely used and reliable thermal modeling method for power electronic cooling systems. To analytically calculate the junction temperature of IGBT module for long-term dynamic operation using fast and easy circuit-level simulation, RC network models provide sufficient accuracy and computational efficiency. 1D RC equivalent thermal models in Foster and Cauer forms, as depicted in Figures 11.22.–11.23, are generally used for studying thermal reliability.

But the 1D Foster models generally provided by the manufacturers fail to represent the accurate geometrical multilayer structure of IGBT module as the individual RC components do not signify the sequence layers. If steady-state power loss (P_{loss}) is applied as input to this mode, the junction temperature will increase exponentially according to (11.44).

$$T(t) = T_{ambient} + P_{loss} \cdot \sum_{i=1}^{n} R_{thi} \cdot \left[1 - e^{-\frac{t}{\tau_i}} \right] \tag{11.44}$$

where $T_{ambient}$ is the ambient temperature at the heat sink and τ is the thermal time constant given by (11.45).

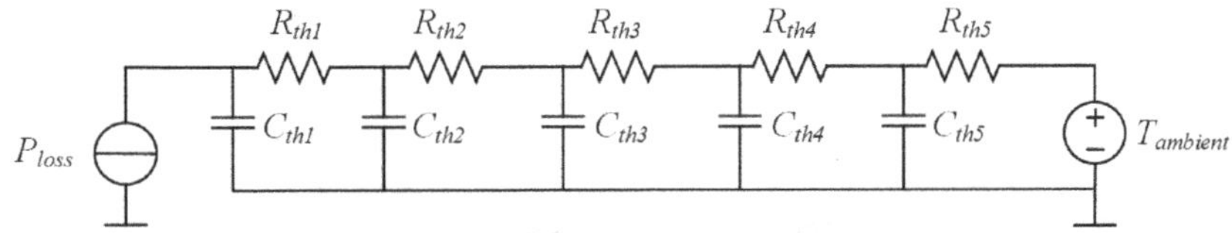

Figure 11.23 An equivalent of Cauer thermal network model

$$\tau = R_{th} \cdot C_{th} \tag{11.45}$$

Under DUT condition a step power loss is applied and a corresponding cooling system is provided for power IGBT modules. Then the junction temperature (T_j) is continuously measured until it reaches its steady-state value. The transient thermal impedance of the power module is calculated using (11.46).

$$Z_{th}(t) = \frac{T_j(t) - T_{ambient}}{P_{loss}} = \sum_{i=1}^{n} R_{thi} \cdot \left[1 - e^{-\frac{t}{\tau_i}} \right] \tag{11.46}$$

A transient thermal impedance curve was obtained and the values of R and C elements of the equivalent circuit in the 1D thermal model were extracted by fitting the measured thermal dynamic curve. The coefficients of transient thermal time constants τ_i and R_{thi} were obtained by numerical analysis of the transient thermal impedance curves and a non-linear least-square data fitting program was used in MATLAB® environment to extract the values of the coefficients up to n (n>=4) number of pairs. Frequency domain Foster thermal network equation is given by (11.47).

$$Z_{th}(s) = \sum_{i=1}^{n} \frac{R_{thi}}{R_{thi} C_{thi} \cdot s + 1} \tag{11.47}$$

Since the Foster RC network is an equivalent mathematical model with no physical meaning, it is unsuitable for temperature analysis at different layers of stacked IGBT module.

However, the Cauer networks are designed from the real physical setup of the semiconductor-based fundamental heat-transfer equivalent RC network of each layer and provide realistic physical regions of heat transferring paths. For each individual layer of the module, an equivalent RC element can be assigned in the Cauer network. Therefore, the nodes of RC networks in the Cauer model indicate the real temperature of each layer in IGBT power modules. But the main drawback of the Cauer model is that the correct mapping of the thermal spreading on the individual layer is not possible.

In Cauer networks, one or several R_{th} and C_{th} pairs represent physical information and equivalent thermal properties of each layer. The thermal resistors and thermal capacitors of each layer can be calculated by (11.48) and (11.49).

$$R_{th} = \frac{l}{\lambda \cdot A} \tag{11.48}$$

$$C_{th} = c_P \cdot \rho \cdot l \cdot A \tag{11.49}$$

where l is the length at heat flow direction of each layer, ρ is the mass density, and A is the area of heat dissipation perpendicular to the direction of heat flow, as illustrated in Figure 11.24. The equation for thermal equivalent transfer function using the Cauer network is given by (11.50).

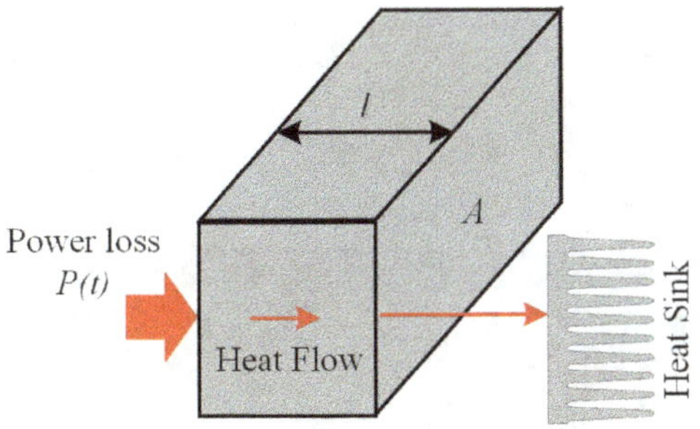

Figure 11.24 Simplified structure for the illustration of (11.48) and (11.49)

$$Z_{th}(s) = \cfrac{1}{sC_1 + \cfrac{1}{R_1 + \cfrac{1}{sC_2 + \cdots + \cfrac{1}{R_n}}}} \tag{11.50}$$

The conventional methods for determining average power dissipation in IGBT modules at rated currents involve the use of curves and information from the data sheet. The rises in temperature of the heated chips are measured using a 1D lumped RC thermal impedance model. Such model has the limitation to represent the precise temperature at different underlying layers, whereas it provides only equivalent junction to case response for the chips only. Moreover, those models do not show the temperature distribution at different positions in every layer.

11.4.2 Design and FEM analysis of power modules in ANSYS

1D lumped thermal resistance (*R*) and thermal capacitance (*C*) models using Cauer or Foster networks are generally provided by the manufacturers, which are very fast and rough for estimating the distributed junction temperatures. Whereas multichip differential temperature coupling effect of the IGBT power modules can be designed accurately with 3D thermal networks along with adjustable power losses, cooling system, and boundary conditions.

With high-computational speed, the temperature distribution can be calculated by the finite element method (FEM) for the sophisticated geometrical shape of semiconductor devices and cooling systems. However, the co-simulation of the thermal model using a commercial tool such as ANSYS with circuit simulation tool becomes furthermore difficult for advanced discontinuous modulation techniques. Therefore, the electro-thermal parameters (*RC*) extracted from numerical simulation of structural assembly can be utilized for chip-wise multilayer 3D thermal model to predict the operating point of semiconductor chips within allowable range of temperature. A satisfactory power loss and temperature profile of high-power IGBT modules can guarantee the reliable and long-term operation of the converter with reduced cost. Interdependent multilayer thermal coupling effects with 3D thermal mode provide

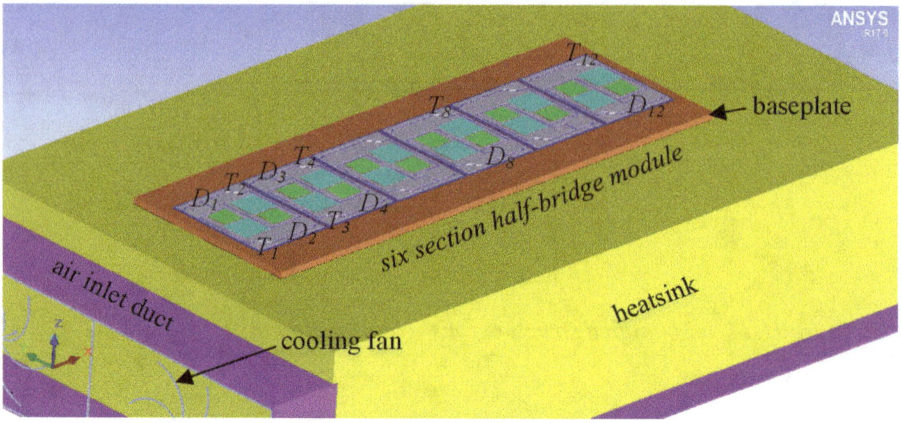

Figure 11.25 Geometrical structure of power module in ANSYS/Icepak

better reliability study for long-term loss profile and corresponding thermal stability. To investigate the functionality of the 3D thermal model a FEM analysis is required to be performed in ANSYS/Icepak to extract the lumped RC parameters, which is used in the circuit simulation tool to examine the long-term effectiveness of the thermal model. An accurate geometrical structure of IGBT module is generated using ANSYS/Icepak, which contains six half-bridge sections in parallel, as depicted in Figure 11.25.

The Geometry tool in ANSYS is used to create the accurate geometrical structure of all separate layers, heat sink, and cooling system of the power IGBT module, shown in Figure 11.26.

A data linker is used to link up the Geometry and Icepak tool in ANSYS project window so that all the geometrical data can be transferred to Icepak. In each section, two IGBT chips and two diode chips are assembled on a stacked multilayer structure. TIM is applied to install the whole IGBT module on an extruded heat sink. All stacked multilayers are designed in ANSYS/Icepak, as illustrated in Figure 11.27. However, the variation of thermal conductivity and specific heat capacity of the materials with the change of operating temperature is needed to be considered. Figure 11.28(a) and (b) represents temperature-dependent thermal properties of different material at stacked multilayer structure inside the package of IGBT module.

The curve data of temperature-dependent properties (λ, c_p) of several materials are utilized and imported in FEM simulation using graph editing tool, as depicted in Figure 11.28(c). Cooling mechanism and the properties of the heat sink affect the thermal behavior of IGBT modules. A forced-air cooling system with a radial blower is designed for FEM simulation, which is depicted in Figure 11.29.

The intersection of the pressure drop curve of the heat sink and fan characteristics curve indicates the operating point and the volumetric flow rate of the air, as depicted in Figure 11.30. Besides, the thermal resistance of heat sink is also a function of volumetric flow rate. All of the mentioned factors are needed to be considered to accurately estimate the thermal behavior of power modules via FEM analysis.

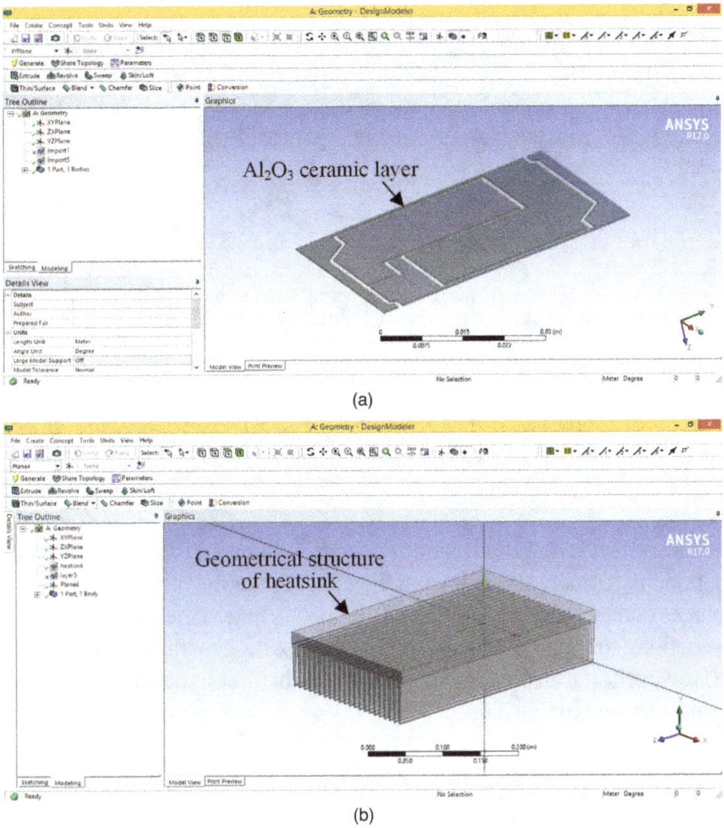

Figure 11.26 *ANSYS Geometry tool windows to create geometrical structure of different objects: (a) formation of* Al_2O_3 *ceramic layers of IGBT module and (b) formation of heat sink*

11.4.3 3D thermal model of IGBT modules and thermal coupling

A novel 3D thermal model is essentially required to precisely estimate the junction temperature of silicon chips to prevent over thermo-mechanical stress, partial discharge, current constriction, thermal runaway, and eventually failures. Change of thermal properties of materials at every layer with the variation of operating temperature is needed to be considered to extract accurate lumped parameters.

The model is called "3D thermal model," because it not only indicates junction (*j*) to case (*c*) thermal coupling among chips but also thermal coupling between junction (*j*) to several underlying layers at surrounding chips. Soldering distinct chips with different operating temperatures on the same layer causes high thermal coupling among the chips and the variation of temperature on the containing surface will be higher. To accurately estimate the self-heating and thermal-coupling impact of other chips, five parallel coupling locations are considered in this 3D model, which are also shown in Figure 11.31.

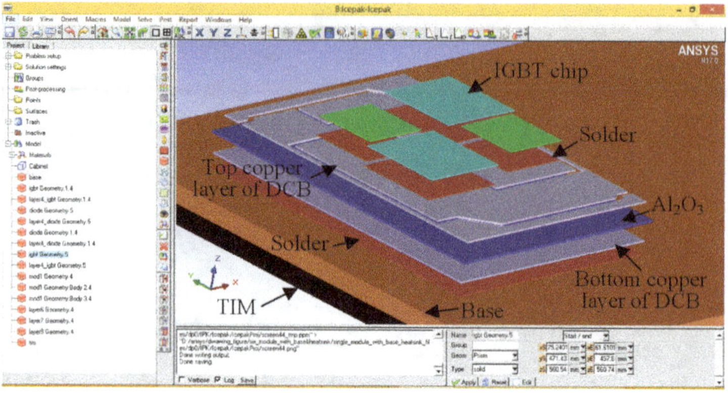

Figure 11.27 Stacked multilayer structure for power IGBT module in ANSYS/ Icepak

The self-heating effect is designed using conventional lumped RC networks from junction to ambient cooling temperature point and the effect of thermal coupling is modeled using voltage sources connected in series with RC lumps. Whereas the voltage sources are controlled by coupling thermal impedance and power losses generated from other neighboring chips. However, in transient operation the self-thermal impedance (Z_{th}) between two parallel points can be expressed as

$$\begin{aligned} Z_{th(j-l_1)}^{T_{25}}(t) &= \frac{T_j(t)-T_{l_1}(t)}{P_{T_2}} \\ &= \sum_{i=1}^{n} R_{th_i}\left(1-e^{-\frac{t}{\tau_{th_i}}}\right) \end{aligned} \tag{11.51}$$

$$\tau_{th_i} = R_{th_i} \times C_{th_i} \tag{11.52}$$

where $T_j(t)$ and $T_{l_2}(t)$ are the transient temperatures of vertical monitoring points at two parallel layers, P_{T_2} is the total power loss of IGBT chip T_2, R_{th_i} is the thermal resistance, C_{th_i} is the thermal capacitance, τ_{th_i} is the time constant, and n is the number of exponential terms to fit Z_{th} well to the transient thermal impedance curve with minimum errors.

A detailed 3D thermal model is shown in Figure 11.32, where thermal coupling effect among chips, which are soldered on the same DCB layer is considered. To analyze the distribution of temperature on every underlying layer, nine monitoring points are considered for each chip as shown in Figure 11.33. At IGBT chip-2, the equation of induced coupling thermal impedance (Z_c) between layers j and l_1 for the power loss in diode-2 can be written as

$$Z_{c(j-l_1)}^{T_{25}@D_2} = \frac{T_{25}^{j}-T_{25}^{l_1}}{P_{diode-2}} \tag{11.53}$$

The 3D equivalent thermal impedance model can be summarized as an impedance matrix (Z_{th}), where both self (Z_{ii}) and coupling thermal impedances (Z_{ij}) are

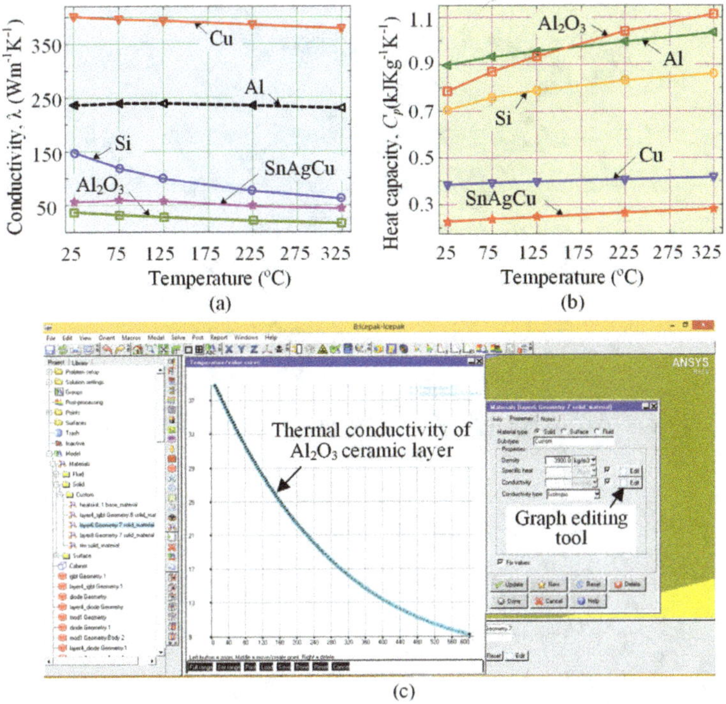

Figure 11.28 Temperature-dependent properties of several materials used in IGBT module: (a) thermal conductivity, (b) specific heat capacity of materials at different operating temperatures, and (c) ANSYS/Icepak window to assign thermal conductivity of solid ceramic material Al_2O_3

considered at junction (j) and several underlying layers to study the impact of thermal coupling on the chips and to find the temperature rise at several layers.

The equation of junction temperatures at monitoring points considering the thermal impedance matrix is given by (11.54).

$$\begin{bmatrix} T_{jx1}^{T_x/D_x} \\ T_{jx2}^{T_x/D_x} \\ \vdots \\ T_{jxy}^{T_x/D_x} \end{bmatrix} = \begin{bmatrix} A_1 & \cdots & B_1 \\ A_2 & \cdots & B_2 \\ \vdots & \cdots & \vdots \\ A_y & \cdots & B_y \end{bmatrix} \cdot \begin{bmatrix} P_{T_x/D_x} \\ \vdots \\ P_{T_n/D_n} \end{bmatrix} + \begin{bmatrix} T_a \\ T_a \\ T_a \\ T_a \end{bmatrix} \tag{11.54}$$

where

$$A_1 = Z_{th(j-l_1)}^{T_{x1}/D_{x1}} + Z_{th(l_1-l_2)}^{T_{x1}/D_{x1}} + Z_{th(l_2-b)}^{T_{x1}/D_{x1}} + Z_{th(b-h)}^{T_{x1}/D_{x1}}$$

$$A_2 = Z_{th(j-l_1)}^{T_{x2}/D_{12}} + Z_{th(l_1-l_2)}^{T_{x2}/D_{x2}} + Z_{th(l_2-b)}^{T_{x2}/D_{x2}} + Z_{th(b-h)}^{T_{x2}/D_{x2}}$$

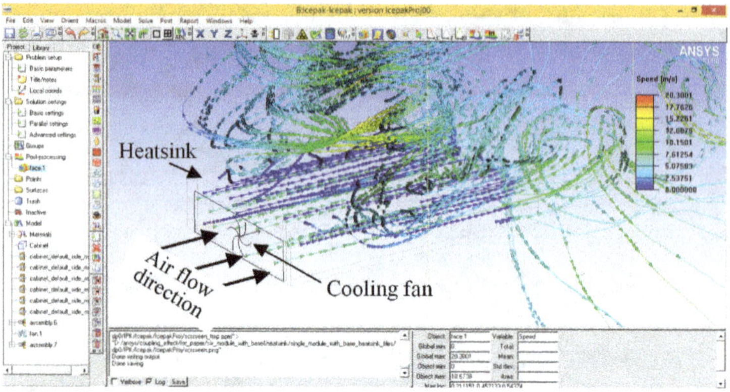

Figure 11.29 ANSYS/Icepak post-processing window to indicate forced-air cooling and flow direction

$$A_y = Z_{th(j-l_1)}^{T_{xy}/D_{xy}} + Z_{th(l_1-l_2)}^{T_{xy}/D_{xy}} + Z_{th(l_2-b)}^{T_{xy}/D_{xy}} + Z_{th(b-h)}^{T_{xy}/D_{xy}}$$

$$B_1 = Z_{c(j-l_1)}^{T_{x1@T_n/D_n}/D_{x1@T_n/D_n}} + Z_{c(l_1-l_2)}^{T_{x1@T_n/D_n}/D_{x1@T_n/D_n}} + Z_{c(l_2-b)}^{T_{x1@T_n/D_n}/D_{x1@T_n/D_n}} + Z_{c(b-h)}^{T_{x1@T_n/D_n}/D_{x1@T_n/D_n}}$$

$$B_2 = Z_{c(j-l_1)}^{T_{x2@T_n/D_n}/D_{x2@T_n/D_n}} + Z_{c(l_1-l_2)}^{T_{x2@T_n/D_n}/D_{x2@T_n/D_n}} + Z_{c(l_2-b)}^{T_{x2@T_n/D_n}/D_{x2@T_n/D_n}} + Z_{c(b-h)}^{T_{x2@T_n/D_n}/D_{x2@T_n/D_n}}$$

$$B_y = Z_{c(j-l_1)}^{T_{xy@T_n/D_n}/D_{xy@T_n/D_n}} + Z_{c(l_1-l_2)}^{T_{xy@T_n/D_n}/D_{xy@T_n/D_n}} + Z_{c(l_2-b)}^{T_{xy@T_n/D_n}/D_{xy@T_n/D_n}} + Z_{c(b-h)}^{T_{xy@T_n/D_n}/D_{xy@T_n/D_n}}$$

where $T_{jxy}^{T_x/D_x}$ is the junction temperature, P_{T_n/D_n} is the power losses of IGBT/diode chip, T_a is the reference ambient temperature, A_l represents self thermal impedance components, and B_y represents coupling impedance components. Therefore,

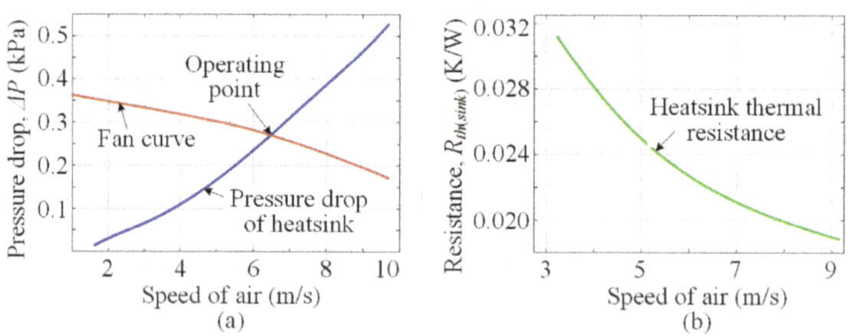

Figure 11.30 Properties of heat sink: (a) pressure drop vs. speed of air flowing through heat sink and fan characteristics and (b) thermal resistance as a function of the speed of air

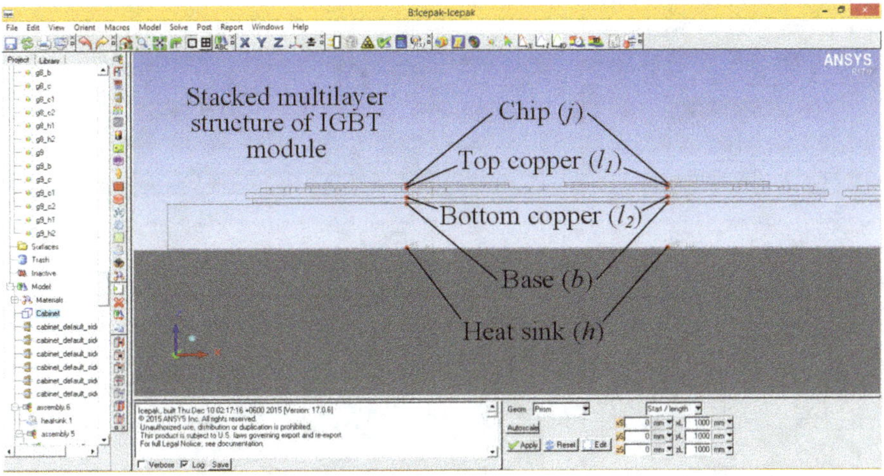

Figure 11.31 ANSYS/Icepak window to indicate parallel monitoring points for analyzing thermal coupling

considering 3D thermal impedance equation the junction temperature $T_{j12}^{D_1}$of Diode chip-1 can be written as

$$
\begin{aligned}
T_{j12}^{D_1} = {} & (Z_{th(j-l_1)}^{D_{12}} + Z_{th(l_1-l_2)}^{D_{12}} + Z_{th(l_2-b)}^{D_{12}} + Z_{th(b-h)}^{D_{12}})P_{D_1} + (Z_{c(j-l_1)}^{D_{12}@T_1} + Z_{c(l_1-l_2)}^{D_{12}@T_1} + Z_{c(l_2-b)}^{D_{12}@T_1} + \\
& Z_{c(b-h)}^{D_{12}@T_1})P_{T_1} \\
& (Z_{c(j-l_1)}^{D_{12}@T_2} + Z_{c(l_1-l_2)}^{D_{12}@T_2} + Z_{c(l_2-b)}^{D_{12}@T_2} + Z_{c(b-h)}^{D_{12}@T_2})P_{T_2} + (Z_{c(j-l_1)}^{D_{12}@D_2} + Z_{c(l_1-l_2)}^{D_{12}@D_2} + Z_{c(l_2-b)}^{D_{12}@D_2} + \\
& Z_{c(b-h)}^{D_{12}@D_2})P_{D_2} \\
& + \ldots. + (Z_{c(j-l_1)}^{D_{12}@T_n/D_n} + Z_{c(l_1-l_2)}^{D_{12}@T_n/D_n} + Z_{c(l_2-b)}^{D_{12}@T_n/D_n} + Z_{c(b-h)}^{D_{12}@T_n/D_n})P_{T_n/D_n} + T_a
\end{aligned}
\tag{11.55}
$$

where P_{T_n/D_n} is the power loss in IGBT or diode chips, $Z_{th(j-l_1)}^{D_{12}}$, $Z_{th(l_1-l_2)}^{D_{12}}$, $Z_{th(l_2-b)}^{D_{12}}$, and $Z_{th(b-h)}^{D_{12}}$ are self impedances, and $Z_{c(j-l_1)}^{D_{12}@T_1}$, $Z_{c(l_1-l_2)}^{D_{12}@T_1}$, $Z_{c(l_2-b)}^{D_{12}@T_1}$ and $Z_{th(j-l_1)}^{D_{12}}$ are coupling impedances for parallel monitoring points.

For high-power IGBT modules 3D thermal model provides better coupling effect and precision in transient temperature measurement at junction to several underlying layers. To validate the accuracy of 3D thermal model, an accurate geometrical structure of commercially available IGBT module is designed in ANSYS/Icepak. A half-bridge module has six sections connected in parallel and each section contains two IGBT and two diode chips, as shown in Figure 11.22. An extruded heat sink with temperature-dependent thermal property is also considered for FEM simulation. Radial blower is designed for forced-air cooling mechanism and the characteristics curve, as shown in Figures 11.27 and 11.28, respectively. Using

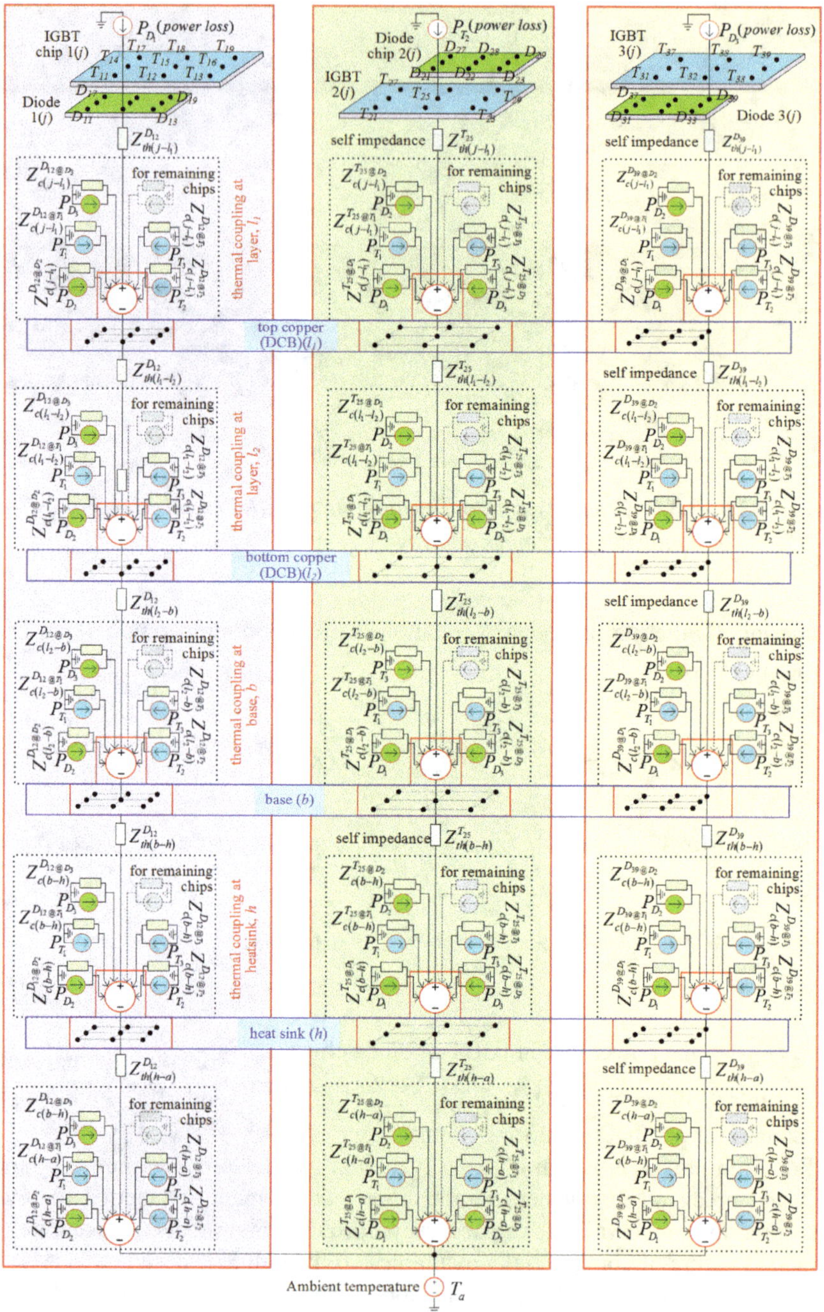

Figure 11.32 3D thermal model of T_{2y} and D_{2y} from junction (j) to ambient (a) with coupling effect T_{1y}, D_{2y}, T_{3y} and D_{3y} to validate the loss profile within IGBT module

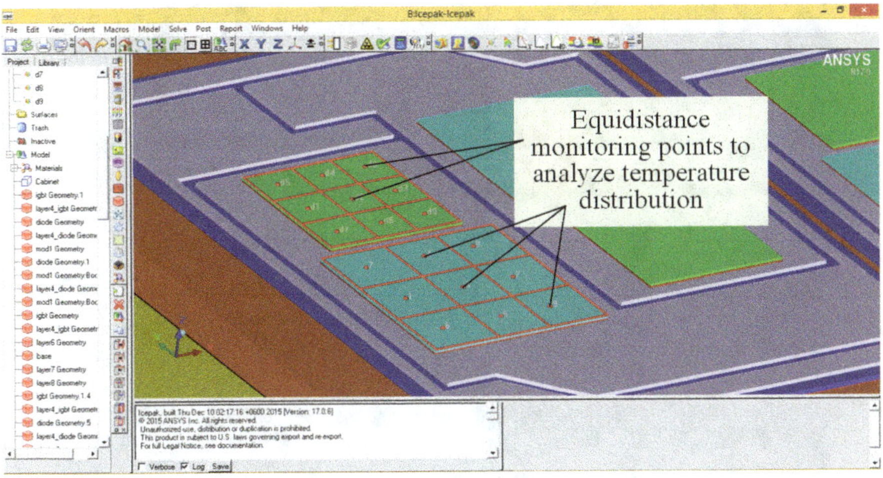

Figure 11.33 ANSYS/Icepak window indicating equidistance temperature monitoring points to analyze thermal distribution

analytical equations the total power loss of each half-cell module at nominal storage temperature (125 °C) is calculated and this loss profile is selected as the input variable of FEM verification. To calculate the junction temperature accurately using 3D model of any IGBT/diode chip (T_x/D_x) at any position, the mutual coupling effect of all previous chips (T_0/D_0 to $T_{(x-1)}/D_{(x-1)}$) and all upcoming chips ($T_{(x+1)}/D_{(x+1)}$ to T_{12}/D_{12}) is needed to be considered. The effect of thermal coupling is demonstrated in Figure 11.34.

Temperature-dependent thermal conductivity and specific heat capacity of materials are considered with the structural geometry of power modules. In this model, not only the thermal couplings among chips are considered but the thermal couplings between the layers beneath the silicon chips are also considered, which is represented in Figure 11.31. To extract lumped RC parameters, a steady-state average power loss is applied in each IGBT and diode chip separately. Therefore, the transient temperature responses at intended monitoring points are investigated to calculate both the self and the coupling thermal impedances. Using curve fitting approach, an optimal number of RC foster networks is formed to satisfy transient thermal impedance with minimum error. The cooling mechanism and the properties of heat sink affect the thermal behavior of power modules, which is also considered in this model. Including all of those factors, the values of RC lumps are then extracted precisely. However, the long-term thermal stability analysis via ANSYS/Icepak is a very time-consuming issue. Therefore, the extracted lumped parameters from FEM analysis are further utilized to implement the 3D thermal model in circuit simulation tools such as MATLAB/Simulink to calculate the temperature of IGBT modules for long-term converter operation. That will reduce the simulation time with adequate accuracy compared to that of the FEM-based simulation. The calculated junction temperature of IGBT chip is found to be 127 °C from 3D

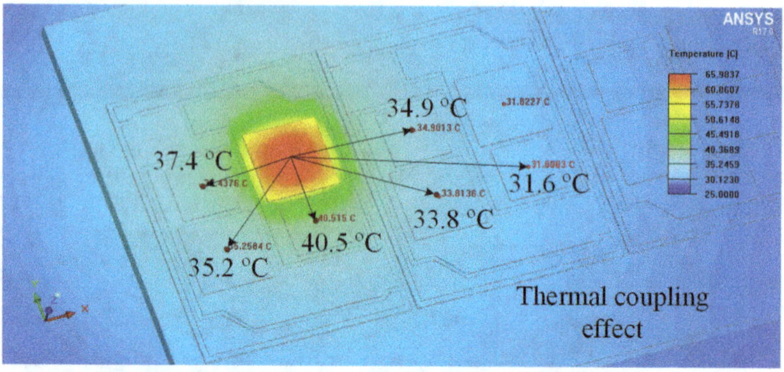

(a)

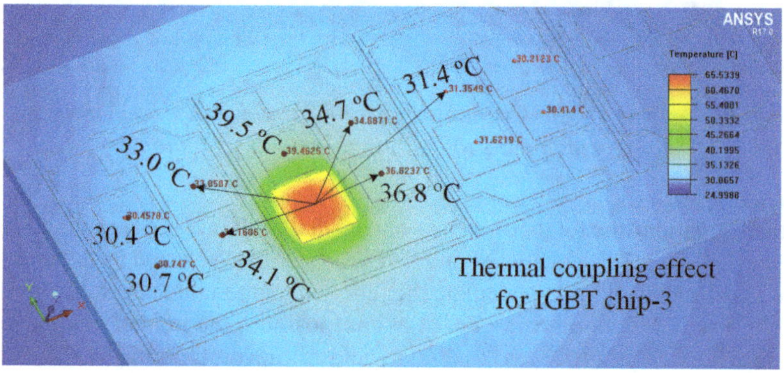

(b)

Figure 11.34 *Effect of thermal coupling: (a) thermal coupling due to the heating of IGBT chip-2 and (b) coupling effect for the heating of IGBT chip-3*

thermal model, which does not cross the junction temperature of 131 °C found from FEM analysis shown in Figure 11.35. An excellent match of junction temperature between analytical calculation and FEM analysis demonstrates a good accuracy of 3D thermal model. Afterward, the parallel monitoring points can be set for every layer and the number of equidistance temperature distribution monitoring points can be increased to analyze more coupling effects and better temperature variation on the surface of the chips and layers.

11.5 Conclusion

This chapter presents a systematic development of thermal model for high-power semiconductor modules as well as thermal reliability study of power electronic converters in EVs or HEVs. The adopted method presents a better coupling effect and precision during temperature measurement at junctions and several layers lying

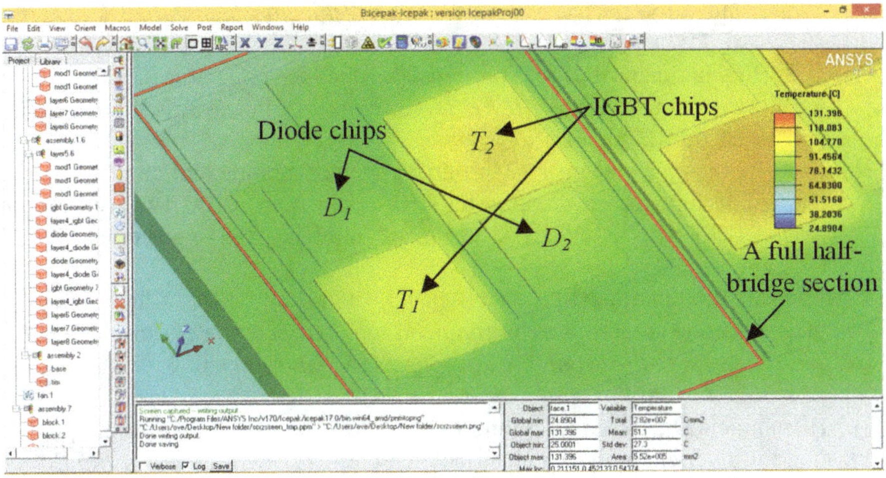

Figure 11.35 *Overall temperatures of IGBT and diode chips of a half-bridge section with forced-air cooling*

underneath it in IGBT power modules. Here, the temperature-dependent thermal conductivity and specific heat of materials are considered in addition to the structural geometry of power modules. In this model, not only the thermal couplings among chips are considered but also the thermal couplings between the layers beneath the silicon chips are also considered. The analytical equations to calculate the average and RMS currents of IGBTs and APDs of power converters with reduced switching topology are very advantageous in the design procedure. The discrete form of equations representing the analytical method to calculate the device turn-on and turn-off transition points at each and every sampling duration and corresponding switching energy loss with data sheet approximation is the precise method for switching power loss calculation, where the numerical results agree well with circuit-level simulation. 3D thermal model of power modules with interdependent multilayer thermal coupling effect is the optimum solution for reliability study of the power electronic converters. The extracted lumped RC parameters from FEM analysis can be utilized in 3D thermal model for long-term reliability study of power converters. Though the boundary condition may differ for separate cooling systems, the 3D thermal model can be used for any surroundings and cooling systems. The high-speed circuit-level thermal reliability analysis using 3D thermal model provides long-term reliability study with sufficient accuracy.

References

[1] Saponara S., Lee C.H.T., Wang N.X., Kirtley J.L. 'Electric drives and power Chargers: recent solutions to improve performance and energy efficiency for

hybrid and fully electric vehicles'. *IEEE Vehicular Technology Magazine*. 2020, vol. 15(1), pp. 73–83.

[2] Ji B. 'In situ diagnostics and prognostics of solder fatigue in IGBT modules for electric vehicle drives'. *IEEE Transactions on Power Electronics*. 2015, vol. 30(3), pp. 1535–43.

[3] Ning P., Liang Z., Wang F., Marlino L. 'Power module and cooling system thermal performance evaluation for HEV application'. 2012 Twenty-Seventh Annual IEEE Applied Power Electronics Conference and Exposition (APEC); Orlando, FL; 2012. pp. 2134–9.

[4] Johnson R.W., Evans J.L., Jacobsen P., Thompson J.R.R., Christopher M. 'The changing automotive environment: high-temperature electronics'. *IEEE Transactions on Electronics Packaging Manufacturing*. 2004, vol. 27(3), pp. 164–76.

[5] Hirschmann D., Tissen D., Schroder S., De Doncker R.W. 'Reliability prediction for inverters in hybrid electrical vehicles'. *IEEE Transactions on Power Electronics*. 2007, vol. 22(6), pp. 2511–7.

[6] Ye J., Yang K., Ye H., Emadi A. 'A fast electro-thermal model of traction inverters for electrified vehicles'. *IEEE Transactions on Power Electronics*. 2017, vol. 32(5), pp. 3920–34.

[7] Emadi A., Young Joo Lee., Rajashekara K. 'Power electronics and motor drives in electric, hybrid electric, and plug-in hybrid electric vehicles'. *IEEE Transactions on Industrial Electronics*. 2008, vol. 55(6), pp. 2237–45.

[8] Rajashekara K. 'Parallel between more electric aircraft and electric\/hybrid vehicle power conversion technologies'. *IEEE Electrification Magazine*. 2014, vol. 2(2), pp. 50–60.

[9] Lai J.-S., Nelson D.J. 'Energy management power converters in hybrid electric and fuel cell vehicles'. *Proceedings of the IEEE*. 2007, vol. 95(4), pp. 766–77.

[10] Lee Y., Khaligh A., Emadi A. 'Advanced integrated bidirectional AC/DC and DC/DC converter for plug-in hybrid electric vehicles'. *IEEE transactions on vehicular technology*. 2009, vol. 58(8), pp. 3970–80.

[11] Rastogi S.K., Sankar A., Manglik K., Mishra S.K., Mohanty S.P. 'Toward the vision of all-electric vehicles in a decade [energy and security]'. *IEEE Consumer Electronics Magazine*. 2019, vol. 8(2), pp. 103–7.

[12] Tran V.T., Islam M.R., Muttaqi K.M., Sutanto D. 'An on-board V2X electric vehicle charger based on amorphous alloy high-frequency magnetic-link and SiC power devices'. Proceedings of the 2019 IEEE Industry Applications Society Annual Meeting (IAS); Baltimore, USA, Sept. 29–Oct. 3; 2019.

[13] Tu H., Feng H., Srdic S., Lukic S. 'Extreme fast charging of electric vehicles: a technology overview'. *IEEE Transactions on Transportation Electrification*. 2019, vol. 5(4), pp. 861–78.

[14] Anderson J.A., Haider M., Bortis D., Kolar J.W., Kasper M., Deboy G. 'New synergetic control of a 20kW isolated VIENNA rectifier front-end EV battery charger'. 2019 20th Workshop on Control and Modeling for Power Electronics (COMPEL); Toronto, ON, Canada; 2019. pp. 1–8.

[15] Muttaqi K.M., Islam M.R., Sutanto D. 'Future power distribution grids: integration of renewable energy, energy storage, electric vehicles, superconductor, and magnetic bus'. *IEEE Transactions on Applied Superconductivity*. 2019, vol. 29(2), pp. 1–5.

[16] Tran V.T., Islam M.R., Muttaqi K.M., Sutanto D. 'An efficient energy management approach for a solar-powered EV battery charging facility to support distribution grids'. *IEEE Transactions on Industry Applications*. 2019, vol. 55(6), pp. 6517–26.

[17] Khan S.A., Islam M.R., Guo Y., Zhu J. 'A new isolated multi-port converter with multi-directional power flow capabilities for smart electric vehicle charging stations'. *IEEE Transactions on Applied Superconductivity*. 2019, vol. 29(2), pp. 1–4.

[18] Williamson S.S., Emadi A., Rajashekara K. 'Comprehensive efficiency modeling of electric traction motor drives for hybrid electric vehicle propulsion applications'. *IEEE Transactions on Vehicular Technology*. 2007, vol. 56(4), pp. 1561–72.

[19] Mukherjee S., Kumar Giri S., Kundu S., Banerjee S. 'A generalized discontinuous PWM scheme for three-level NPC traction Inverter with minimum switching loss for electric vehicles'. *IEEE Transactions on Industry Applications*. 2019, vol. 55(1), pp. 516–28.

[20] Mahfuz-Ur-Rahman A.M., Islam M.M., Islam M.R. 'Performance analysis of modulation techniques in multilevel inverters for direct grid connected photovoltaic arrays'. 2016 9th International Conference on Electrical and Computer Engineering (ICECE); Dhaka; 2016. pp. 66–9.

[21] Islam M.M., Rahman M.A., Islam M.R. 'Power loss and thermal impedance modeling of multilevel power converter with discontinuous modulation'. *IEEE Transactions on Energy Conversion*. 2020, vol. 36(1), pp. 36–47.

[22] Islam M.M., Mahfuz-Ur-Rahman A.M., Islam M.R. 'Precise loss evaluation method of CHBMC with reduced switching strategy for direct integration with medium voltage smart microgrids'. 2016 9th International Conference on Electrical and Computer Engineering (ICECE); Dhaka; 2016. pp. 62–5.

[23] Islam M.R., Mahfuz-Ur-Rahman A.M., Islam M.M., Guo Y.G., Zhu J.G. 'Modular medium-voltage grid-connected converter with improved switching techniques for solar photovoltaic systems'. *IEEE Transactions on Industrial Electronics*. 2017, vol. 64(11), pp. 8887–96.

[24] Mahfuz-Ur-Rahman A.M., Islam M.R., Fahim T.A., Islam M.M., Muttaqi K.M., Sutanto D. 'Performance analysis of symmetric and asymmetric multilevel converters'. 2017 IEEE Region 10 Humanitarian Technology Conference (R10-HTC); Dhaka; 2017. pp. 383–6.

Index

www.ingramcontent.com/pod-product-compliance
Lightning Source LLC
LaVergne TN
LVHW011842200926
848484LV00002B/390
* 9 7 8 1 8 3 9 5 3 1 7 8 1 *